Organotin Compounds:
New Chemistry and Applications

Organotin Compounds:
New Chemistry and Applications

J. J. Zuckerman, EDITOR

University of Oklahoma

A symposium sponsored by the

Division of Inorganic Chemistry,

at the 171st Meeting of the

American Chemical Society,

New York, N. Y.,

April 6–7, 1976.

ADVANCES IN CHEMISTRY SERIES **157**

AMERICAN CHEMICAL SOCIETY

WASHINGTON, D. C.   1976

Library of Congress CIP Data

Organotin compounds.
(Advances in chemistry series; 157)

Includes bibliographical references and index.

1. Organotin compounds—Congresses.
I. Zuckerman, Jerold J., 1936-   II. American
Chemical Society. Division of Inorganic Chemistry. III.
Series.

QD1.A355	[QD412.S7]	540′.8s
[547′.05]		76-54338
ISBN 0-8412-0343-1	ADCSAJ 157 1-299(1976)	

FOREWORD

ADVANCES IN CHEMISTRY SERIES was founded in 1949 by the American Chemical Society as an outlet for symposia and collections of data in special areas of topical interest that could not be accommodated in the Society's journals. It provides a medium for symposia that would otherwise be fragmented, their papers distributed among several journals or not published at all. Papers are refereed critically according to ACS editorial standards and receive the careful attention and processing characteristic of ACS publications. Papers published in ADVANCES IN CHEMISTRY SERIES are original contributions not published elsewhere in whole or major part and include reports of research as well as reviews since symposia may embrace both types of presentation.

CONTENTS

Preface ix

1. Organotin Chemistry: Past, Present and Future 1
   G. J. M. van der Kerk

2. Homolytic Reactions Involving Organotin Compounds 26
   Alwyn G. Davies

3. Synthetic and Mechanistic Aspects of Organostannylanionoin
   Chemistry 41
   Henry G. Kuivila

4. Organotin Phosphines, Arsines, Stibines, and Bismuthines: Starting
   Materials for New Catalysts 57
   Herbert Schumann, Joachim Held, Wolf-W. Du Mont,
   Gisbert Rodewald, and Bernd Wöbke

5. Di- and Trivalent Trimethylsilyl-Substituted Tin Amides and
   Related Compounds Such as $Sn[N(SiMe_3)_2]_2$ or $_3$ 70
   Michael F. Lappert and Philip P. Power

6. Organotin Alkoxides and Amines: New Chemistry and
   Applications 82
   Jean-Claude Pommier and Michael Pereyre

7. Reactions of Electrophilic Reagents with Tin Compounds
   Containing Organofunctional Groups 113
   James L. Wardell

8. Synthesis of Novel Substituted Alkyltin Halides 123
   R. E. Hutton and V. Oakes

9. The Use of Estertin Stabilizers in PVC 134
   D. Lanigen and E. L. Weinberg

10. Structure and Bonding in Organotins by Gamma Resonance
    Spectroscopy 155
    Rolfe H. Herber and Michael F. Leahy

11. Organotins in Agriculture 167
    Melvin H. Gitlitz

12. Some Recent Chemistry Related to Applications of Organotin
    Compounds 177
    R. C. Poller

13. The Influence of Organotin Compounds on Mitochondrial
    Functions 186
    W. N. Aldridge

14. Bioorganotin Chemistry: Biological Oxidation of Organotin
Compounds 197
Richard H. Fish, Ella C. Kimmel, and John E. Casida

15. Triorganotin Compounds as Ionophores and Inhibitors of Ion
Translocating ATPase 204
M. J. Selwyn

16. Photoelectron Spectroscopy as a Tool in Organotin Chemistry 227
Y. Limouzin and J. C. Maire

17. The Optical Stability of Organotin Compounds 249
Marcel Gielen, Cornelis Hoogzand, Serge Simon, Yves Tondeur,
Ivan Van Den Eynde, and Michel Van De Steen

18. The Structural Chemistry of Some Organotin Oxygen-Bonded
Compounds 258
Phililp G. Harrison

19. Chiral Pentacoordinate Triorganotin Halides 275
G. Van Koten and J. G. Noltes

Index 291

PREFACE

Organotin chemistry is among the very strongest areas in the inter-disciplinary organometallic field. Basic studies in organotin chemistry are stimulated by the success with which a large number of modern physical techniques can be applied to organotin compounds. Tin possesses, for example, two spin of one-half isotopes, tin-117 and tin-119, which become important in nuclear magnetic resonance, 10 stable isotopes (the largest of any element) which allow the easy identification of tin-bearing fragments in the mass spectrometer, one of the easiest to record Mössbauer resonances from the tin-119m nuclide, and easily assignable tin–carbon stretching frequencies in the infrared and Raman. Tin has available two stable oxidation states—tin(II) and tin(IV)—with contrasting chemistries, and a wide variety of structural types, which among the tin(IV) derivatives alone encompass four-, five-, six-, seven-, and eight-coordination at tin in neutral, cationic and anionic species, with intra- and intermolecular association to give dimers and higher oligomers with one-, two-, and three-dimensional lattices in the solid state. The synthesis of fluxional and pseudorotating organotin derivatives brings tin chemistry into the fourth dimension.

Equally important are the well-developed commercial applications of organotin chemicals. Organotins have become a leading commercial organometallic first through their use in poly(vinyl chloride) stabilization, and now as biocides, where their success is based upon their extremely high performance/unit weight ratio and their degradation by chemical action into non-toxic inorganic tin compounds. Today organotins are used widely as agricultural fungicides and miticides, industrial biocides and surface disinfectants, and as anthelminthics and marine antifouling agents. Industrial production of organotins now exceeds 25,000 tons per year in the free world with a selling price of over $150 million.

This volume brings together a group of American and European scientists working on different aspects of the organotin field. These include leading figures of organotin chemistry along with younger investigators working actively in this exciting area. Chemists from industrial and governmental as well as academic laboratories are included, and both basic and applied research is discussed. Although the organometallic aspects of tin chemistry are well represented in the contents, the biological effects of organotin compounds are also stressed. Because of the delays imposed by correspondence with overseas authors, some of the logical sequencing of chapters was sacrificed for speed in production.

My own contribution to the symposium program, entitled "Organic and Organometallic Derivatives of Tin(II)," was delivered in New York City. The talk described the efforts in our research group over several years in the synthesis and spectroscopic study of carbon-containing compounds of tin in its subvalent state. My intention of subsequently preparing a manuscript for inclusion in this volume, however, went awry in my relocation from the State University of New York at Albany to the University of Oklahoma to take up the chairmanship of the Department of Chemistry. Rather than see the volume delayed to include my late contribution, I shall instead content myself to bask in the reflected glory of its international team of authors.

Acknowledgments

The symposium on which this volume is based received generous financial support from the ACS Division of Inorganic Chemistry, as well as from the ACS Experimental Program Development Fund, the National Science Foundation, the Petroleum Research Fund Special Educational Opportunities Program, and the following commercial suppliers of organotin chemicals: Argus Chemical Corp., subsidiary of Witco Chemical Corp., Cincinnati Milacron Chemicals, Inc., and M & T Chemicals, Inc. This support is gratefully acknowledged.

University of Oklahoma   J. J. ZUCKERMAN
Norman, Okla. 73019
November 1976

Organotin Chemistry: Past, Present, and Future

G. J. M. van der KERK

Organic Chemistry Institute, TNO, Utrecht, The Netherlands

A review is given of the development and of the present scope or organotin applications. The striking versatility of these applications is emphasized, and it is concluded that the two main application fields of organotin compounds, viz. as PVC stabilizers and as agricultural and general biocides, are likely to expand considerably in the future. The simultaneous and impressive development of fundamental organotin chemistry has been illustrated by stressing one particular aspect, viz., organotin hydride chemistry.

A t Utrecht we have been engaged since 1950 in an explorative program of organotin research in close cooperation with the International Tin Research Council and the Tin Research Institute at London. From the beginning, this program aimed at a dual approach; the simultaneous development of fundamental organotin chemistry and a search for new applications of organotin compounds.

Tin as a metal—either as such or in the form of alloys—and in its chemical compounds, has an astonishing amount of usefulness. Characteristically, in the majority of its applications, only small amounts of tin are needed to see its effect. This is generally true for the organotin compounds which, during the past 25 years, have developed into extremely important industrial commodities. A further characteristic of tin is that it is unsurpassed by any other metal in the multiplicity of its organic applications. These involve such widely divergent fields as stabilizers for polyvinyl chloride, industrial catalysts, industrial and agricultural biocides, and wood-preserving and anti-fouling agents to mention only the most important applications. A third characteristic of tin, particularly true for the organotin compounds, is that in each of its present organic applications it has to compete with quite different materials, either old or new. However important the present-day organotin applications may be, it should be realized that so far no single large-size application has been found in which a particular type of organotin compound is absolutely unique as is, for instance, true for the lead tetraalkyls as anti-knock agents.   This implies that cost/performance relations are and will remain of paramount significance for the continuation of the present and the development of future organotin applications.

The final cost of using organotin compounds depends strongly on the price of tin, but other costs, among them manufacturing costs, are significant.   Since the industrial chemist has to accept the price of tin as it is, he has only two means at his disposal for making organotins competitive for any given application: (a) to find the most satisfactory process for manufacturing the required compound, and (b) to establish what particular compound gives optimal performance in any given application.

The first organotin compound was prepared by Frankland in 1849, and from that start organotin chemistry developed as a regular, though not very exciting, research subject for almost 100 years.   This is very evident from the relevant chapter in Krause and Von Grosse's monumental book of 1937 "Die Chemie der Metall-Organischen Verbindungen."

In retrospect, the patents granted in 1940 and 1943 to V. Ungve (12), describing the utility of certain dialkyltin derivatives as heat stabilizers for PVC, were a landmark, although their full industrial significance did not become apparent until 10 to 15 years later.   When we started our systematic organotin program at Utrecht in 1950, the annual industrial world production of organotin compounds was less than 50 tons.   In 1960 it was 2000 tons, in 1965, 5000 tons, and in 1969 it had risen to about 14,000 tons.   For 1975 a production of 25,000 tons, with a selling value of around $150 million, may be a conservative estimate.

Simultaneous with the enlargement and extension of organotin applications, new and improved methods have been developed for the industrial manufacture of the required types of organotin compounds.   The Utrecht work has contributed substantially to the present practical significance of quite a variety of organotin compounds.   In addition, our group has been actively engaged in further development of fundamental organotin chemistry.   This paper reviews, in a condensed form, the manufacturing processes and the established applications of organotin compounds.   A few cautious predictions are ventured as well.   In addition, I intend to recall a few important fundamental developments in organotin chemistry since 1950 and to deal with one particular aspect in some detail, viz. organotin hydride chemistry.   This choice is justified by two considerations. First, the field has been opened at Utrecht and was next extended both in Utrecht and elsewhere to one of the most fertile research areas in organotin chemistry. Secondly, one of the chapters in this book deals with the synthesis of novel substituted alkyltin halides in which the formation and reactivity of tin–hydrogen bonds play an essential part.   According to the authors, R. E. Hutton and V. Oakes (Chapter 8), the compounds obtained have considerable industrial potential as new types of PVC stabilizers.   I felt particularly challenged by this relationship between an originally purely fundamental approach and what may become a significant new applicational development.

A few preliminary remarks can be made regarding the toxicity problems connected with the applications of organotin compounds. In this respect I shall not deal with toxicology proper (3). The several basic types of organotin compounds (R_4Sn, R_3SnX, R_2SnX_2, and $RSnX_3$), as well as the inorganic forms of tin (SnX_2 and SnX_4) represent a toxicity pattern of great divergence. Under practical conditions the inorganic tin compounds are known to have very low to negligible toxicity. For organotin compounds, great differences exist between the basic types but also with regard to the nature of R. It is generally accepted that the basic types of organotin compounds are subject to the following generalized pattern of physical, chemical, and/or biochemical degradation

$$R_4Sn \rightarrow R_3SnX \rightarrow R_2SnX_2 \rightarrow RSnX_3 \rightarrow SnX_4 \text{ or } SnX_2, \text{ respectively}$$

which ultimately leads to inorganic forms of tin. Until recently, there was very little direct evidence for the actual course and rates of such degradation processes under environmental conditions. The overall toxicity picture for any compound is, however, dependent both on its own toxicity and on the toxicity of the degradation products formed under the conditions of its application. To fill this gap, a joint program was started some years ago at the Institute for Organic Chemistry TNO at Utrecht under the final responsibility of the Tin Research Institute. This program—the Organotin Environmental Project or ORTEP—is supported by about 10 major organotin-producing companies all over the world.

The Manufacture of Organotin Compounds

So far, the structural types R_3SnX and R_2SnX_2 comprise the largest part of practical organotin applications, those of the type $RSnX_3$ being much smaller. The compounds R_4Sn are scarcely used as such but are important industrial intermediates in the manufacture of the other types. In industrial manufacturing processes, X stands for chlorine which in the compounds used in practice is replaced by a variety of other substituents, but I shall not deal in detail with these structural variants. As far as practical applications are concerned, R mostly denotes simple, unsubstituted alkyl or aryl groups.

Four methods are in current use for the manufacture of the basic products, three of which—the old Grignard process or its Barbier version, the newer alkylaluminum process, and the much less important modified Würtz process—all lead to compounds R_4Sn:

$$SnCl_4 + 4RMgCl \rightarrow R_4Sn + 4MgCl_2$$

$$3SnCl_4 + 4R_3Al \rightarrow 3R_4Sn + 4AlCl_3$$

$$SnCl_4 + 4RCl + 8Na \rightarrow R_4Sn + 8NaCl$$

A second step, the disproportionation reaction with $SnCl_4$, is required to convert these into the practically important organotins with a lower degree of organic substitution:

$$R_4Sn \xrightarrow{SnCl_4} \left\{ \begin{array}{l} R_3SnCl \\ R_2SnCl_2 \\ RSnCl_3 \end{array} \right.$$

The so-called direct reaction allows the one-step preparation of compounds R_2SnX_2 but this process is restricted to compounds of that type and to R = alkyl:

$$Sn + 2RX \rightarrow R_2SnX_2$$

This, of course, represents an example of an oxidative addition reaction in which tin, in the oxidation state zero is transformed into an organotin(IV) species (presumably via the highly reactive oxidation state two). For phenyltin compounds, so far only the Grignard route is available. Both the Grignard and the alkylaluminum process are carried out on a large scale, a characteristic of the latter process being that it is only practical for primary manufacturers of trialkylaluminum compounds.

Because of the considerable recent growth of the dialkyltins for PVC stabilizers, there has been a strong industrial interest during the past five years in the direct reaction. For the higher alkyls (butyl and octyl) this reaction requires specific catalysts, e.g., quaternary ammonium or phosphonium salts or inorganic antimony compounds to lower the reaction temperature and thus to avoid extensive dehydrohalogenation. In Utrecht it was found that certain organoantimony compounds allow the conversion of metallic tin with butyl or octyl chloride into compounds R_2SnCl_2 with yields of over 90%. A more recent development is the introduction of dimethyltin compounds as PVC stabilizers. For the manufacture of dimethyltin dichloride, the direct reaction between tin and methyl chloride is the method of choice. The reason is, of course, that here much higher reaction temperatures can be used, and complex catalysts are not required since the dehydrohalogenation is much less important and the tin–carbon bond in methyltin compounds is very thermostable.

During the past years a certain interest has developed in monoalkyltin compounds as synergists for the usual dialkyltin PVC stabilizers. Although monoalkyltins can be made from R_4Sn by way of redistribution, tri- and dialkyltin chlorides are inevitable side products. In Utrecht a very convenient oxidative addition reaction has been developed for the manufacture of monoalkyltin trihalides in high yields:

$$SnCl_2 + RX \xrightarrow[\text{catalyst}]{} RSnCl_2X \qquad X = Cl, Br$$

$$R_3Sb \qquad\qquad R = alkyl \,(C_1-C_{18})$$

A highly intriguing synthetic development is reported in this volume by workers of Akzo Chemie U.K. Ltd.   At the same time, this development could mean an important extension of the range of available PVC stabilizers with the so-called estertins.   Clearly, it is suggested that conditions have been found to form the elusive chlorostannanes $HSnCl_3$ and H_2SnCl_2 and to bring these into reaction with activated olefins, e.g.:

$$H_2SnCl_2 + 2CH_2{=}CH{-}COOR \rightarrow (ROOCCH_2CH_2)_2SnCl_2$$

Although this type of compound has been prepared by Noltes about 20 years ago as follows:

$$(nC_3H_7)_2SnH_2 + 2CH_2{=}CHCOOCH_3 \rightarrow (nC_3H_7)_2Sn(CH_2CH_2COOCH_3)_2$$

$$(nC_3H_7)_2Sn(CH_2CH_2COOCH_3)_2 + Br_2 \rightarrow$$
$$2nC_3H_7Br + (CH_3OOCCH_2CH_2)_2SnBr_2$$

his method could, of course, never have developed into an industrially attractive process or product.

Applications of Organotin Compounds

At present the following applications—arranged according to diminishing commercial importance—are well established:

(a) Stabilization of PVC by dialkyltin and, to a much lesser extent, by mono-alkyltin compounds.

(b) The use of tributyltin compounds, in particular tributyltin oxide TBTO, as industrial biocides in materials protection and as surface disinfectants.

(c) The use of triphenyltin compounds and of tricyclohexyl- and trisneo-phyltin compounds as agricultural fungicides and as agricultural acaricides, respectively.

(d) A number of smaller and quite diverse applications in which all types of organotin compounds are involved.

These application fields will be briefly reviewed in this sequence.

Use of Organotin Compounds as PVC Stabilizers.   During the past 15 years the growth of PVC manufacture has been phenomenal: in 1962 the annual world production was 0.5 million ton, and in 1972 this figure had increased to 7.5 million tons.   In addition, there has been a shift from flexible to rigid PVC products, the latter requiring much higher processing temperatures.   There is a consensus that dialkyltin compounds are the best general-purpose stabilizers for rigid PVC especially if colorlessness and transparency are required.   Growth of organotins has been particularly strong in PVC for packaging films and bottles and for roofing.

Early development, based on the original work of Yngve, was in dibutyltin compounds.   Tremendous efforts were made in improving the overall performance through variation of the groups bound to the dibutyltin moiety.   Via the dilaurate, the maleate and several other variations, the present peak was reached in certain mercapto derivatives, particularly those derived from the octyl and isooctyl esters of thioglycolic acid.

In 1957 Luijten and Pezarro (4), on the basis of results of Kaars Sijpesteijn regarding the biocidal activity of tri- and dialkyltin compounds, proposed di-n-octyltin compounds as PVC stabilizers with anticipated mammalian nontoxicity. The dioctylins were found to be excellent stabilizers and their nontoxicity was proved by Barnes in Great Britain and by Klimmer in Germany. During the past 10 years there has been a selective growth of the dioctyltins, particularly in PVC for food-packaging purposes.

More recently another class of dialkyltin compounds, the dimethyltins, has made its appearance on the stabilizer market. Because of their extreme thermostability the dimethyltin stabilizers allow the use of high working temperatures and concomitant high working speeds which is of particular advantage in the manufacture of PVC pipings, bottles, and films. On the basis of the simple direct manufacturing process discussed above and of reported excellent performance, it is claimed that the application of dimethyltin stabilizers has economic advantages. If this is true and if the toxicological and environmental properties of the dimethyltins are acceptable, it would seem that these compounds are a valuable extension of the earlier PVC stabilizers.

An important recent development is based on the observation that the addition of small amounts (5–10%) of monoalkyltin derivatives to the usual dialkyltin formulations has a synergistic effect on stabilizing effectiveness. A particular effect of the monoalkyltins, though generally far less effective stabilizers than the dialkyltins, seems to be the prevention of so-called early yellowing. Such synergistic mixtures thus allow the manufacture of PVC articles of perfect colorlessness and clarity. The newest stars in the stabilizers sky are the estertins, a subject covered in Chapter 9 by E. L. Weinberg.

The Use of Tributyltin Compounds as Biocides. The use of tributyltin compounds, particularly bis(tributyltin)oxide, as biocides is presently a widespread and strongly expanding field of applications which stems entirely from early work at Utrecht by Luijten and Kaars Sijpesteijn. The original work, started in 1950, pointed primarily to the high antifungal and antibacterial activities of certain triorganotin compounds, particularly tributyl- and triphenyltin derivatives, but the much wider implications were immediately recognized (5, 6). The present broad applications of TBTO as an industrial biocide, in materials protection, as an antifouling agent, and as a surface disinfectant will be reviewed briefly. Some other types of biocidal triorganotin compounds are finding their main outlet in agricultural applications and will be discussed in the next section.

The main uses of TBTO now comprise wood preservation, antifouling, and the disinfection of circulating industrial cooling water. Until recently, there was an annual world use of around 12,000 tons of mercury in the form of organomercurials for a multitude of biocidal applications. These applications are now under hard scrutiny and will have disappeared completely within a few years.

Except for the agricultural uses, several other applications of organomercurials may be partly taken over by TBTO in the near future.

Among the first publications on the preservation of wood against fungal attack by means of triorganotin compounds were those of Hof and Luijten (TNO) (7) and of Fahlstrom (8). Since then, wood preservation using TBTO as or among the active ingredients has become common practice. Tributyltin compounds are characterized by their high activity and broad antifungal spectrum. Their leachability by water is extremely low and they have the advantage of being colorless and non-corrosive. Amounts of 0.5–2 kg of TBTO per m^3 of wood are quite effective not only against fungal decay but against the attack by marine borers as well: shipworms (*Teredo*) and gribble (*Limnoria*). Much higher concentrations are required to protect wood against wood-boring insects such as the common furniture beetle and particularly termites, and here combinations with other active ingredients like insecticides are required. An excellent review on TBTO-based wood preservatives has been given by Richardson at the 1970 Annual Convention of the British Wood Preservers Association (9).

The low aqueous leachability of TBTO is attributed to its high affinity for cellulose in particular. As a consequence, however, its lack of penetration into deeper layers of the treated wood poses some problems. To a certain extent, these have been solved by special impregnation techniques and also by combining organotins with other biocidal agents that have better penetrating properties. During the past few years our group at Utrecht, in cooperation with the Wood Research Institute TNO at Delft, has developed a different approach. The hydrocarbon-like compound hexabutylditin($Bu_3SnSnBu_3$) which is very soluble in non-polar hydrocarbon solvents was found to have much better wood-penetrating properties than TBTO and an equal wood-preserving capacity. We believe that hexabutylditin offers considerable promise as a new wood-preserving agent provided that a satisfactory method can be developed for its technical manufacture.

Triorganotin compounds, in particular TBTO and tributyltin fluoride, are finding increased use in marine antifouling paints. An important factor is the leaching rate of the active ingredient which must be low enough to give long-range protection and yet allow the active agent to be released in sufficient concentration. An interesting development realized by Cardarelli (B. F. Goodrich) (cf. 10, 11) is the incorporation of TBTO in elastomeric coatings based on natural or nitrile rubber. The rationale behind this development is that such coatings may be made much thicker than coatings of antifouling paints. The reservoir of toxicant is thus enlarged, and a much longer fouling-free lifetime is reached. A prerequisite is, of course, a sufficient rate of migration of the toxicant within the elastomeric coating. This rate is very dependent on the nature of the anionic substituent in the compounds Bu_3SnX, TBTO giving the best performance. A growing application is the use of TBTO for slime control in paper mills which until recently was dominated by the organomercurials. The amount of TBTO added to the mill water may be very small since only incomplete inhibition of microbial growth needs to be achieved. Higher concentrations of TBTO are being applied in recirculating cooling water or in water for the secondary recovery of oil.

Another interesting application where organotins may replace organo-mercurials is paint preservation, although a few complications have to be resolved. For instance, the antimicrobial spectrum of triorganotins, though considerable, is not so wide as that of the organomercurials. To reach an equivalent degree of protection, new formulations have to be developed which, in certain cases, must contain other active ingredients as well. One notable deficiency of TBTO is its modest activity against gram-negative bacteria. It has been found in Utrecht that tripropyltin compounds have a wider antibacterial spectrum and are rather active against gram-negative bacteria as well.

Potentially very important is the use of TBTO and of triphenyltin compounds in fighting the vector of the wide-spread and very debilitating tropical Bilharzia disease. These compounds are highly active against the snails which serve as intermediate hosts for the parasitic worms causing the disease. In this application, vast volumes of surface water must be supplied with continuous but extremely low concentrations of the triorganotin compounds (far below 1 ppm). Here again a breakthrough may have been achieved by incorporating the organotin compounds into small rubber pellets which secrete the active compound continuously in the very low concentrations required (Cardarelli, B. F. Goodrich) (12). In this way it might be possible to cope with the high fish toxicity of these organotins.

A modest but important use of certain tributyltin-containing formulations is in hospital and veterinary disinfectants. Similar formulations are applied to protect textiles against fungal and bacterial attack, both in the industrial and the hygienic sector (sanitizing). In reconsidering the biocidal properties of TBTO, one cannot get away from the conclusion that its biocidal applications are likely to expand strongly in the future both as a result of extending present uses and of developing new ones. Much will depend here on the outcome of the work within ORTEP concerning the metabolic fate of organotin compounds under environmental conditions.

The Agricultural Applications of Organotin Compounds. Our original observations regarding the very high antifungal activity of the lower trialkyltin and the triphenyltin compounds raised the expectation that these compounds might be generally useful protectant agricultural fungicides. Because of their broad antifungal spectrum it was anticipated that they would be suitable to combat a variety of fungal plant diseases. This expectation has not been realized.

Independent work of Härtel (Farbwerke Hoechst, Germany) (13, 14) has shown that in the laboratory, trialkyl- and particularly tributyltins are better fungicides than triaryltin compounds but that the reverse is true in the field. This has been ascribed to the lower stability and the higher volatility of the former. Moreover, triaryltin compounds are less phytotoxic than trialkyltin compounds.

The final result has been that now certain triphenyltin formulations—containing either triphenyltin hydroxide or acetate—have become important agricultural fungicides. Their importance is not connected with their general usefulness but with their specific effectiveness against two economically important plant diseases viz. late blight of potatoes caused by *Phytophthora infectans* and leaf spot in sugar beets, caused by *Cercospora beticola*. In these applications they have almost completely ousted the formerly dominating inorganic copper compounds. Later on it was found that also a number of important tropical plant diseases, viz. those in coffee, rice, ground nuts, banana and pecan can be controlled successfully.

A further extension of the triphenyltin compounds as agricultural fungicides was found in their combination with manganese ethylenebisdithiocarbamate (Maneb). A particular advantage of triphenyltin formulations is that so far no development of field resistance has been observed. The problem of toxic residues from field sprays with triphenyltin compounds has been very thoroughly investigated. An important feature is the relatively short half life of these compounds on the foliage under field conditions (3–4 days). Moreover, the compounds do not penetrate into the plant and their action is thus purely protective.

Another agricultural development of great potential interest is based on the more recent observation that certain rather unusual triorganotin compounds have considerable acaricidal activity. Well known at present is the compound tricyclohexyltin hydroxide developed by Dow Chemical Co. and M & T Chemicals. This compound is very effective against spider mites in fruit orchards and has been found to act as well against varieties of spider mites which had developed resistance toward the usual acaricides based on organic phosphorus compounds and carbamates. Another promising compound with a similar application field is marketed by Shell. It is the tris(neophyl)tin oxide:

$$\left[\left(\text{\large\textcircled{}}-\overset{\displaystyle CH_3}{\underset{\displaystyle CH_3}{\overset{|}{\underset{|}{C}}}}-CH_2\right)_3 SnO\right]$$

In summary, it may be said that the agricultural applications of organotin compounds so far are restricted to a comparatively small, though very important, number of plant diseases and pests. Moreover, on the basis of the more recent developments and because tin compounds in several cases are fully active against resistant varieties, it may be expected that a further modest growth of the uses of organotins in agriculture is likely.

Miscellaneous Applications. In addition to the main fields of application discussed so far, there are a number of quite divergent uses of organotin compounds. Although the total tonnage involved may be modest, these uses illustrate in an impressive way the versatility and the potential of organotin applications. I shall briefly discuss these uses on the basis of the types of organtin compounds involved, in the sequence R_4Sn, R_3SnX, R_2SnX_2, $RSnX_3$.

R_4SN.   Contrary to what is true for lead, the tin compounds of this type so far have attained only very modest commercial importance.   The old application of tetraphenyltin as a heat stabilizer for polychlorinated aromatics used as transformer oils is no longer of practical importance.   An interesting use, both of tetraalkyltins and of tetraphenyltin, is still the one indicated in 1958 by Carrick (15).   These compounds in combination with aluminum chloride and certain transition metal chlorides (e.g., VCl_4 or $TiCl_4$) yield extremely active catalytic systems for the polymerization of ethylene and other alkenic compounds.   The tin compounds act in situ as alkylating agents for aluminum chloride, and in using such systems the separate handling of the extremely susceptible and hazardous organoaluminum compounds is avoided.   It should be realized that the interaction of compounds R_4Sn with aluminum chloride results in the formation of partially alkylated aluminum compounds $RAlCl_2$ and R_2AlCl, never in the fully alkylated species R_3Al.   As a consequence, in the catalytic systems under consideration, transition metals should be present which exert their maximum catalytic activity in combination with incompletely alkylated aluminum species.

R_3SNX.   There is not much to add to what has already been said about this type of compound which so far has only found biocidal applications, but this is, of course, a field widely open for extension and expansion.

R_2SNX_2.   This class of compounds, in addition to their main use as PVC stabilizers, exhibits an astonishing variety of quite different applications which are not very large but nevertheless significant.   An old use of dibutyltin dilaurate, in particular in the USA, is the treatment of chickens as a cure against intestinal worm infections.   One application of about 100 mg is sufficient.   Although it seems that this use is now declining, around 1960 it involved an annual consumption of about 150 tons of dibutyltin dilaurate.   The compounds R_2SnX_2 find extensive use as catalysts and as components of catalytic systems.   Most important is the application of certain dibutyltin compounds as catalysts in the manufacture of flexible polyurethane foams and in the cold-cure of silicone rubbers.   Other potential applications under this heading are the uses as esterification catalysts and as catalysts in the curing of epoxy resins.

An interesting application is the use of dimethyltin and dibutyltin dichloride in the surface treatment of glass.   The compounds are vaporized, and the vapor is brought into contact with the hot glass surface to form a thin and transparent layer of SnO_2.   Very thin layers considerably improve the resistance of the glass surface against abrasion and chemical corrosion.   Somewhat thicker layers give surface heat-reflectance properties.   Still thicker layers are electrically conductive, and the treated glass can be used for window heating.   Very thick layers give decorative iridescence.

$RSNX_3$.   Outside of PVC stabilization, no practical applications have been realized for this class of compound.   Many years ago we suggested, on the basis of very encouraging experiments, their use as waterproofing agents for paper, textiles, and, remarkably, for brick walls (*16*), but so far no practical applications have evolved in this respect.   Interesting and rather surprising is the claim, made in a recent patent application by Albright and Wilson (*17*), that monobutyl- and monooctyltin compounds are effective wood preserving agents notwithstanding their negligible in vitro antifungal activity.   It is suggested that these compounds effectively block places within the wood which are vulnerable to fungal attack. Now that a very convenient synthesis for this type of compound is available, allowing a wide variation of the alkyl group, it would seem that a closer study of their possibilities is justified.

Conclusion

I have attempted to impress on the reader a few applicational aspects of organotin chemistry which are somewhat alien to the average academic worker in the field of organometallic research.   In Utrecht we have tried to balance the fundamental and the applied aspects; in fact, we have always aimed at integrating these into one approach.   The second part of my paper deals with a few fundamental developments in organotin chemistry since 1950.

Some Aspects of Fundamental Organotin Chemistry Since 1950

Since about 1940 and especially since 1950 an enormous activity has been going on in the field of organometallic and coordination chemistry.   For tin, this revival of interest started in 1950 and today organotin chemistry is one of the most intensively explored areas of main-group organometallic chemistry.   In the foregoing I have emphasized the multiplicity of the practical applications of organotin compounds.   To do justice to the variety and importance of fundamental organotin developments is virtually impossible.   Many of these developments will be coverd in depth in this volume and, as a consequence, there is little reason for me to review the whole field in a shallow way.   Rather, I prefer to select just one subject that follows.

The Versatility of Organotin Hydride Chemistry.   The selection of this subject as a representative research theme in organotin chemistry involves some pride.   The extremely fruitful field of organotin hydride chemistry was started at Utrecht in 1955 by J. G. Noltes.   Although the several types of organotin hydrides and even stannane, SnH_4, had been described early in this century, a really suitable synthesis did not become available before 1947 when Schlesinger et al. (*18*) introduced lithium aluminum hydride as an agent for reducing organotin halides to the corresponding hydrides.   In his exploration of the synthetic possibilities of these compounds, Noltes discovered the following types of reactions (cf. *16*):

(a) The addition of tin–hydrogen bonds to carbon–carbon double and triple bonds

$$R_3SnH + CH_2\!=\!CH\!-\!R' \to R_3Sn\!-\!CH_2\!-\!CH_2\!-\!R'$$

$$R_3SnH + CH\!\equiv\!C\!-\!R' \to R_3Sn\!-\!CH\!=\!CH\!-\!R'$$

This type of reaction, which was later extended to include unsaturated carbon–hetero atom and hetero–hetero atom bonds, is now generally known as hydro-stannation.

(b) The reductive cleavage of carbon–halogen bonds

$$R_3SnH + R'X \xrightarrow[\text{X = halogen}]{} R_3SnX + R'H$$

Also this reaction was found later on to be much more general:

$$-\overset{|}{\underset{|}{Sn}}-H + A\!-\!B \longrightarrow -\overset{|}{\underset{|}{Sn}}-A + HB$$

A–B may a.o. represent a metal–element bond.   Reactions of this type are hy-drostannolysis.

(c) The catalytic decomposition of tin–hydrogen bonds by amines with the formation of tin–tin bonds.

$$2R_3SnH \xrightarrow{\text{amines}} R_3Sn\!-\!SnR_3 + H_2$$

These basic observations caused widespread interest and activity in organotin hydride chemistry until the present day when many groups are engaged in this subject.   In this section I deal in particular with some aspects of hydrostannation and hydrostannolysis reactions.   In the hydrostannation reactions a free-radical mechanism predominates, but ionic mechanisms have been established as well. For the hydrostannation of non-activated or weakly activated carbon–carbon double bonds, Neumann et al. (cf. *19*) and Kuivila et al. (cf. *20*) arrived at the following free-radical chain mechanism:

$$R_3SnH + R''. \longrightarrow R_3Sn. + R''H$$

β adduct

The almost exclusive formation of terminal β adducts depends on the relative stabilities of the free radical intermediates I and II.

In Utrecht, Leusink (*21, 22*) proved that the hydrostannation of activated carbon–carbon double bonds follows a different course, e.g.:

$$R_3SnH \; + \; CH_2{=}CH{-}CN \; \longrightarrow \; \underset{\underset{\alpha\ adduct}{\underset{|}{SnR_3}}}{CH_3{-}CH{-}CN} \; + \; \underset{\beta\ adduct}{R_3SnCH_2CH_2{-}CN}$$

$$\text{acrylonitrile}$$

He demonstrated that, as in the former case, the β adduct is formed by a free radical mechanism but that the formation of the α adduct involves an ionic mechanism. The rate of formation of the α adduct, but not of the β adduct, increases strongly with increasing polarity of the solvent; the presence of radical initiators selectively promotes the formation of the β adduct. The rate of the ionic hydrostannation is also strongly dependent on the nature of R with both electronic and steric factors playing a part. Altogether, the observed facts are in accord with a nucleophilic attack of the organotin hydride hydrogen on the β carbon atom as the rate-determining step in the ionic reaction:

$$R_3SnH \; + \; H_2C{=}CH{-}CN \xrightarrow{slow} \left[\begin{array}{c} H_2C{=}CH{-}CN \\ H{-}\!\!\diagdown \\ SnR_3 \end{array} \right] \longrightarrow$$

$$\longrightarrow \; R_3Sn^+ \; + \; H_3C{-}\overset{-}{CH}{-}CN \xrightarrow{fast} \underset{\underset{\alpha\ adduct}{\underset{|}{SnR_3}}}{H_3C{-}CH{-}CN}$$

Similar results were obtained in the study of the more complicated hydrostannations of substituted acetylenes (*23, 24*), but these will not be discussed here.

The studies of Leusink (*23, 24*) have shown that hydrostannation may follow exclusively an ionic mechanism (e.g., with iso(thio)cyanates or with strongly electrophilic acetylenes), a combination of an ionic and a radical mechanism (e.g., with strongly electrophilic alkenes and electrophilic acetylenes), or exclusively a radical mechanism (e.g., with nucleophilic acetylenes and nucleophilic or weakly electrophilic alkenes).

In contrast, studies at Utrecht of Creemers (*25, 26, 27*) showed that with very few exceptions hydrostannolysis reactions follow an ionic course, in which the organotin hydride hydrogen acts as an electrophile, e.g.:

$$R_3SnH + R'_3Sn-N\!\!\!\begin{array}{c}\diagup\\ \diagdown\end{array} \xrightarrow[\text{slow}]{} \left[\begin{array}{c} \overset{\delta^+}{R'_3Sn}-\overset{\delta^-}{N}\!\!\!\begin{array}{c}\diagup\\ \diagdown\end{array} \\ \overset{\delta^+}{\underset{|}{H}} \\ \underset{\delta^-}{SnR_3} \end{array}\right] \longrightarrow$$

transition state

$$\longrightarrow R'_3Sn-\overset{+}{\underset{H}{N}}\!\!\!\begin{array}{c}\diagup\\ \diagdown\end{array} + R_3\overset{-}{Sn} \xrightarrow[\text{fast}]{} R'_3Sn.SnR_3 + HN\!\!\!\begin{array}{c}\diagup\\ \diagdown\end{array}$$

In a somewhat simplified form the transition states in the ionic reactions of tin–hydrogen bonds can be pictured as follows:

$$\begin{array}{c} R\diagdown \\ R-Sn---H----E \\ R\diagup \end{array} \qquad\qquad \begin{array}{c} R\diagdown \\ R-Sn---H---N \\ R\diagup \end{array}$$

E = an electrophile
N = a nucleophile

In the ionic reactions of organotin hydrides the importance of these transition states is not only determined by the nature of E and N, but by the electron-donating or -attracting properties of the substituents R at the tin atom as well. It could be demonstrated that in these reactions hydrogen transfer is the rate-determining step. Both hydrostannation and hydrostannolysis reactions have led to interesting developments, a few of which will be discussed.

Some Reactions Involving Hydrostannation. Hydrostannation has allowed the preparation of a great variety of functionally-substituted organotin compounds in which the functional groups are placed either in α position or in β position or beyond with regard to the tin atom. Starting from mono-, di-, or trihydrides, mono, di, or trifunctionally-substituted organotin compounds were obtained. The scope of these reactions could be widened considerably since the functional groups placed in β position or beyond proved accessible to a variety of nucleophilic secondary reactions without cleavage of the tin–carbon bonds. A few examples may illustrate these possibilities:

$$R_3SnH \begin{cases} \xrightarrow{CH_2=CH_2-CN} R_3SnCH_2CH_2CN \begin{cases} \xrightarrow{LiAlH_4} R_3SnCH_2CH_2CH_2NH_2 \\ \xrightarrow[H_2O]{C_6H_5MgBr} R_3SnCH_2CH_2-\underset{\underset{O}{\|}}{C}-C_6H_5 \end{cases} \\ \xrightarrow{CH_2=CH-COOCH_3} R_3SnCH_2CH_2COOCH_3 \begin{cases} \xrightarrow{LiAlH_4} R_3SnCH_2CH_2CH_2OH \\ \xrightarrow[H_2O]{CH_3MgI} R_3SnCH_2CH_2-\underset{\underset{CH_3}{|}}{\overset{\overset{CH_3}{|}}{C}}-OH \end{cases} \end{cases}$$

The 1:1 reactions of organotin dihydrides with bifunctionally-unsaturated organic and organometallic molecules provided a synthesis of a variety of organotin polymers and tin-containing ring compounds. In particular the latter are of interest.

The 1:1 reaction of diphenyltin dihydride with divinylorganometallics afforded the novel 1-stanna-4-sila- and 1-stanna-4-germanacyclohexane ring systems (*28*):

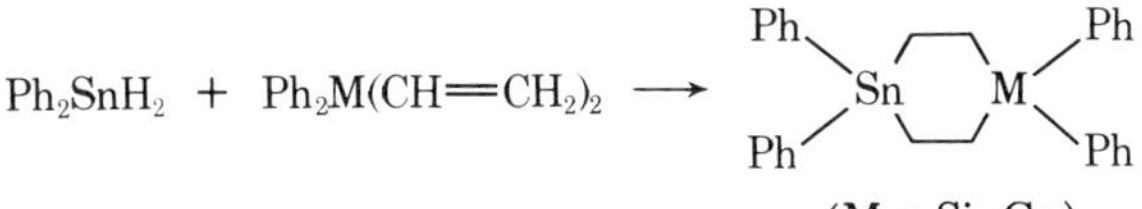

$$Ph_2SnH_2 \ + \ Ph_2M(CH{=}CH_2)_2 \ \longrightarrow$$

(M = Si, Ge)

From diphenyltin dihydride and phenylacetylene a 1,3,5-tristannacyclohexane derivative was obtained (*29*).

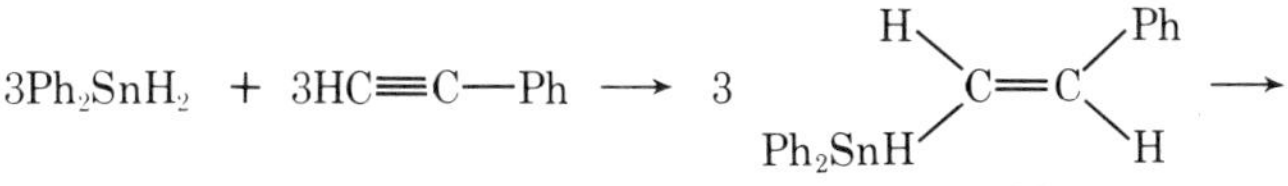

$$3Ph_2SnH_2 \ + \ 3HC{\equiv}C{-}Ph \ \longrightarrow \ 3$$

trans adduct

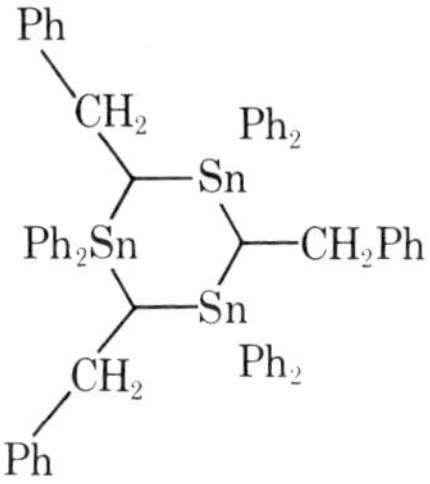

a 1,3,5-tristannacyclohexane

not (as might be expected):

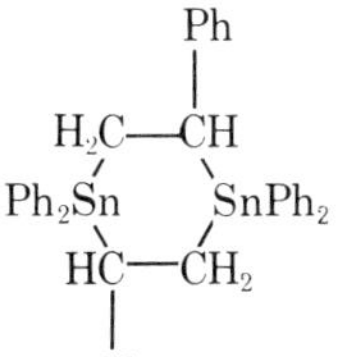

a 1,4-distannacyclohexane

Particularly interesting ring compounds were formed in the reactions of dialkyl- and diaryltin dihydrides with organic dienes and diynes (*30*).

RING COMPOUNDS FROM o-DIVINYLBENZENE AND FROM o-DIETHYNYLBENZENE

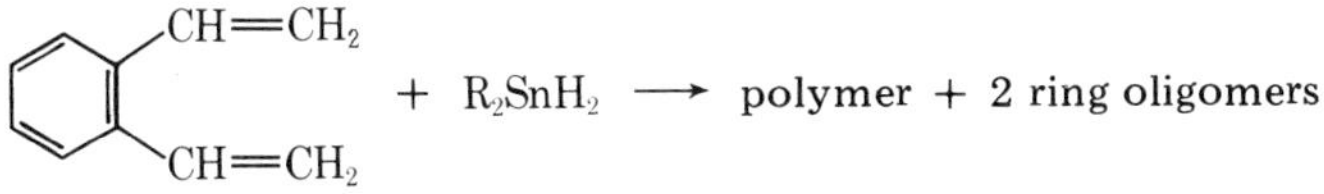

$$+ \ R_2SnH_2 \ \longrightarrow \ polymer \ + \ 2 \ ring \ oligomers$$

o-divinylbenzene

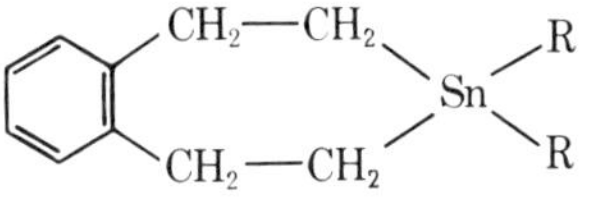

1:1 adduct
seven-membered ring
a benzotetrahydrostannepin

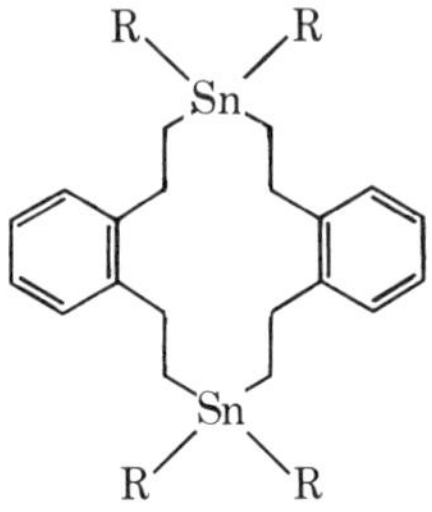

2:2 adduct
14-membered ring

o-diethynylbenzene

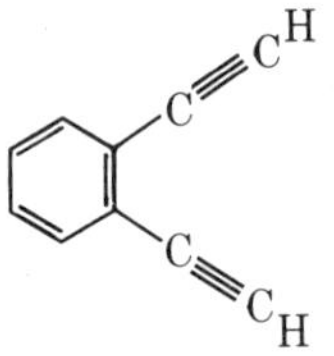

$+ R_2SnH_2 \longrightarrow$ polymer $+$ 2 ring oligomers

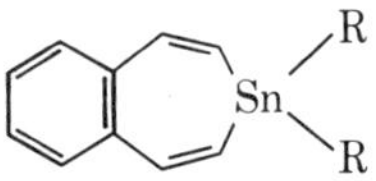

1:1 adduct
seven-membered ring
a benzostannepin

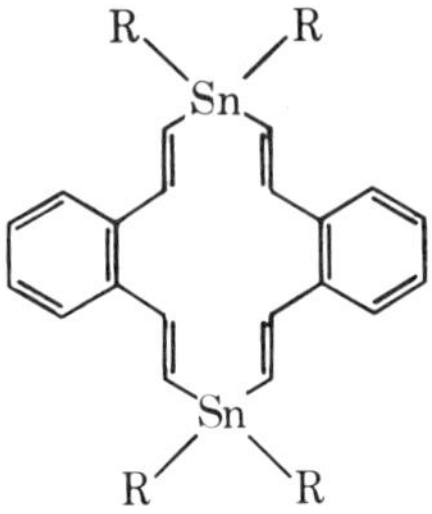

2:2 adduct
14-membered ring

A fascinating development, also from a theoretical viewpoint, was the conversion of 3,3-dimethyl-3-benzostannepin into a cyclic boron compound, 3-phenyl-3-benzoborepin (*31*):

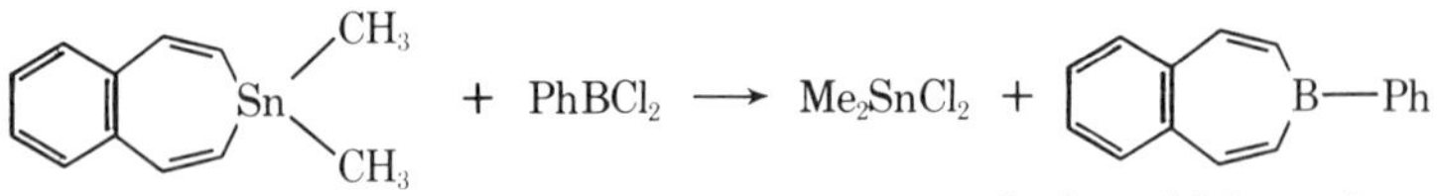

$+ PhBCl_2 \longrightarrow Me_2SnCl_2 +$

3,3-dimethyl-3-benzostannepin

3-phenyl-3-benzoborepin

This compound contains, in a seven-membered ring, three conjugated double bonds and an electron sink. It thus obeys Hückel's closed shell $(4n + 2)$ rule for aromaticity, being isoelectronic with the benzotropylium cation. In fact, ^{1}H NMR and UV spectroscopic measurements clearly indicated aromatic character for this novel ring system which more recently could be confirmed by theoretical calculations using the simple Hückel–MO method (*32*). This aromatic character followed from the UV spectra connected with the following reaction equilibrium (Figure 1):

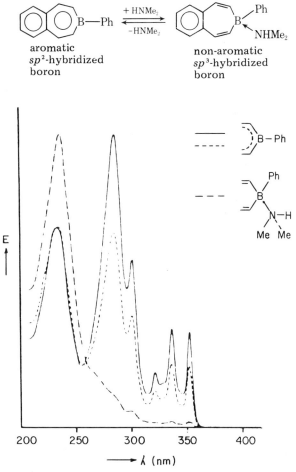

*Figure 1.   UV spectra of a solution of 3-phenyl-3-ben-
zoborepin in cyclohexane (solid line and dotted line);
excess dimethylamine (dashed line).*

The solid line represents the spectrum of a solution of 3-phenyl-3-benzoborepin in cyclohexane.   The addition of an excess of dimethylamine causes a profound change (the striped line).   All specific absorptions beyond 250 nm disappear, and the spectrum becomes much like that of 3,3-dimethyl-3-benzostannepin with absorptions typical for a vinyl–metal compound.   Upon evaporating the solution in vacuo and redissolving the solid residue in cyclohexane, the resulting solution yields a UV spectrum which is qualitatively identical with that of the uncoordinated starting material (the dotted line).

Although it was possible to prepare tin compounds of the type

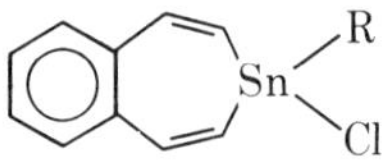

no spectral characteristics were found which would indicate aromatization in the sense

It must be assumed that sp^2 hybridization of tin, required for aromaticity, does not occur in this type of compound but that sp^3—or even sp^3d—hybridization is preferred.

RING COMPOUNDS FROM ALIPHATIC DIYNES. Tin-containing unsaturated rings were obtained when organotin dihydrides were allowed to react with diynes of the type $HC{\equiv}C{-}(CH_2)_n{-}C{\equiv}CH$ (n = 1 and 2). For n = 2 a seven-membered ring was formed (33):

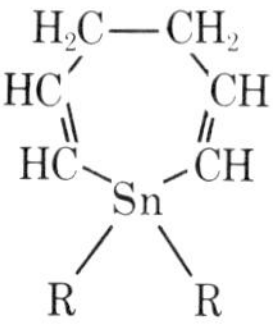

In relation to the last subject in the foregoing section, a most interesting product was obtained by Ashe and Shu from dibutyltin dihydride and 1,4-pentadiyne (34):

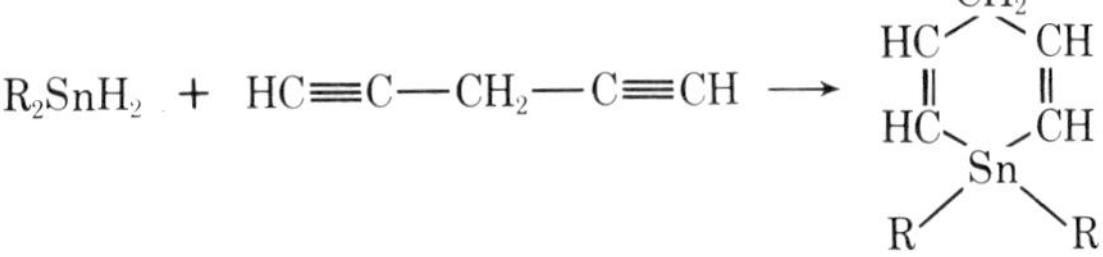

a stannacyclohexadiene

The treatment of this compound with phenylborium dibromide yielded the corresponding boron derivative:

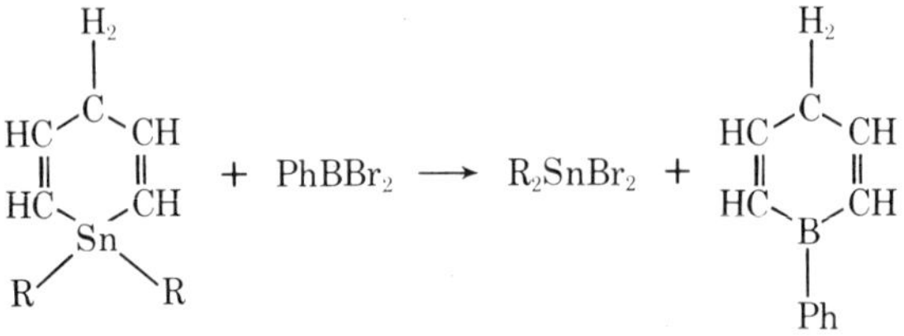

1-phenyl-1,4-dihydroborabenzene

Deprotonation of this cyclic borium compound with *tert*-butyllithium yielded
the aromatic 1-phenylborabenzene anion:

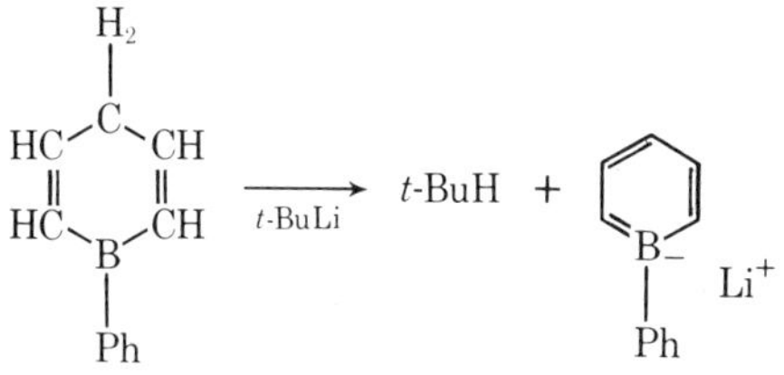

The conversion of this anion into highly interesting transitional metal sandwich
compounds will not be discussed.

Quite recently in Utrecht the following reaction sequence was performed
(32). Hexa-1,5-diyne-3-ene, consisting of a mixture of the cis and trans forms,
was allowed to react with diethyltin dihydride and the reaction products with
methylborium dibromide. The volatile, low molecular weight final products
were frozen over in a high vacuum. Spectroscopic evidence was obtained for
the presence of the two compounds, 1-methylborepin and a 1-methylborir-
ane:

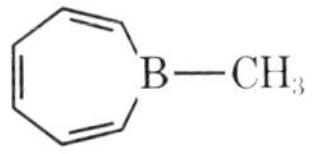

1-methylborepin	1-methylborirane
aromatic	not aromatic

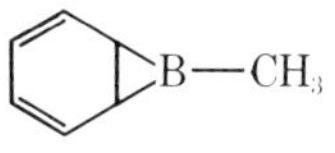

So far, the amounts formed were too small for rigid chemical analysis.

Some Reactions Involving Hydrostannolysis. The hydrostannolysis of
carbon–halogen bonds has become of considerable synthetic value in preparative
organic chemistry, in particular for the selective mono-dehalogenation of geminal
dihalides >CCl_2, e.g.:

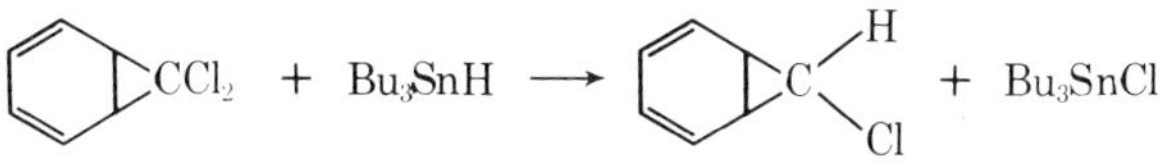

cis- and trans

Since organotin hydrides are not easily available reagents everywhere, it was an
important improvement of the original procedure of Noltes et al. (35) when it
was found that such reactions can succeed with catalytic amounts of a triorganotin
halide and a stoichiometric amount of the generally available reagents $LiAlH_4$
(36) and $NaBH_4$ (37).

Hydrostannation has become of particular importance in organometallic
chemistry for establishing tin–metal bonds (Scheme 1). For tin, the combination
of hydrostannation, hydrostannolysis, and catalytic decomposition of the Sn–H

bond allowed the synthesis of catenated polystannanes (38, 39). Scheme 1, which is given without further explanation, illustrates the versatility of this combination.

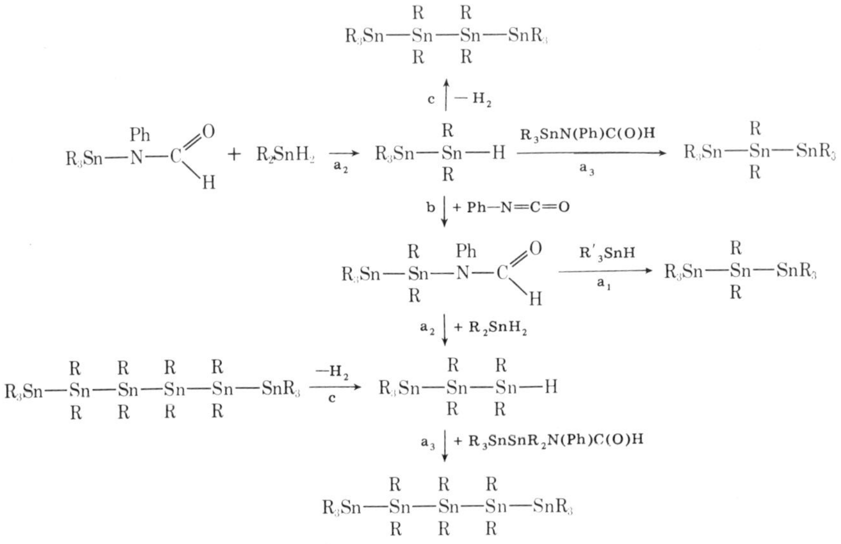

$$R_3Sn\!-\!\underset{R}{\overset{R}{Sn}}\!-\!\underset{R}{\overset{R}{Sn}}\!-\!SnR_3$$

$$c \uparrow -H_2$$

$$R_3Sn\!-\!N\!-\!\underset{H}{\overset{Ph}{C}}\!\!\!\overset{O}{\diagup} + R_2SnH_2 \xrightarrow{a_2} R_3Sn\!-\!\underset{R}{\overset{R}{Sn}}\!-\!H \xrightarrow[a_3]{R_3SnN(Ph)C(O)H} R_3Sn\!-\!\underset{R}{\overset{R}{Sn}}\!-\!SnR_3$$

$$b \downarrow + Ph\!-\!N\!=\!C\!=\!O$$

$$R_3Sn\!-\!\underset{R}{\overset{R}{Sn}}\!-\!\underset{\diagdown H}{\overset{Ph}{N}}\!-\!C\!\!\!\overset{O}{\diagup} \xrightarrow{R'_3SnH}_{a_1} R_3Sn\!-\!\underset{R}{\overset{R}{Sn}}\!-\!SnR_3$$

$$a_2 \downarrow + R_2SnH_2$$

$$R_3Sn\!-\!\underset{R}{\overset{R}{Sn}}\!-\!\underset{R}{\overset{R}{Sn}}\!-\!\underset{R}{\overset{R}{Sn}}\!-\!\underset{R}{\overset{R}{Sn}}\!-\!SnR_3 \xrightarrow[c]{-H_2} R_3Sn\!-\!\underset{R}{\overset{R}{Sn}}\!-\!\underset{R}{\overset{R}{Sn}}\!-\!H$$

$$a_3 \downarrow + R_3SnSnR_2N(Ph)C(O)H$$

$$R_3Sn\!-\!\underset{R}{\overset{R}{Sn}}\!-\!\underset{R}{\overset{R}{Sn}}\!-\!\underset{R}{\overset{R}{Sn}}\!-\!SnR_3$$

Scheme 1. a_1, a_2, and a_3 denote hydrostannolysis steps, and b and c denote hydrostannation and catalytic decomposition, respectively.

Hydrostannation offers a simple method as well for making compounds containing tin–transition metal bonds, e.g.:

$$Ti(NMe_2)_4 + 4Ph_3SnH \longrightarrow (Ph_3Sn)_4Ti + 4HNMe_2$$

$$Ti(NMe_2)_4 + 3Ph_3SnH \longrightarrow (Ph_3Sn)_3TiNMe_2 + 3HNMe_2$$

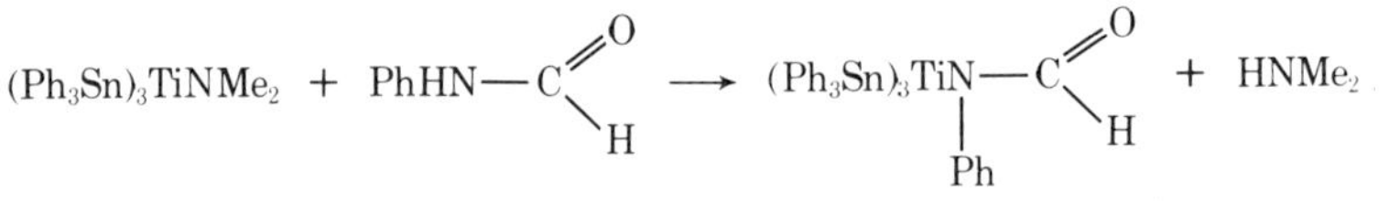

$$(Ph_3Sn)_3TiNMe_2 + PhHN\!-\!C\!\!\!\overset{O}{\underset{\diagdown H}{\diagup}} \longrightarrow (Ph_3Sn)_3TiN\!-\!\underset{Ph}{\overset{|}{C}}\!\!\!\overset{O}{\underset{\diagdown H}{\diagup}} + HNMe_2$$

formanilide

$$2(Ph_3Sn)_3TiN\!-\!\underset{Ph}{\overset{|}{C}}\!\!\!\overset{O}{\underset{\diagdown H}{\diagup}} + Ph_2SnH_2 \longrightarrow (Ph_3Sn)_3Ti\!-\!\underset{Ph}{\overset{\overset{Ph}{|}}{Sn}}\!-\!Ti(SnPh_3)_3 +$$

$$2PhHN\!-\!C\!\!\!\overset{O}{\underset{\diagdown H}{\diagup}}$$

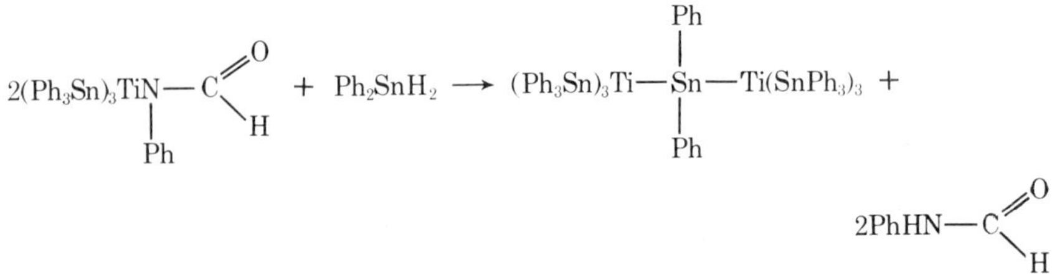

Corresponding reactions were realized with zirconium. The hydrostannolysis of certain types of metal–carbon bonds, in particular of magnesium and zinc, became interesting to us because here a link became apparent with the organic chemistry of bivalent tin. Compounds containing tin bound to Group II elements are readily obtained from the reaction of triphenyltin hydride with coordination complexes of Group II metal alkyls or alkyl halides (40). The product of the following reaction was studied in more detail:

$$Ph_3SnH + EtMgBrEt_3N \rightarrow Ph_3SnMgBr \cdot Et_3N + EtH$$

It showed the behavior to be expected for a compound containing a triphenyltin group bound to a strongly electropositive metal:

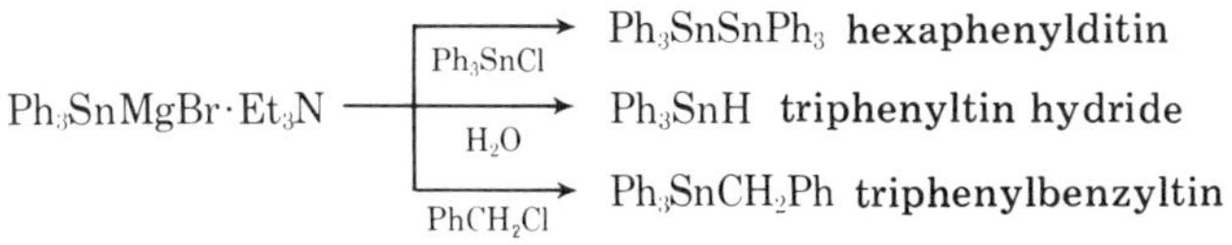

$$Ph_3SnMgBr \cdot Et_3N \longrightarrow
\begin{cases}
\xrightarrow{Ph_3SnCl} & Ph_3SnSnPh_3 \text{ hexaphenylditin} \\
\xrightarrow{H_2O} & Ph_3SnH \text{ triphenyltin hydride} \\
\xrightarrow{PhCH_2Cl} & Ph_3SnCH_2Ph \text{ triphenylbenzyltin}
\end{cases}$$

Treatment of the liquid triphenyltin–magnesium bromide/triethylamine complex [which still contains coordinated diethyl ether and whose molecular weight was not determined] at 50° C in vacuo yielded solid unsolvated triphenyltin–magnesium bromide which, according to a cryoscopic molecular-weight determination in benzene, is a dimer in solution. This reaction appeared to be reversible:

$$2Ph_3SnMgBr \cdot Et_3N \underset{+Et_3N}{\overset{-Et_3N}{\rightleftarrows}} (Ph_3SnMgBr)_2$$
$$\quad\quad I \quad\quad\quad\quad\quad\quad\quad\quad II$$

Removing the complexing agent, triethylamine, causes a striking change in the chemical reactions of the tin–magnesium compound:

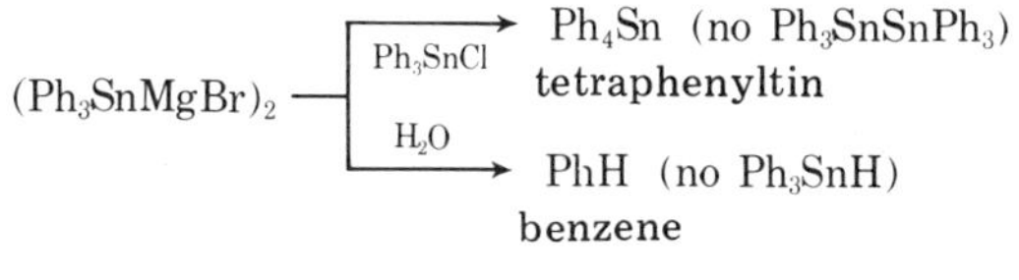

$$(Ph_3SnMgBr)_2 \longrightarrow
\begin{cases}
\xrightarrow{Ph_3SnCl} & Ph_4Sn \text{ (no } Ph_3SnSnPh_3) \\
 & \text{tetraphenyltin} \\
\xrightarrow{H_2O} & PhH \text{ (no } Ph_3SnH) \\
 & \text{benzene}
\end{cases}$$

These reactions indicate that upon removal of the complexing ligand, the compound reacts as if a phenylmagnesium rather than a triphenyltinmagnesium species is present. We interpreted these results in terms of a phenyl-group transfer from tin to magnesium which occurs upon transformation of the triethylamine complex I into the ligand-free species II. Assuming for compound I a dimeric structure as well (for which there is some evidence from the x-ray structural proof for the dimeric structure of EtMgBr · Et_3N in which bromine–magnesium bridges occur), the following reaction course, involving a 1,2-phenyl shift from tin to magnesium, seems plausible:

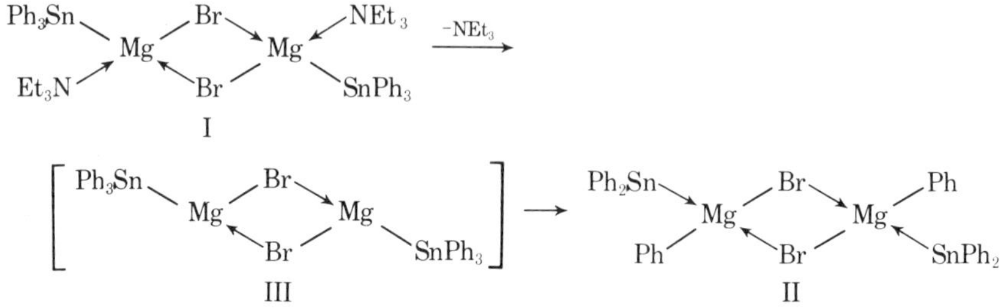

In the ligand-free intermediate III, the magnesium has become an electrophilic center, and the shift of a phenyl group as a phenyl anion from tin to magnesium leaves the tin behind in the formal oxidation state two.   According to this view, compound II contains monomeric diphenyltin—a bivalent organotin species—which evidently is stabilized by acting as an electron-donating ligand for magnesium.   This picture completely agrees with the chemical reactions of compound II, but it is neither supported nor denied by the results of ^{119m}Sn Mössbauer measurements (41).

Direct evidence for the occurrence of a Ph_2Sn species in compound II followed from two insertion reactions:

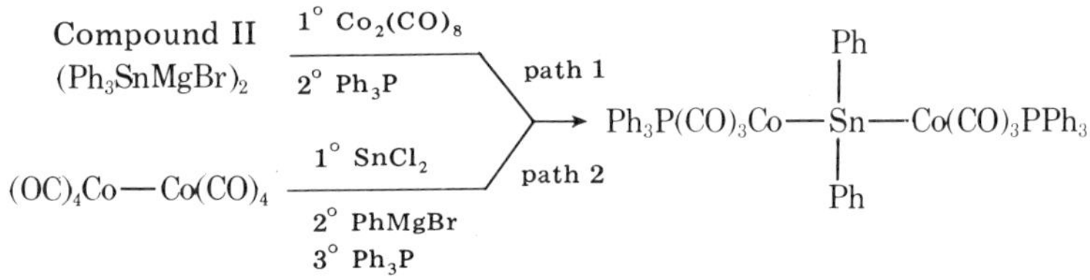

Pathway 2, realized in 1966 by Bonati et al. (42), and pathway 1 resulted in identical products.   This work led, inter alia, to the view that the so far elusive dialkyl- and diaryltins, i.e., bivalent organotin compounds, might be stabilized by coordination with strongly electron-deficient ligands.   This can be an important lead for developing the currently explored field of the organic chemistry of bivalent tin.   So far, the usual types of electron-donating ligand molecules have been found unsuitable for that purpose.

To conclude, I report a very recent finding which, in a way, stands in contrast to the results mentioned above (43).   Earlier attempts to hydrostannolyze bis-(cyclopentadienyl)tin by means of triphenyltin hydride were unsuccessful, probably because of insufficient acidity of the hydride hydrogen.   With a much stronger acidic metal–hydrogen bond, viz. the one occurring in pentacarbonyl-manganese hydride, a hydrometalolysis reaction occurred, but the result was quite unexpected.   The following reaction was expected:

$$(C_5H_5)_2Sn + 2HMn(CO)_5 \xrightarrow{benzene} Sn(Mn(CO)_5)_2 + 2C_5H_6$$
$$MW\ 509$$

Instead, a completely different product was obtained:

$$2(C_5H_5)_2Sn + 4HMn(CO)_5 \rightarrow C_{20}H_2O_{20}Mn_4Sn_2$$
$$MW\ 1020$$

The orange crystalline product had about twice the expected molecular weight and contained hydrogen. The analysis fits very well with the molecular formula given, and the Mössbauer spectrum revealed that the tin was in the tetravalent rather than in the divalent state. The mass-spectroscopic investigation showed a parent peak at 1020 rather than at 1018, the latter being expected for dimeric tin(II)-bis(manganesepentacarbonyl). These results pointed to the following structure for the compound obtained:

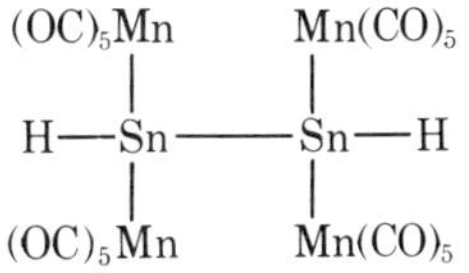

$$(OC)_5Mn \qquad Mn(CO)_5$$
$$| \qquad\qquad |$$
$$H—Sn————Sn—H$$
$$| \qquad\qquad |$$
$$(OC)_5Mn \qquad Mn(CO)_5$$

tetra(manganesepentacarbonyl)ditin dihydride

Chemical evidence for the presence of tin–hydrogen bonds in the molecule was obtained from the formation of chloroform and of red-colored tetra(manganesepentacarbonyl)ditin dichloride in the reaction with carbon tetrachloride. In the reaction with acrylonitrile, the expected hydrostannation product was formed:

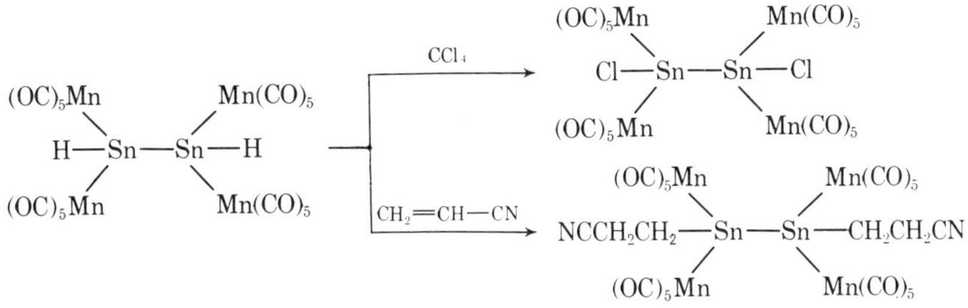

The proposed structure was confirmed by x-ray analysis, but I will not deal with this in detail. The formation of a ditin dihydride in the protolysis of Cp_2Sn by $HMn(CO)_5$ is not readily explained, although of course the formation of Sn(IV) compounds from Sn(II) compounds is not at all surprising.

At the end of my selection of topics in organotin hydride chemistry—in which I have leaned heavily on the Utrecht work—I realize that I may be blamed for quite some bias. What I have tried to do, however, is to present some basic concepts which I think could be fairly well covered by results from the Utrecht work. Personally, I have always found a tremendous challenge in the combination of applied and fundamental approaches or rather in their integration. For that reason, I am particularly pleased that we may learn from this volume about the first industrial application of the reactivity of the tin–hydrogen bond.

Looking back, I am astonished by the tremendous multitude of applied and fundamental developments which have emerged from the world-wide study of organotin chemistry in the past 25 years. Looking ahead, I venture to predict that this field is still a gold mine for the industrial chemist and the academic chemist as well.

Literature Cited

1. Yngve, V., U.S. Patent **2,219,463**; 1940.
2. Yngve, V., U.S. Patent **2,307,092**, 1943.
3. Luijten, J. G. A., Klimmer, O. R., "Eine toxikologische Würdigung der Organozinnverbindungen" ("A toxicological evaluation of organotin compounds"), Tin Research Institute, Fraser Road, Perivale, Greenford, Middlesex, Great Britain, 1974.
4. Luijten, J. G. A., Pezarro, S., *Brit. Plast.* (1957) **30**, 183.
5. van der Kerk, G. J. M., Luijten, J. G. A., *J. Appl. Chem.* (1954) **4**, 314.
6. van der Kerk, G. J. M., Luijten, J. G. A., *J. Appl. Chem.* (1956) **6**, 56.
7. Hof, T., Luijten, J .G. A. *Timber Technol* (1959) **67**, 83.
8. Fahlstrom, G. B. *Proc. Am. Wood-Preservers' Ass.* (1958) **54**, 178.
9. Richardson, B. A. *Brit. Wood Preservers Ass. Ann. Meetg.* (1970).
10. Evans, C. J. *Tin and Its Uses.* (1970) **85**, 3.
11. Evans, C. J., *Tin and Its Uses* (1973) **96**, 7. Evans, C. J., Smith, P. J. *J. Oil. Col. Chem. Ass.* (1975) **58**, 160. (Tin Research Institute Publication n. **505**)
12. Cardarelli, N. T. *Tin and Its Uses* (1972) **93**, 16.
13. Härtel, K. *Tin and Its Uses* (1958) **43**, 9.
14. Härtel, K., *Agr. Vet. Chem.* (1962) **3**, 19.
15. Carrick, W. L. *J. Am. Chem. Soc.* (1958) **80**, 6455.
16. van der Kerk, G. J. M., Noltes, J. G., Luijten, J. G. A. *Angew. Chem.* (1958) **70**, 298.
17. Albright & Wilson Ltd., *Ger. Offenl.* **2,351,188** (13-10-1972–9-5-1974).
18. Finholt, A. E., Bond Jr., A. C., Wilzbach, K. E., Schlesinger, H. J., *J. Am. Chem. Soc.* (1947) **69**, 2692.
19. Neumann, W. P., "The Organic Chemistry of Tin," Wiley, London, 1970.
20. Kuivila, H. G., "Advances in Organometallic Chemistry," Vol. I, pp. 47–87, Academic, New York, 1964.
21. Leusink, A. J. "Hydrostannation. A mechanistic Study," Doctoral Thesis, State University of Utrecht, 1966.
22. Leusink, A. J., Noltes G. J. *Tetrahedron Lett.* (1966) 335.
23. Leusink, A. J., Budding, H. A. *J. Organometal Chem.* (1968) **11**, 533.
24. Leusink, A. J., Budding, H. A., and Drenth, W., *J. Organometal. Chem.* (1968) **13**, 163.
25. Creemers, H. M. J. C., "Hydrostannolysis. A General Method for Establishing Tin–Metal Bonds," Doctoral Thesis, State University of Utrecht, 1967.
26. Creemers, H. M. J. C., Noltes, J. G. *Rec. Trav. Chim. Pays-Bas* (1965) **84**, 590.
27. Creemers, H. M. J. C., Verbeek, T., Noltes, J. G. *J. Organometal. Chem.* (1967) **8**, 469.
28. Henry, M. C., Noltes, J. G. *J. Am. Chem. Soc.* (1960) **82**, 558.
29. Leusink, A. J., Noltes, J. G. *J. Organometal. Chem.* (1969) **16**, 91.
30. Leusink, A. J., Budding, H., Noltes, J. G. *J. Organometal. Chem.* (1970) **24**, 375.

31. Leusink, A. J., Drenth, W., Noltes, J. G., van der Kerk, G. J. M. *Tetrahedron Lett.* (1967) **14**, 1263.
32. van der Kerk, S. M., unpublished data.
33. Noltes, J. G., van der Kerk, G. J. M. *Rec. Trav. Chim. Pays-Bas* (1962) **81**, 41.
34. Ashe, A. J., Shu, P. *J. Am. Chem. Soc.* (1971) **93**, 1804.
35. van der Kerk, G. J. M., Noltes, J. G., Luijten, J. G. A. *J. Appl. Chem.* (1957) **7**, 356.
36. Kuivila, H. G., Menapace, L. W. *J. Org. Chem.* (1963) **28**, 2165.
37. Corey, E. J., Suggs, J. W. *J. Org. Chem.* (1975) **40**, 2554.
38. Creemers, H. M. J. C., Noltes, J. G., van der Kerk, G. J. M. *Rec. Trav. Chim. Pays-Bas* (1964) **83**, 1284.
39. Creemers, H. M. J. C., Noltes, J. G. *Rec. Trav. Chim. Pays-Bas* (1965) **84**, 382.
40. Creemers, H. M. J. C., Noltes, J. G., van der Kerk, G. J. M., *J. Organometal. Chem.* (1968) **14**, 217.
41. Harrison, P. G., Zuckerman, J. J., Noltes, J. G. *J. Organometal. Chem.* (1971) **31**, C23.
42. Bonati, F., Cenini, S., Morelli, A., Ugo, R. *J. Chem. Soc. A* (1966) 1052.
43. Bos, K. D., Bulten, E. J., Noltes, J. G., Spek, A. L. *J. Organometal. Chem.* (1974) **71**, C52.

RECEIVED May 3, 1976.

2

Homolytic Reactions Involving Organotin Compounds

ALWYN G. DAVIES

University College London, 20 Gordon Street, London WC1H OAJ, England

Free radicals may react at a variety of sites in an organotin compound. (1) Reaction at the tin center can displace on alkyl radical, e.g. t-BuO· + Pr₃SnCl → t-BuOSnPr₂Cl + Pr·. The rates of the reactions can be rationalized in terms of electronic and steric effects in a five-coordinate intermediate. (2) Reaction at an atom bonded to tin (particularly H or Sn) displaces a tin-centered radical, e.g. t-BuO· + HSnBu₃ → t-BuOH + ·SnR₃. These organotin radicals will, for example, abstract halogen from an alkyl halide, or add to an alkene, and their reactions are useful in organic synthesis. (3) Reaction at hydrogen on the α- or β-carbon atoms occurs more readily than in simple alkenes, and a variety of electronic interactions have been invoked to account for the enhanced reactivity.

As the field of organometallic chemistry has developed in recent years it has become apparent that it provides an important new context for the operation of homolytic reactions. The principal reasons for this are twofold. First, the metal–carbon, metal–hydrogen, and metal–metal bonds are often rather weak, and, second, the metals carry low-energy p or d orbitals which can accommodate an attacking radical to establish a bimolecular homolytic transition state or intermediate. These two factors together facilitate both unimolecular and bimolecular homolytic reactions. The homolytic chemistry of organic compounds of tin has probably been studied more extensively than that of the compounds of any other metal, and it is too vast to cover comprehensively in a short review. Thus, this paper concentrates on principles rather than practice and mentions only briefly the very important synthetic methods which are based on these principles, and that have been reviewed elsewhere.

Although some organotin compounds will undergo unimolecular photolysis or thermolysis under mild conditions, radicals are usually derived from organotins by the bimolecular reaction with some other radical. The various sites at which this radical can react are illustrated in Reactions 1–7 where the reactants are labelled with the approximate bond dissociation energies. All the reactions which are listed are exothermic, although it is not necessarily the most exothermic reaction possible which in fact occurs.

Reactions at Tin Centers

$$t\text{-BuO}\cdot + Pr_2SnCl\text{---}Pr \longrightarrow t\text{-BuO}\text{---}SnPr_2Cl + Pr \qquad \Delta H \text{ kcal mol}^{-1} \quad (1)$$
$$\qquad\qquad 61 \qquad\qquad\qquad 77 \qquad\qquad\qquad -16$$

$$Br\cdot + Et_3Sn\text{---}Et \longrightarrow Br\text{---}SnEt_3 + Et\cdot \qquad\qquad\qquad\qquad (2)$$
$$\qquad 61 \qquad\qquad\qquad 83 \qquad\qquad\qquad\qquad -22$$

$$(CH_2CO)_2N\cdot + Bu_3Sn\text{---}Bu \dashrightarrow (CH_2CO)_2N\text{---}SnBu_3 + Bu\cdot \qquad\qquad (3)$$
$$\qquad\qquad 61 \qquad\qquad\qquad\qquad ? \qquad\qquad\qquad\qquad ?$$

Reactions Producing Organotin Radicals

$$t\text{-BuO}\cdot + H\text{---}SnBu_3 \longrightarrow t\text{-BuO}\text{---}H + \cdot SnBu_3 \qquad\qquad\qquad (4)$$
$$\qquad\qquad 82 \qquad\qquad\qquad 104 \qquad\qquad -22$$

$$t\text{-Bu O}\cdot + Me_3Sn\text{---}SnMe_3 \longrightarrow t\text{-BuO}\text{---}SnMe_3 + \cdot SnMe_3 \qquad\qquad (5)$$
$$\qquad\qquad 62 \qquad\qquad\qquad\quad 77 \qquad\qquad\qquad -15$$

Reactions not Involving a Bond to Tin

$$t\text{-BuO}\cdot + H\text{---}CH_2CH_2SnEt_3 \longrightarrow t\text{-BuO}\text{---}H + \cdot CH_2CH_2SnEt_3 \qquad (6)$$
$$\qquad\qquad <99 \qquad\qquad\qquad\qquad 104 \qquad\qquad\qquad >-5$$

$$Ph_2\dot{C}\text{---}\dot{O} + CH_3\text{---}CH\text{---}SnEt_3 \longrightarrow Ph_2\dot{C}\text{---}OH + CH_3\dot{C}HSnEt_3 \qquad (7)$$
$$\qquad\qquad\qquad\qquad\quad | $$
$$\qquad\qquad\qquad\qquad\quad H$$
$$\qquad\qquad <99 \qquad\qquad\qquad <104 \qquad\qquad\qquad\qquad ?$$

These elementary reactions can occur in the context of a chain or a non-chain process (*1*). For example, *tert*-butyl hypochlorite reacts exothermically with tripropyltin chloride by a chain process in which *tert*-butoxyl and propyl radicals are the chain carriers (Reactions 9 and 10).

$$t\text{-BuOCl} + Pr_3SnCl \longrightarrow t\text{-BuOSnPr}_2Cl + PrCl \qquad\qquad (8)$$
$$t\text{-BuO}\cdot + Pr_3SnCl \longrightarrow t\text{-BuOSnPr}_2Cl + Pr\cdot \qquad\qquad (9)$$
$$Pr\cdot + t\text{-BuOCl} \longrightarrow PrCl + t\text{-BuO}\cdot \qquad\qquad\qquad (10)$$

On the other hand, di-*tert*-butyl peroxide reacts only when it is photolyzed to give *tert*-butoxyl radicals, and the propyl radical which is displaced (Reaction 12) will not now propagate a chain. If such reactions are carried out in the cavity

of an ESR spectrometer, the spectra of the alkyl radicals which are involved can
be observed.

$$t\text{-BuOOBu-}t \xrightarrow{h\nu} 2\ t\text{-BuO}\cdot \tag{11}$$

$$t\text{-BuO}\cdot + \text{Pr}_3\text{SnCl} \longrightarrow t\text{-BuOSnPr}_2\text{Cl} + \text{Pr}\cdot \tag{12}$$

By measuring the yield of products from the former reaction (by NMR or
GLC for example), or of the radical intermediates from the latter reaction (by
ESR), under appropriate conditions, the rate constants for the elementary steps
involving the organotin compounds can be derived (2). The values quoted in
this paper have been determined by one or other of these procedures.

Reactions Involving an Organotin Radical

Organotin radicals can be generated in solution by the reactions exemplified
in Reactions 13–18.

$$\text{Bu}_3\text{SnN-}i\text{-Pr}_2 \xrightarrow{h\nu} \text{Bu}_3\text{Sn}\cdot + \cdot\text{N-}i\text{-Pr}_2 \tag{13}$$

$$(\text{Bu}_2\text{Sn})_n \xrightarrow{h\nu} \text{Bu}_2\dot{\text{S}}\text{n(SnBu}_2)_{n-2}\dot{\text{S}}\text{nBu}_2 \longrightarrow n\text{-Bu}_2\text{Sn:} \tag{14}$$

$$\text{Me}_3\text{SnSnMe}_3 \xrightarrow{h\nu} 2\text{Me}_3\text{Sn}\cdot \tag{15}$$

$$t\text{-BuO}\cdot + \text{Me}_3\text{SnSnMe}_3 \longrightarrow t\text{-BuOSnMe}_3 + \cdot\text{SnMe}_3 \tag{16}$$

$$t\text{-BuO}\cdot + \text{HSnMe}_3 \longrightarrow t\text{-BuOH} + \cdot\text{SnMe}_3 \tag{17}$$

$$\text{Cl}_3\text{C}\cdot + \text{CHMe}{=}\text{CHCH}_2\text{SnBu}_3 \longrightarrow \text{Cl}_3\text{C}{-}\text{CHMeCH}{=}\text{CH}_2 + \cdot\text{SnBu}_3 \tag{18}$$

Table I. ESR Spectra

Radical	Source	g
Me$_3$Sn$\cdot$	Me$_3$SnH + t-BuO$\cdot$	2.017
	Me$_6$Sn$_2$ + t-BuO$\cdot$	
	Me$_3$SnCl + Na	2.0163
	Me$_4$Sn + γ	2.0154
		2.018
Me$_2$$\dot{\text{S}}$nSnMe$_3$	Me$_6$Sn$_2$ + γ	
Me$_2$$\dot{\text{S}}$nCl	Me$_3$SnCl + γ	2.0113
		ca. 2.10
Me$\dot{\text{S}}$nCl$_2$	Me$_2$SnCl$_2$ + γ	2.0009
Et$_3$Sn$\cdot$	Et$_4$Sn + γ	ca. 2.0
Bu$_3$Sn$\cdot$	Bu$_4$Sn + γ	ca. 2.0
i-Bu$_3$Sn$\cdot$	i-Bu$_3$SnH + UV	ca. 2.0
Ph$_3$Sn$\cdot$	Ph$_3$SnH + UV	ca. 2.0
	Ph$_4$Sn + UV	
[(Me$_3$Si)$_2$CH]$_3$Sn$\cdot$	R$_2$Sn: + UV	2.0094
[(Me$_3$Si)$_2$N]$_3$Sn$\cdot$	R$_2$Sn: + UV	1.9912

Aminotin compounds are photolyzed in solution (Reaction 13) to give strong ESR spectra of the appropriate dialkylaminyl radicals (3), and the nitrogen-containing products (from Me_3SnNEt_2) show a CIDNP effect in the proton NMR spectrum (4). The photolysis of cyclic oligomeric dialkyltins (Reaction 14) provides a convenient source of the dialkylstannylenes (R_2Sn:) which are apparently formed via the spontaneous decomposition of the polystanna α,ω-di-radicals (5). The hexaalkylditins are similarly photolyzed to the trialkylstannyl radicals, though in rather low quantum yield, but this can be improved if di-*tert*-butyl peroxide is added to the system, when the bimolecular Reaction 16 is superimposed on the unimolecular Reaction 15 (3, 35). For preparative purposes the most useful source of trialkyltin radicals is the abstraction of hydrogen from a trialkyltin hydride with a radical such as t-BuO·, (Reaction 17), and Lehnig has observed both ^1H- and ^{119}Sn-CIDNP effects during such reactions (6). The conjugative substitution of Reaction 18 is involved in the chain reaction which occurs between organic halides and allyltin compounds (7, 8, 9, 10).

A further source of trialkyltin radicals is provided by the benzpinacol derivatives (Reaction 19) which dissociate into ketyl radicals at room temperature, and above 60° dissociate into benzophenone and trialkyltin radicals (11). In many reactions, however, the trialkyltin group is transferred directly to the reagent (e.g., O_2, I_2, alkyl halides) without existing as a free intermediate.

$$Me_3SnOCPh_2CPh_2OSnMe_3 \rightleftharpoons 2Me_3SnO\dot{C}Ph_2 \longrightarrow Me_3Sn\cdot + OCPh_2 \quad (19)$$

The most direct evidence for the formation of trialkyltin radicals is their detection by ESR, and data on the radicals which have been observed to date are given in Table I. Unfortunately, apart from the last two atypical examples, only

of Organotin Radicals

$a(Sn)/G$	$a(H)/G$	*Ref.*
	2.75(9H)	(15)
		(16)
1610		(17)
(1983)	2.76(9H)	(12)
1550		(18)
1171	4.5(6H)	(18)
and 250		
		(12)
1780		(18)
		(12)
1550		(12, 19)
1550		(18)
		(20)
1550		(18)
		(20)
1776	2.1(3H)	(21)
	$a(N)10.9(3N)$	(22)

the Me$_3$Sn· radical has been observed in fluid solution, probably because the slower tumbling of the larger radicals, combined with their strongly anisotropic g factors, gives rise to very broad signals.   The spectra of these larger radicals have been detected only by irradiating solid samples with UV light, γ-rays or x-rays, or by the rotating cryostat technique, and the kinetics of the reactions of the radicals cannot be studied under these conditions.

The large value of the tin hyperfine coupling has been interpreted as implying that the R$_3$Sn· radicals have a pyramidal structure: for example, the Me$_3$Sn· radical has been proposed to be distorted by about 14° from the planar structure (12).   Though the conclusion may well be correct, the argument should be used with caution in view of the continuing controversy about the structure of R$_3$C· radicals for which many more data are available (13); on the other hand, the evidence for the pyramidal structure of the R$_3$Si· and R$_3$Ge· radicals is convincing (14).

Rate constants for the self reaction of trialkyltin radicals to give hexaalkylditins, are given in Table II.   The values which are obtained for the simple radicals (ca. 2 × 10^9 1 mole^{-1} sec^{-1}) are typical of those for small radicals which combine without chemical complications and reflect the viscosity-limited rate of diffusion of radicals in the solvent (2).   The bottom two examples in the table illustrate the now-familiar persistence which is conferred by introducing bulky ligands around the radical center; concentrated solutions of these radicals can be prepared which give excellent ESR spectra.

The principal reactions of the trialkyltin radicals are summarized in Reactions 20–32 together with a few leading references.

$$
\begin{array}{lccr}
 & & \textit{Ref.} & \\
\text{R}_3\text{Sn· + R}_3\text{Sn·} \xrightarrow{k} \text{R}_3\text{SnSnR}_3 & & (15,23) & (20) \\
\end{array}
$$

$$
\begin{array}{lcr}
\text{O}_2 \longrightarrow \text{R}_3\text{SnOO·} & (24) & (21) \\
\text{C=C} \longrightarrow \text{R}_3\text{SnCC·} & (25) & (22) \\
\text{C}\equiv\text{C} \longrightarrow \text{R}_3\text{SnC=C·} & (25) & (23) \\
\text{R'N=CHR} \longrightarrow \text{R}_3\text{SnNR'ĊHR'} & (26) & (24) \\
\text{O=CR'}_2 \longrightarrow \text{R}_3\text{SnOĊR'}_2 & (27) & (25) \\
\text{S=CR'(OR')} \longrightarrow \text{R}_3\text{SnSĊR'(OR')} & (28) & (26) \\
\end{array}
$$

$$
\begin{array}{lcr}
\text{XR'} \xrightarrow{k} \text{R}_3\text{SnX + R'·} & (29) & (27) \\
\text{R'OOR'} \xrightarrow{k} \text{R}_3\text{SnOR' + ·OR'} & (30) & (28) \\
\text{R'SSR'} \xrightarrow{k} \text{R}_3\text{SnSR' + ·SR'} & (31) & (29) \\
\text{R'}_2\text{NNR'}_2 \longrightarrow \text{R}_3\text{SnNR'}_2 + ·\text{NR'}_2 & (32) & (30) \\
\text{R'ON=NOR'} \longrightarrow \text{R}_3\text{SnOR' + N}_2 + ·\text{OR'} & (33) & (31) \\
\text{R'}_2\text{NN=NNR'}_2 \longrightarrow \text{R}_3\text{SnNR'}_2 + \text{N}_2 + ·\text{NR'}_2 & (34) & (32) \\
\end{array}
$$

The symbol k above a reaction arrow implies that rate constants for the reaction have been determined; these values are important in that they provide the primary data against which other rate constants can be determined by competition methods.   The most important of these reactions are those involving

Table II.   Self-Reaction of Organometallic Radicals
$(2\ k_t/1\ \text{mole}^{-1}\ \text{sec}^{-1}\ \text{at}\ 298°\text{K})$

		Ref.			*Ref.*
$Me_3Sn\cdot$	3.1×10^9	*(16)*	$Me_3C\cdot$	8.1×10^9	*(16)*
$Bu_3Sn\cdot$	1.4×10^9	*(23)*	$Me_3Si\cdot$	5.5×10^9	*(16)*
$Ph_3Sn\cdot$	2.8×10^9	*(23)*	$Me_3Ge\cdot$	3.6×10^9	*(16)*
Bu_2SnH	1.6×10^9	*(23)*			
			$(Me_3Si)_3C\cdot$	$t_{1/2}$ 2 days	*(16)*
$[(Me_3Si)_2CH]_3Sn\cdot$	$t_{1/2}$ 1 year				*(22)*
$[(Me_3Si)_2N]_3Sn\cdot$	$t_{1/2}$ 90 days				*(22)*

alkyl halides (Reaction 27) and alkenes (Reaction 22), and these two reactions are now considered in more detail.

Reaction of Organotin Radicals with Alkyl Halides. Trialkyltin radicals, generated from the corresponding hexaalkylditins, react with alkyl halides (particularly bromides) to display strong ESR spectra of the appropriate alkyl radicals (*35*) (e.g., Figure 1). This is a very convenient method for generating various types of organic radicals for ESR studies.

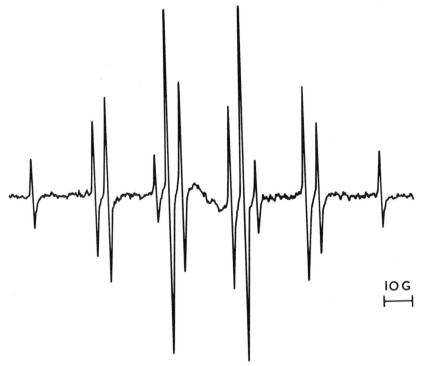

10 G
⊢——⊣

Figure 1.   ESR spectrum of the ethyl radical obtained from the photolysis of di-tert-butyl peroxide and hexamethylditin in the presence of ethyl bromide (3): $CH_3CH_2Br + Me_3Sn\cdot \rightarrow Me_3SnBr + CH_3CH_2\cdot$

Table III.   Rate Constants $(k_1/\text{l mole}^{-1}\text{ sec}^{-1}$ at $25°\text{C})$ for the Reaction $R_3Sn\cdot + R'X \rightarrow R_3SnX + R'\cdot$

$R_3Sn\cdot$	R'	X	k
$Bu_3Sn\cdot$	$c\text{-}C_5H_{11}$	Cl	2.0×10^3
$Bu_3Sn\cdot$	$c\text{-}C_5H_{11}$	Br	2.2×10^7
$Bu_3Sn\cdot$	Me	I	2.5×10^9
$Bu_3Sn\cdot$	$n\text{-}C_5H_{11}$	Cl	8.5×10^2
$Bu_3Sn\cdot$	$c\text{-}C_6H_{11}$	Cl	2.0×10^3
$Bu_3Sn\cdot$	$t\text{-}Bu$	Cl	1.6×10^4
$Bu_3Sn\cdot$	$PhCH_2$	Cl	6.4×10^5
$Me_3Sn\cdot$	$t\text{-}Bu$	Cl	5.9×10^3
$Bu_3Sn\cdot$	$t\text{-}Bu$	Cl	1.6×10^4
$Ph_3Sn\cdot$	$t\text{-}Bu$	Cl	2.0×10^4

If the organotin radicals are derived instead from a trialkyltin hydride, a chain reaction is set up leading to the formation of an alkane and a trialkyltin halide (Reactions 33 and 34).

$$R_3Sn\cdot + R'X \xrightarrow{k_1} R_3SnX + R'\cdot \qquad (33)$$

$$R'\cdot + R_3SnH \xrightarrow{k_2} R'H + R_3Sn\cdot \qquad (34)$$

This reaction was discovered by Noltes and van der Kerk in 1957 (36), interpreted as a free radical process by Kuivila in 1962 (29, 37), and the kinetics analyzed by Ingold in 1968 (23, 38). Alkyl chlorides are much less reactive toward the tin radicals than are bromides or iodides, and different kinetic equations apply, allowing the rate constants for the production (k_1) and the removal (k_2) of the alkyl radicals to be determined.

Representative rate constants (k_1) for the reaction of trialkyltin radicals with alkyl halides (Reaction 33) are given in Table III (23). This illustrates the low reactivity of the alkyl chlorides and the fact that the reactivity follows the predicted order of stability (e.g., $1°R\cdot < 2°R\cdot < 3°R\cdot$) of the alkyl radicals which are liberated.

Table IV.   Rate Constants $(k_2/\text{l mole}^{-1}\text{ sec}^{-1}$ at $25°\text{C})$ for the Reaction $R'\cdot + R_3SnH \rightarrow R'H + R_3Sn\cdot$

R	k
$t\text{-}Bu\cdot + Bu_3SnH$	7.4×10^5
$n\text{-}C_6H_{13}\cdot + Bu_3SnH$	1.0×10^6
$Me\cdot + Bu_3SnH$	5.8×10^6
$t\text{-}Bu\cdot + Bu_3SnD$	2.7×10^5
$c\text{-}C_6H_{11}\cdot + Bu_3SnD$	4.4×10^5
$t\text{-}Bu\cdot + Me_3SnH$	2.9×10^5
$t\text{-}Bu\cdot + Ph_3SnH$	3.1×10^6

Rate constants for the abstraction of hydrogen from trialkyltin hydrides by alkyl radicals (Reaction 33) are given in Table IV (*23*). These values vary little with the nature of the attacking radical R′·, and they have been used as the primary standards against which other rate constants can be determined by competition methods (*39*).

Reactions of Organotin Radicals with Alkenes. The hydrostannation of an olefin by an organotin hydride was reported by van der Kerk, Luijten, and Noltes in 1956 (*40*), and the free radical mechanism was first clearly defined by Neumann in 1961 (*41*), where the propagating steps are:

$$R_3Sn\cdot \ + \ C{=}C \ \longrightarrow \ R_3SnCC\cdot \tag{35}$$

$$R_3SnCC\cdot \ + \ R_3SnH \ \longrightarrow \ R_3SnCCH \ + \ R_3Sn\cdot \tag{36}$$

These reactions have been exploited in the preparation of new types of organotin compounds (*42*). More recently, the intermediate β-stannylalkyl radicals have been studied by ESR spectroscopy (*43, 44, 45, 46, 47, 48, 49, 50*).

If a radical X· is added to an olefin, the adduct appears to prefer either of two conformations in which the C–X bond eclipses (I), or is staggered with respect to (II), the axis of the $2p$ orbital containing the unpaired electron. These two conformations can be distinguished by the values of the ESR hyperfine coupling constants caused by the β-hydrogen atoms. The eclipsed conformation (I) places

I   II

$a(H_\beta)$ 10 – 18 G $a(H_\beta)$ 25 – 30 G

these hydrogen atoms close to the nodal plane of the p orbital, as shown in the Newman projection, and $a(H\beta)$ is low, whereas in the staggered conformation (II), the β-CH bonds are more nearly parallel to the axis of the p-orbital, where they are better able to hyperconjugate with the unpaired electron and to show a larger coupling constant.

The values for $a(H\beta)$ and their positive temperature coefficients (44) (Table V) show that whereas the adducts of alkyl radicals prefer to exist in the staggered conformations, the adducts formed from organotin (or silicon or germanium)

Table V.   ESR. Hyperfine Coupling Constants for β-Substituted Ethyl Radicals at ca. $-100\ °C$

	$a(H\beta)/G$	$C\alpha - C\beta$ barrier/kcal mole^{-1}
$Me_3CCH_2CH_2\cdot$	24.71	
$Me_3SiCH_2CH_2\cdot$	17.68	1.2
$Me_3GeCH_2CH_2\cdot$	16.57	1.6
$Me_3SnCH_2CH_2\cdot$	15.84	2.0

radicals prefer to adopt the eclipsed conformations, and, from the variation of $a(H\beta)$ with temperature, the barriers to rotation about the $C\alpha$ and $C\beta$ bond can be derived (44).

Reactions at a Distant Site in a Ligand

The preference for the eclipsed conformation in the $R_3SnCH_2CH_2\cdot$ radical is usually ascribed to hyperconjugation between the p-electron and the carbon–tin bond with perhaps some contribution from p_π–d_π homoconjugation (43, 44, 45, 46, 47, 48, 49, 50).

Whatever the origin of the effect, it does appear that the presence of a β-stannyl substituent confers an enhanced reactivity on a CH group (see Table VI) toward abstracting tert-butoxyl radicals (1, 43, 44, 45).   A CH group in the α position to tin is even more reactive (see Table VI) (43); this may in part be caused

Table VI. Approximate Rate Constants $(1 \text{ mole}^{-1} \text{ sec}^{-1}$ at $40°C)$
for the Reactions
$$t\text{-}BuO\cdot + R_3'MCH_2CH_2R \rightarrow t\text{-}BuO\cdot + R_3'MCHCH_2R + R_3'MCH_2CHR$$

$R_3' \overset{\alpha}{M}CH_2 \overset{\beta}{C}H_2R$	α Reactivity	β Reactivity
$R_3'CCH_2CH_2R$	7×10^3	6.1×10^2
$R_3'SnCH_2CH_2R$	1.7×10^5	3×10^4

by conjugative delocalization of the p electron into the vacant $5d$ orbital on tin, but a more significant factor is probably that the inductive electron release of the tin accommodates the electronegative nature of the attacking alkoxyl radical.

Reactions at the Tin Center

Reactions involving bimolecular homolytic substitution (S_H2) reactions at a tin center form part of a much wider field which includes a variety of metals and of attacking radicals. The present scope of the reaction is as follows (*51, 52, 53*):

$$X\cdot + MR_n \longrightarrow XMR_{n-1} + R\cdot \qquad (37)$$

$X = ROO\cdot, R'O\cdot, R_3'SiO\cdot, PhCO_2\cdot, R'S\cdot, R_2'N; (CH_2CO)_2N\cdot, R_2'CO^T$
$M = Li, Mg, Zn, Cd, Hg, B, Al, Tl, Si, Sn, Pb, S, P, As, Sb, Bi.$

Most work has been carried out on compounds of boron and of phosphorus. Homolytic substitution at tin was proposed by Razuvaev in 1961 (*54*), and the reactions which have been reasonably well established are shown in Reactions 38 to 44.

	Ref.	
$t\text{-}BuO\cdot + R_3Sn\text{—}SnR_3 \longrightarrow t\text{-}BuOSnR_3 + \cdot SnR_3$	(*16,35*)	(38)
$(CH_2CO)_2N\cdot + R_3Sn\text{—}SnR_3 \longrightarrow (CH_2CO)_2NSnR_3 + \cdot SnR_3$	(*55*)	(39)
$t\text{-}BuO\cdot + R_nSnX_{4-n} \longrightarrow t\text{-}BuOSnR_{n-1}X_{4-n} + R\cdot$	(*56,57*)	(40)
$R_2\dot{C}O + R_nSnX_{4-n} \longrightarrow R_2\dot{C}OSnR_{n-1}X_{4-n} + R\cdot$	(*56,57*)	(41)
$Br\cdot + R_4Sn \longrightarrow BrSnR_3 + R\cdot$	(*58*)	(42)
$(CH_2CO)_2N\cdot + R_4Sn \longrightarrow (CH_2CO)_2NSnR_3 + R\cdot$	(*59,60*)	(43)

$$t\text{-}BuO\cdot + R_2Sn \text{(ring)} \longrightarrow t\text{-}BuOSnR_2CH_2CH_2CH_2CH_2\cdot \qquad (61) \quad (44)$$

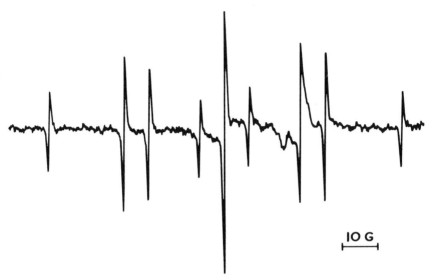

10 G

Figure 2. ESR spectrum of the propyl radical obtained by the photolysis of di-tert-butyl peroxide in the presence of tripropyltin chloride: $t-BuO \cdot + Pr_3SnCl \rightarrow t-BuOSnPr_2Cl + CH_3CH_2CH_2 \cdot$

There is an interesting dichotomy in the reactivities of the various radicals toward the tetraalkyltins and the trialkyltin halides. Whereas bromine radicals (58) and succinimidyl radicals (59) react at tin in tetraalkyltins, they appear not to react with trialkyltin halides; on the other hand, alkoxyl radicals and the related ketone triplets (56, 57) bring about an S_H2 reaction at tin in trialkyltin halides (Figure 2) but abstract hydrogen from the alkyl groups of tetralkyltins. This difference in behavior may in part be explained by the conflicting electronic demands of the reagents. All four are electronegative species which should prefer to react at tin in R_4Sn rather than in R_3SnX where X is an electron-attracting group such as chloride. On the other hand, all four presumably wish to make use of the vacant $5d$ orbitals on the tin to form the 5-coordinate transition state or intermediate for the S_H2 reactions, and these orbitals are lower-lying when the tin carries an electronegative group such as Cl. These conflicting demands appear to be finely balanced, and the balance can favor the tetraalkyltins or the trialkyltin halides for different reagents. The special behavior of the stannacyclopentanes (Reaction 44) in undergoing the S_H2 reaction at tin at least 1000

Table VII.   Relative Leaving Abilities of Alkyl Radicals

$Me_3SnR + Br \cdot$	*i*-Pr 6.95; Et 3.7; Pr, Bu 3.2; PhCH$_2$ 2.2; Me 1.0
$Et_3SnR + Br \cdot$	*i*-Pr 1.5; Et 1.2; Pr, Bu 1.1; Me 1.0
$n\text{-}Bu_2SnBu_2\text{-}s + (CH_2CO)_2N \cdot$	*s*-Bu 2.1; *n*-Bu 1
$n\text{-}Bu_3SnCH_2Ph + (CH_2CO)_2N \cdot$	$PhCH_2 \cdot \gg n\text{-}Bu$
$n\text{-}Bu_2Bu\text{-}t\text{-}SnCl + t\text{-}BuO \cdot$	$n\text{-}Bu \gg t\text{-}Bu$

times more rapidly than the acyclic tetraalkyltins, must be associated with the ring strain in the cyclic compound.

The leaving ability of different groups in the reactions of mixed organotin compounds follows the expected sequence (3°R > 2°R > 1°R) with the remarkable exception that di-*n*-butyl-*tert*-butyltin chloride reacts with *tert*-butoxyl radicals with the selective displacement of the *n*-butyl radical (*see* Table VII).

The overall reactivities of different organotin compounds toward S_H2 reactions at tin, which are shown in Table VIII, are less easy to interpret. Although

Table VIII. Absolute or Relative Rate Constants for S_H2 Reactions at Tin

	s-Bu	*i-Pr*	*n-Bu*	*n-Pr*	*Et*	*Me*
$R_4Sn + (CH_2CO)_2N\cdot$						
k(1 mole^{-1} sec^{-1} at 25°C)	8×10^2		4×10^3	4×10^3	4×10^4	1.4×10^5
$R_4Sn + Br\cdot$, k(rel) at 25°C		7.9	4.3	4.2	5.5	1.0

the reactivity toward the succinimidyl radical, $1°R_4Sn < 2°R_4Sn$, follows a steric sequence, the reactivity toward Br· appears to follow the reverse order, $2°R_4Sn > 1°R_4Sn$. Clearly, much more work is needed in this area.

Three pieces of evidence combine to suggest that these S_H2 reactions at tin may proceed through an intermediate rather than a transition state. First, Boué et al. (58) showed that although, for example, *i*-PrSnMe$_3$ reacted with bromine radicals with a selectivity $k_{i-Pr}/k_{Me} = 5.23$, the selectivity rose to 6.95 if tetraethyltin was added to the system. They proposed that the tetraethyltin formed a radical complex which could proceed to products, or could act as a more selective brominating agent toward the trimethylisopropyltin (Reaction 45) (58):

$$R_4Sn + Br\cdot \longrightarrow R_4\dot{S}nBr \longrightarrow R_3SnBr + R\cdot$$
$$\Big\downarrow R'_4Sn$$
$$\longrightarrow R_4Sn + R'_3SnBr + R'\cdot \qquad (45)$$

Second, we found that the presence of an organotin halide altered the selectivity of the attack of *tert*-butoxyl radicals, derived from *tert*-butyl hypochlorite, on 2,3-dimethylbutane, from a ratio of k_{CH}/k_{CH_3} of ca. 40 to ca. 300. Again, this would be compatible with the formation of a radical complex which is more selective in its reactivity (56, 57):

$$R_3SnCl + \cdot OBu\text{-}t \longrightarrow R_3\dot{S}nCl(OBu\text{-}t) \longrightarrow R\cdot + R_2SnCl(OBu\text{-}t)$$
$$\Big\downarrow Me_2CHCHMe_2$$
$$\longrightarrow R_3SnCl + Bu\text{-}t\text{-}OH + Me_2\dot{C}CHMe_2 +$$
$$\dot{C}H_2CHMeCHMe_2 \qquad (46)$$

Third, in the reaction of ketone triplets with organotin halides, quenching is complete, those triplets which escape chemical reaction being quenched physically. This type of behavior is usually ascribed to the formation of a complex (an exciplex) which can proceed to products, or can revert to the unchanged substrate and singlet ketone (*56, 57*).

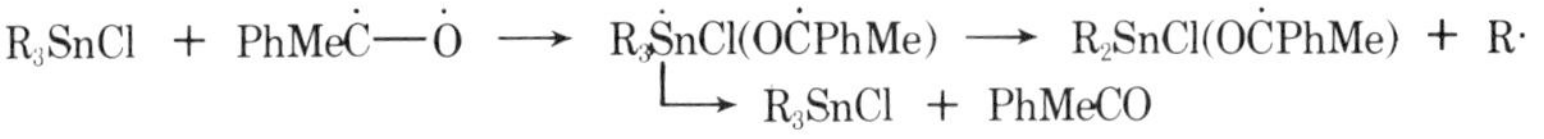
R3SnCl + PhMeĊ—Ȯ ⟶ R3ṠnCl(OĊPhMe) ⟶ R2SnCl(OĊPhMe) + R·
⟶ R3SnCl + PhMeCO

(47)

On this basis, the reactivity of the stannacyclopentanes can be accommodated in terms of the formation of a trigonal bipyramidal intermediate in which the initial strain is relieved by the ring bridging equatorial and apical positions and which then decomposes by the breaking of an apically-directed bond (*61*).

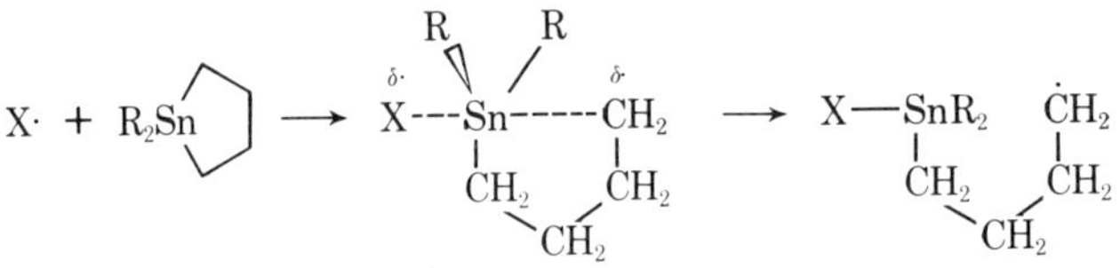
X· + R2Sn
X---Sn-----CH2
R R
δ· δ·
CH2 CH2
CH2
X—SnR2 ĊH2
CH2 CH2
CH2

(48)

The unusual selectivity which is observed in the reaction of di-*n*-butyl-*tert*-butyltin chloride might then be ascribed to the *tert*-butyl ligand occupying a less sterically congested equatorial site in the trigonal bipyramid so that apical departure involves the less bulky *n*-butyl ligand (*62*).

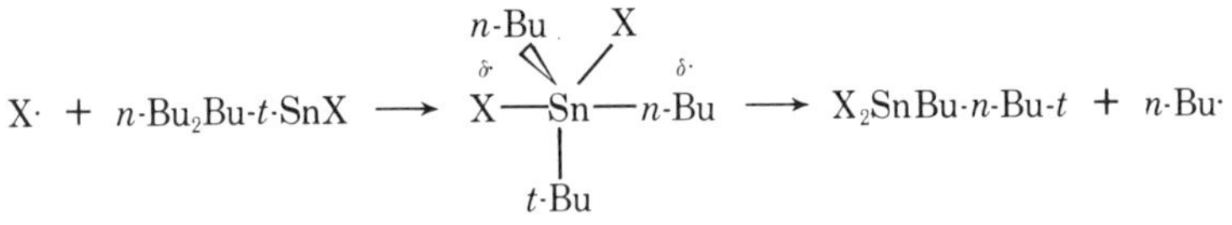
X· + n-Bu2Bu-t-SnX
n-Bu X
δ· δ·
X—Sn—n-Bu
t-Bu
X2SnBu-n-Bu-t + n-Bu·

(49)

Summary

Figure 3 illustrates the principal reactions of free radicals with organotin compounds which we have discussed in the previous sections, and includes approximate second-order rate constants for the attack of *tert*-butoxyl radicals on representative substrates.

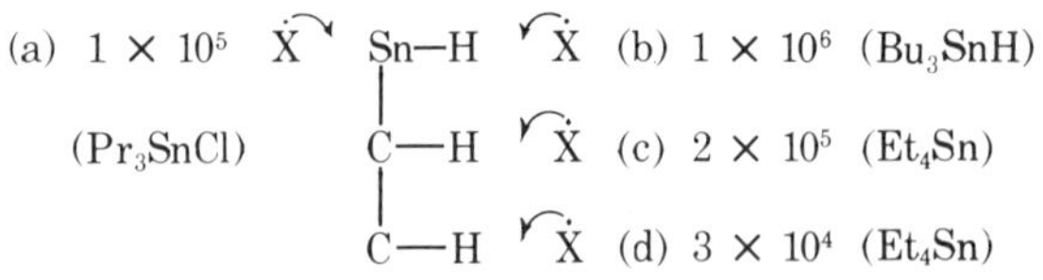
(a) 1 × 10^5 Ẋ Sn—H Ẋ (b) 1 × 10^6 (Bu3SnH)
(Pr3SnCl) C—H Ẋ (c) 2 × 10^5 (Et4Sn)
C—H Ẋ (d) 3 × 10^4 (Et4Sn)

Figure 3. Approximate rate constants for the attack of tert-*butoxyl radicals on organotin compounds (1 mole^{-1} sec^{-1} at 25°C).*

We have seen that only a few examples of attack at the tin center (process a) are as yet recognized, although such reactions do form part of a much wider and well-established field involving the organic derivatives of other metals. Evidence is accumulating that these reactions at a tin center may proceed through a 5-coordinate radical complex.

Reaction at an atom bonded to tin is well established, and with the organotin hydrides (process b) it occurs to the exclusion of all other processes and provides the basis of some important synthetic methods. Attack at tin bonded to tin is also common, and similar processes can be predicted for many other compounds containing tin–metal bonds.

Attack at hydrogen on an alpha carbon atom (process c) occurs more readily than at hydrogen on a beta carbon atom (process d). The former reaction is probably facilitated by the inductive effect of the metal, perhaps by p_π–d_π conjugation, and the latter by carbon–tin hyperconjugation and perhaps p_π–d_π homoconjugation.

Literature Cited

1. Davies, A. G., Scaiano, J. C., *J. Chem. Soc. Perkin Trans. II* (1973) 1777.
2. Ingold, K. U., "Free Radicals," J. Kochi, Ed. Vol. 1, Wiley-Interscience, New York, 1973.
3. Muggleton, B., Doctoral Thesis, London, 1975.
4. Lehnig, M., *Tetrahedron Lett.* (1974) 3323.
5. Neumann, W. P., Schwarz, A., *Angew. Chem., Int. Ed. Engl.* (1975) **14**, 812.
6. Lehnig, M., *Chem. Phys.* (1975) **8**, 419.
7. Kosugi, M., Kuino, K., Takayama, K., Migita, T., *J. Organomet. Chem.* (1973) **56**, C11.
8. Grignon, J., Pereyre, M., *J. Organomet. Chem.* (1973) **61**, C33.
9. Grignon, J., Servens, C., Pereyre, M., *J. Organomet. Chem.* (1975) **96**, 225.
10. Schröer, U., Neumann, W. P., *J. Organomet. Chem.* (1976) **105**, 183.
11. Hillgärtner, H., Neumann, W. P., Schroeder, B., *Justus Liebigs Ann. Chem.* (1975) 586.
12. Lloyd, R. V., Rogers, M. T., *J. Am. Chem. Soc.* (1973) **95**, 2459.
13. Symons, M. C. R., *Tetrahedron* (1973) 207.
14. Krusic, P. J., Meakin, P., *J. Am. Chem. Soc.* (1976) **98**, 228.
15. Sakurai, H., "Free Radicals," J. Kochi, Ed. Vol. 2, Wiley-Interscience, New York, 1973.
16. Watts, G. B., Ingold, K. U., *J. Am. Chem. Soc.* (1972) **94**, 491.
17. Bennett, J. E., Howard, J. A., *Chem. Phys. Lett.* (1972) **15**, 322.
18. Fieldhouse, S. A., Lyons, A. R., Starkie, H. C., Symons, M. C. R., *J. Chem. Soc., Dalton Trans.* (1974) 1966.
19. Lyons, A. R., Symons, M. C. R., *J. Am. Chem. Soc.* (1971) **93**, 7330.
20. Schmidt, U., Kabitzke, K., Merkau, K., Neumann, W. P., *Chem. Ber.* (1965) **98**, 3827.
21. Davidson, P. J., Hudson, A., Lappert, M. F., Lednor, P. W., *Chem. Commun.* (1973) 829.
22. Cotton, J. D., Cundy, C. S., Harris, D. H., Hudson, A., Lappert, M. F., Lednor, P. W., *Chem. Commun.* (1974) 651.
23. Carlsson, D. J., Ingold, K. U., *J. Am. Chem. Soc.* (1968) **90**, 7047.
24. Bennett, J. E., Howard, J. A., *J. Am. Chem. Soc.* (1972) **94**, 8244.
25. Neumann, W. P., "The Organic Chemistry of Tin," Interscience, London, 1970.
26. Lorenz, D. H., Becker, E. I., *J. Org. Chem.* (1963) **28**, 1707.
27. Cooper, J., Hudson, A., Jackson, R. A., *J. Chem. Soc. Perkin Trans. II* (1973) 1933.

28. Barton, D. H. R., McCombie, S. W., *J. Chem. Soc. Perkin Trans. I* (1975) 1574.
29. Kuivila, H. G., *Acc. Chem. Res.* (1968) **1**, 299.
30. Brockenshire, J. L., Ingold, K. U., *Int. J. Chem. Kinetics* (1971) **3**, 157.
31. Spanswick, J., Ingold, K. U., *Int. J. Chem. Kinetics* (1970) **3**, 157.
32. Hollaender, J., Neumann, W. P., Lind, H., *Chem. Ber.* (1973) **106**, 2395.
33. Neumann, W. P., *Angew. Chem.* (1968) **80**, 48.
34. Neumann, W. P., Lind, H., *Chem. Ber.* (1968) **101**, 2837.
35. Cooper, J., Hudson, A., Jackson, R. A., *J. Chem. Soc. Perkin Trans. II* (1973) 1056.
36. van der Kerk, G. J. M., Noltes, J. G., Luijten, J. G. A., *J. Appl. Chem.* (1957) **7**, 356.
37. Kuivila, H. G., Menapace, L. W., Warner, C. R., *J. Am. Chem. Soc.* (1962) **84**, 3584.
38. Carlsson, D. J., Ingold, K. U., *J. Am. Chem. Soc.* (1968) **90**, 1055.
39. Bloodworth, A. J., Davies, A. G., Griffin, M., Muggleton, B., Roberts, B. P., *J. Am. Chem. Soc.* (1974) **96**, 7599.
40. van der Kerk, G. J. M., Luijten, J. G. A., Noltes, J. C., *Chem. Ind.* (1956) 352.
41. Neumann, W. P., Niermann, H., Sommer, R., *Angew. Chem.* (1961) **73**, 768.
42. Noltes, J. G., *J. Organometal. Chem.* (1975) **100**, 177.
43. Krusic, P. J., Kochi, J. K., *J. Am. Chem. Soc.* (1969) **91**, 6161.
44. *Ibid.* (1971) **93**, 846.
45. Kawamura, T., Kochi, J. K., *J. Am. Chem. Soc.* (1972) **94**, 648.
46. Lyons, A. R., Symons, M. C. R., *Chem. Commun.* (1971) 1068.
47. Lyons, A. R., Symons, M. C. R., *J. Am. Chem. Soc.* (1971) **97**, 7330.
48. Symons, M. C. R., *J. Am. Chem.* (1972) **94**, 8589.
49. Symons, M. C. R., *Chem. Phys. Lett.* (1973) **19**, 61.
50. Griller, D., Ingold, K. U., *J. Am. Chem. Soc.* (1973) **95**, 6459.
51. Ingold, K. U., Roberts, B. P., "Free Radical Substitution Reactions," Wiley-Interscience, New York, 1971.
52. Davies, A. G., Roberts, B. P., *Acc. Chem. Res.* (1972) **5**, 387.
53. Davies, A. G., Roberts, B. P. "Free Radicals," J. Kochi, Ed. Vol. 1, Wiley-Interscience, New York, 1973.
54. Ruzuvaev, G. A., D'yachkova, O. S., Vyazankin, N. S., Shchepetkova, O. A., *Doklady Akad. Nauk S.S.S.R.* (1961) **137**, 618.
55. Digiacomo, P. M., Kuivila, H. G., *J. Organomet. Chem.* (1973) **63**, 251.
56. Davies, A. G., Roberts, B. P., Scaiano, J. C., *J. Organomet. Chem.* (1972) **39**, C55.
57. Davies, A. G., Roberts, B. P., Scaiano, J. C., *J. Chem. Soc. Perkin Trans. II* (1972) 2234.
58. Boué, S., Gielen, m., Nasielski, J., *J. Organomet. Chem.* (1967) **9**, 461, 491.
59. Davies, A. G., Roberts, B. P., Smith, J. M., *Chem. Commun.* (1970) 557.
60. Davies, A. G., Roberts, B. P., Smith, J. M., *J. Chem. Soc. Perkin Trans. II* (1972) 2221.
61. Davies, A. G., Parry, G. R., Roberts, B. P., Tse, M.-W., *J. Organomet. Chem.*, (1976) **110**, C33.
62. Davies, A. G., Muggleton, B., Roberts, B. P., Tse, M.-W., Winter, J. N., *J. Organomet. Chem.*, in press.

RECEIVED May 3, 1976.

Synthetic and Mechanistic Aspects of Organostannylanionoin Chemistry

HENRY G. KUIVILA

State University of New York at Albany, Albany, N. Y. 12222

Some of the substantial diversity in the behavior of Group IV anionoids is reviewed and followed by a summary of the results of recent studies on the reactions of organostannylanionoids with organic halides. Simple alkyl halides lead to substitution with primary, substitution and elimination with secondary, and elimination with tertiary halides. Inversion has been shown to occur with sec-butyl halides and retention with cyclopropyl. Strained and bridge-head bromides such as adamantyl and 3-norbornyl produce satisfactory yields of substitution products. Aryl bromides and chlorides react by nucleophilic substitution on halogen (halogen–metal exchange) to form aryl anions as intermediates which may react with haloorganostannane to form substitution products or, with good proton donors, to form reduction products. Functional groups, such as acetyl, methoxycarbonyl, amino, and cyano can be present on the aryl ring and permit good yields of substitution products to be obtained. Carbon tetrachloride and chloroform react primarily by an initial nucleophilic displacement on chlorine as the initial step. Rigid bicyclic and tricyclic organostannanes prepared by these reactions have been used to establish the relationship between the $^3J(^{119}Sn-^{13}C)$ coupling constants and dihedral angle.

This paper is concerned with the chemistry of alkali metal derivatives of organotins which are called anionoids because we do not know as yet under what conditions they may exist as clusters, covalent species, intimate ion pairs, solvent-separated ion pairs, or as free ions. Because carbanionoids may be considered the prototypes of those of Group IV, it is tempting to assume that the chemistry of the higher members of their series will be very similar, varying more in degree than in kind. The only obvious electronic structural variation appearing in silicon, germanium, tin, and lead is that d orbitals of lower energy may become available for involvement in certain situations. Examples in the literature demonstrate that variations in behavior from one metalloid or metal to another, and indeed with a given metalloid as reaction parameters are changed, can bring about profound changes in the course of reaction with a variety of substrates. The discussion here will be limited to organic halides, and will show that, although the reactions are simple and apparently straightforward, they are by no means predictable.

For example, the reactions of triphenylsilyllithium with 1,4-dichloro- and dibromobutanes are shown in Reactions 1 and 2. The major product is the expected 1,4-bistriphenylsilylbutane when the dichloride is used, but only a minimal amount is formed from the dibromide with hexaphenyldisilane being the major product. Another example in which the solvent is the major determinant of the course of the reaction, is shown in Reactions 3 and 4 involving 1,2-dibromobutane and triphenylgermylsodium.

$$2\phi_3\text{SiLi} + \text{Cl(CH}_2)_4\text{Cl} \longrightarrow \phi_3\text{Si(CH}_2)_4\text{Si}\phi_3 \quad (64\%) \tag{1}$$

$$+ \text{Br(CH}_2)_4\text{Br} \longrightarrow (\phi_3\text{Si})_2 \ (74\%) + \phi_3\text{Si(CH}_2)_4\text{Si}\phi_3 \ (3\%) \tag{2}$$

$$2\phi_3\text{GeNa} + \text{Br(CH}_2)_4\text{Br} \xrightarrow{\text{NH}_3} \phi_3\text{Ge(CH}_2)_4\text{Ge}\phi_3 \ (60\%) \tag{3}$$

$$\xrightarrow{\text{Et}_2\text{O}} (\phi_3\text{Ge})_2 \ (62\%) \tag{4}$$

Another set of pertinent observations comes from Willemsens and van der Kerk (4) on the reactions of plumbyllithiums, and of Bulten and Noltes (5) on the reactions of triethylgermylpotassium with polyhalomethanes as shown in Reactions 5, 6, 7, and 8.   Yet another contrast in behavior involves the reaction

$$4\phi_3\text{PbLi} + \text{CCl}_4 \xrightarrow{\text{THF}} [\phi_3\text{Pb}]_4\text{C} \ (61\%) \tag{5}$$

$$3\phi_3\text{PbLi} + \text{CHCl}_3 \xrightarrow{\text{THF}} [\phi_3\text{Pb}]_3\text{CH} \ (66\%) \tag{6}$$

$$4\text{Me}_3\text{PbLi} + \text{CCl}_4 \xrightarrow{\text{NH}_3} [\text{Me}_3\text{Pb}]_4\text{C} \ (78\%) \tag{7}$$

$$\text{Et}_3\text{GeK} + \begin{cases} \text{CCl}_4 \\ \text{CHCl}_3 \\ \text{CBr}_4 \end{cases} \xrightarrow{\text{HMPT}} [\text{Et}_3\text{Ge}]^-_2 \text{ only} \tag{8}$$

of benzyl chloride with trimethylsilylsodium in HMPT (6) and of trimethyls-tannylsodium in tetraglyme (7): the benzyl groups appear predominantly as bibenzyl in the former case, Reaction 9, whereas they appear only as trimethyl-benzylstannane in the latter case, Reaction 10.

$$\text{Me}_3\text{SiNa} + \phi\text{CH}_2\text{Cl} \xrightarrow{\text{HMPA}} \phi\text{CH}_2\text{SiMe}_3 + [\phi\text{CH}_2\text{-}]_2 \tag{9}$$
$$18\% \qquad\qquad 47\%$$

$$\text{Me}_3\text{SnNa} + \phi\text{CH}_2\text{Cl} \xrightarrow{\text{TG}} \phi\text{CH}_2\text{SnMe}_3 + [\phi\text{CH}_2\text{-}]_2 \tag{10}$$
$$75\% \qquad\qquad \text{nil}$$

Reactions of Stannylanionoids with Alkyl and Cycloalkyl Halides

Organostannylanionoids react with great to moderate ease with saturated halides, particularly bromides and iodides, to provide a range of organic substituents on tin. Simple primary and secondary halides (vide infra) react to provide substitution products in good to excellent yields. More interesting are the observations that cycloalkyl halides which might be expected to show little or no reactivity in nucleophilic substitution reactions do in fact react to give satisfactory yields of substitution products. Examples are the reactions of adamantyl- (*8*, *9*), 3-nortricyclyl- (*9*), and 7-norbornyl- (*9*) bromides which react as shown in Reactions 11, 12, and 13.

$$\xrightarrow{\text{Me}_3\text{SnLi, THF}} \quad 1\text{—Ad—SnMe}_3 \ (57\%) \qquad 2\text{—Ad—SnMe}_3 \tag{11}$$

$$\xrightarrow[\text{Me}_3\text{SnNa}]{\text{THF}} \qquad 63\% \qquad + \qquad 37\% \tag{12}$$

$$\xrightarrow[\text{Me}_3\text{SnLi}]{\text{THF}} \qquad (37\%) \tag{13}$$

The reaction with 3-bromonortricyclene is of particular interest because of the formation of both the unrearranged product and the rearranged product, 2-norbornen-5-yltrimethylstannane. Clearly the formation of 1-adamantyl-, 2-adamantyl-, and 3-nortricyclylstannanes should occur with retention in configuration. Other stereochemical results of interest are shown in Reactions 14, 15, and 16.

$$\xrightarrow[\text{THF}]{\text{Me}_3\text{SnLi}} \qquad (30\%) \tag{14}$$

$$\xrightarrow[\text{THF}]{\text{Me}_3\text{SnLi}} \qquad (63\%) \tag{15}$$

$$\xrightarrow[\text{THF}]{\text{Me}_3\text{SnLi}} \qquad 44\% \qquad + \qquad 37\% \tag{16}$$

The axial bromide in Reaction 14 reacts with retention (8) whereas the equatorial tosylate in Reaction 15 (8) reacts with inversion of configuration. The bromocyclopropane shown in Reaction 16 was optically active as were both the trimethylstannyl product and the hydrocarbon (10). Furthermore, cleavage of the trimethyltin group from the cyclopropane as shown yielded hydrocarbon with the same optical rotation as that isolated from the reaction mixture. The clear implication is that the replacement of bromine by tin and by hydrogen in Reaction 16 both occur with complete retention of configuration.

A remarkable set of stereochemical observations was obtained in a study of the reactions of *syn*- and *anti*-7-bromonorbornenes with trimethylstannylalkalis in a series of solvents shown in Table I (11). The solvents tetrahydrofuran

Table I. Stereochemistry of the Reaction of Trimethylstannylanionoids with 7-Bromonorbornenes

		% Retention in Product		
		THF	*DME*	*TG/THF (½)*
Syn	Li	16	3	3
	Na	90	19	9
	K	53	82	4
Anti	Li	96	98	99
	Na	85	83	96
	K	79	92	96

(THF), 1,2-dimethoxyethane (DME), and tetraethylene glycol dimethyl ether (tetraglyme, TG) are arranged in the presumed increasing order of ability to solvate cations. The results indicate that, in the case of the *syn*-bromide, the more free the stannyl anion, the greater the degree of inversion in the reaction, and conversely, the less free, the higher the degree of retention. On the other hand, the degree of dissociation of the stannylanionoid has little effect on the stereochemistry of substitution with the *anti*-bromide, for the major course is retention regardless of the solvent or counterion. In the simpler 2-halobutane series, inversion has been shown to be the primary (and nearly exclusive) stereochemical consequence with triphenyl derivatives of each of the Group IV sodium metalides (lithium in the case of silicon) as seen in Table II. We have found independently that reaction of 2-bromooctane and trimethylstannylpotassium in DME and in THF proceeds with inversion in configuration also, indicating no counterion effect (9). Because the optical rotations of the products used in estimating degrees of inversion presented in Table II are calculated and not experimental values, the absolute figures are probably not exact. However, the trend toward decreased

Table II. Stereochemistry of Reaction of Group IV Anionoids
with 2-Halobutanes

	X	% inversion	% yield
ϕ_3SnNa	Cl	90	46 (DME)
	Br	85	55 (DME)
	I	71	58 (DME)
ϕ_3SiLi	Cl	51	34 (THF)
ϕ_3GeNa	Br	62	39 (NH$_3$)
ϕ_3PbNa	Br	65	26 (NH$_3$, Et$_2$O)

inversion in going from chloride to bromide to iodide in the reactions with tri-phenylstannylsodium and the relatively low yields of products isolated are significant.

In view of the puzzling nature of some of the observations reported above, we have begun a systematic study of the reactions of organostannylanionoids to ascertain their synthetic scope and utility, and to learn as much as possible concerning the mechanisms by which they react with halides. Results obtained with *n*-butyl halides are gathered in Table III, (Sn≡Me₃Sn⁻). The primary halides give only *n*-butyltrimethylstannane as tin-containing product. On the other hand, *tert*-butyl halides yield only hexamethyldistannane, and secondary halides provide both types of product, with the distannane increasing in proportion in the order Cl < Br < I. When the solvent is changed from THF to tetraglyme, all of the secondary halides yield 2-butyltrimethylstannane as the virtually exclusive tin-containing product. The source of the substitution products is the expected S_N2 substitution reaction as we have shown by kinetic experiments, but the source of the distannane is not obvious. Further examination shows that the C$_4$ fragments appear as isobutylene and, apparently, isobutane. A more careful examination with the more conveniently studied *tert*-amyl chloride shows that the overall reaction follows the course shown in Reaction 17. The C$_5$ fragments appear only as the mixture of methylbutenes shown, with 2-methyl-2-butene,

$$Sn\text{—}Na + Me_2C(Et)Cl \longrightarrow Sn\text{—}Sn + H_2 + Me_2C\text{=}CHMe$$

$$+ Me(Et)C\text{=}CH_2 + NaCl \quad (17)$$

Table III. Reactions of *n*-Butyl Halides with
Trimethylstannylsodium in THF at 0°C

	X = Cl		X = Br		X = I	
RX	% RSn	% Sn–Sn	% RSn	% Sn–Sn	% RSn	% Sn–Sn
n-BuX	89	0	86	0.2	78	4
s-BuX	77	3	57	16	33	32
t-BuX	0	60	0	57	0	64

the thermodynamically more stable product, predominating.   More detailed examination shows that hydrogen is also formed in addition to hexamethyldistannane.   In independent experiments these were shown to be artifacts formed in a secondary reaction (Reaction 19) following a putative E2 reaction (Reaction 18).   Treatment of solutions of trimethylstannane with trimethylstannylsodium

$$Sn^- + CH_3CH_2\underset{\underset{Cl}{|}}{\overset{\overset{CH_3}{|}}{C}}CH_3 \longrightarrow SnH + CH_3CH{=}\overset{\overset{CH_3}{|}}{C}CH_3 + Cl^- \qquad (18)$$

$$2Sn{-}H \overset{Sn^-}{\longrightarrow} Sn{-}Sn + H_2 \qquad (19)$$

results in smooth evolution of hydrogen which was formed quantitatively and at a rate dependent on the amount of trimethylstannylanionoid present.   The latter is not consumed.   Thus the reaction is a base-catalyzed decomposition of the stannane.   The low yields reported in Table II may be caused, at least in part, by the intrusion of the elimination reaction which would lead to butenes and hexaphenyldistannane as byproducts.   Other bromides such as benzyl, allyl, cyclobutyl, cyclopentyl, cyclohexyl, isobutyl, and neopentyl react normally and in good yields with trimethylstannylsodium in THF.   Chlorides react more slowly as expected.   However, cinnamyl chloride gives no significant yield of substitution product but rather a mixture of bicinnamyls (13).   Thus, further work is clearly needed to clarify the full mechanistic picture of the reactions of organostannylalkalis with saturated organic halides.

Reactions of Organostannylanionoids with Aryl Halides

We now turn to aryl halides whose reactions with triphenylstannylanionoids have been examined by a number of investigators for synthetic purposes (14). The only example reported using a trimethylstannylanionoid involves trimethylstannylsodium in liquid ammonia–ether in reaction with p-dichlorobenzene which yielded p-bistrimethylstannylbenzene in unspecified yield (15).   We have examined the reactions of a number of aryl halides, especially bromides, and have found that the reactions proceed in high yields and are synthetically useful.   Of particular interest are the results presented in Table IV (16).   None of the products formed could be obtained by the conventional reaction of the aryl Grignard reagent with trimethylchlorostannane.   In the first and third entries the formation of a benzyne would be expected to intrude.   In all of the others the functional group in the para position would be expected to react with the Grignard reagent.   Thus we have here a reaction which can be used to prepare organofunctional aryltins which might be difficult to obtain without recourse to rather elaborate procedures.

Table IV. Reactions of Aryl Bromides with Trimethylstannylsodium in Tetraglyme

	% ArSnMe[a]	% ArH[a]
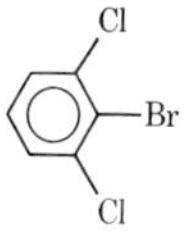 (2,6-dichlorophenyl bromide)	60	—
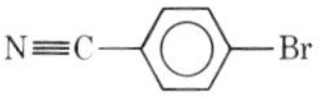 (4-cyanophenyl bromide)	80	—
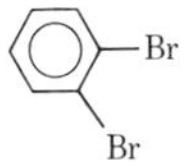 (2-bromophenyl bromide)	42	7 (ϕ SnMe$_3$)
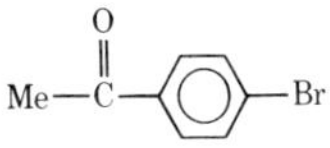 (4-acetylphenyl bromide)	(56)	(43)
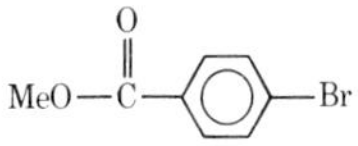 (4-methoxycarbonylphenyl bromide)	(78)	(2)
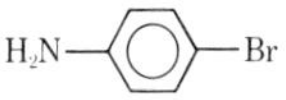 (4-aminophenyl bromide)	(40)	(60)

[a] Yields in parentheses by glpc; others by isolation.

The appearance of product in which the aryl bromine is replaced by hydrogen in the last four entries of Table IV suggests the intermediacy of aryl radicals or anionoids which could abstract hydrogen atoms or protons from the solvent. To probe this point we examined the reaction of o-bromophenyl-3-phenyl-3-propanone with trimethylstannylsodium in tetraglyme as shown in Reaction 20. The direct substitution product, the reduction product, and 1-phenylindene were found in the proportions indicated by the figures outside the parentheses. The formation of phenylindene suggests the intermediacy of the

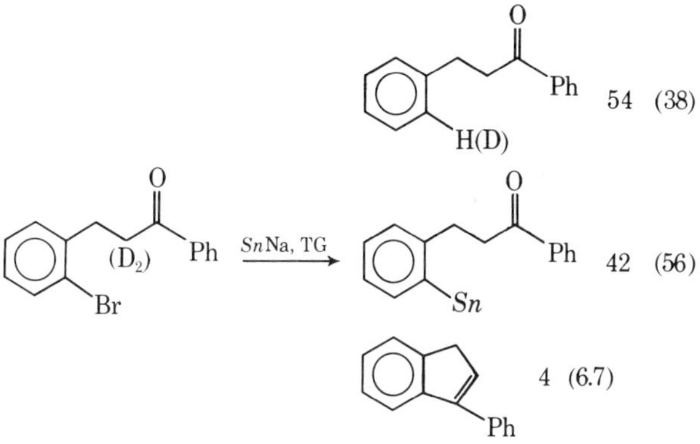

54 (38)

42 (56)   (20)

4 (6.7)

phenylanionoid which undergoes cyclization as shown in Reaction 21a followed by protonation and dehydration in workup. The reduction product also might arise by proton abstraction from solvent.

When the hydrogens alpha to the carbonyl group were replaced by deuteriums, two pertinent observations resulted (indicated by the data in parentheses). First, the proportion of the reduction product dropped from 54% to 38% at the expense of increases in the other two products from 42% to 56% and 4% to 6.7%, respectively. This indicated a kinetic isotope effect in the formation of the reduction product. When its mass spectrum was examined it was found that the benzylic fragment of the phenyl group initially bearing the bromine atom contained three quarters of one deuterium. The only possible source of the deuterium was the carbon alpha to the carbonyl, and this suggests an intramolecular proton abstraction as shown in Reaction 21b to form the enolate anion of the ketone which would provide the aryl-reduced ketone on hydrolytic workup.

In Reaction 20 the arylanionoid suffers from intramolecular trapping by the acidic protons alpha to the carbonyl group. To ascertain whether other acidic species could also function as traps, a series of experiments were conducted using tert-butyl alcohol in this role. Using fixed initial concentrations (ca. 0.1M) of p-bromotoluene and trimethylstannylsodium in tetraglyme, varying amounts of tert-butyl alcohol were added to the halide before the anionoid was added. The product mixture was then analyzed for toluene and p-tolyltrimethylstannane (Figure 1) (17). Even when the ratio of alcohol to trimethylstannylsodium was as large as 20/1, the distribution was the same as that observed at a ratio of 5/1. Thus there is a limit to the proportion of the tolyl groups that can be trapped as the anionoid to form toluene.

One possible interpretation of these observations is that two competing reaction are occurring: one is nucleophilic substitution by the anionoid on carbon of the benzene ring to form arylstannane; the other is nucleophilic substitution on halogen leading to haloorganostannane and aryl anion. This ion can then react with the halostannane to form additional substitution product or, if tert-butyl alcohol is present, it reacts faster in proton abstraction from the hydroxyl group to form the reduction product. This mechanistic dichotomy is consistent with the observation that the reaction is second order; first order in substrate and first order in anionoid.

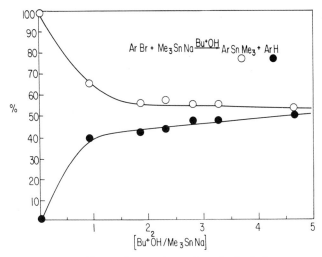

Figure 1. Effect of initial tert-*butyl alcohol concentration on distribution of toluene and* p-*tolyltrimethylstannane; tetraglyme; 0°C*

An alternative mechanism is shown in the Scheme 1 in which the first step in the reaction is a nucleophilic displacement on halogen to form the species shown under the horizontal bracket representing a solvent cage. The species in the cage can couple with rate constant k_c or diffuse apart with rate constant k_d. In the absence of a good proton donor or other electrophilic reagent the aryl anion and halostannane can diffuse back together and undergo reaction to form more of the coupling product. However, in the presence of *tert*-butyl alcohol, proton abstraction occurs (k_r) very rapidly to form the reduction product. The maximum amount of reduction measures the fraction (about one-half; *see* Figure 1) of the coupling product formed after diffusion in the absence of the alcohol, the remainder being formed in the cage before diffusion can occur. If, in the presence of alcohol, the rate of diffusion can be increased, the amount of reduction will increase. To test this, a series of experiments were conducted in mixtures of tetraglyme and DME to vary the viscosity of the medium with a constant ratio of aryl bromide to *tert*-butyl alcohol (1/5). According to a theory developed for caged free radicals developed by Koenig and Deinzer (*18*), the ratio of re-

Scheme I

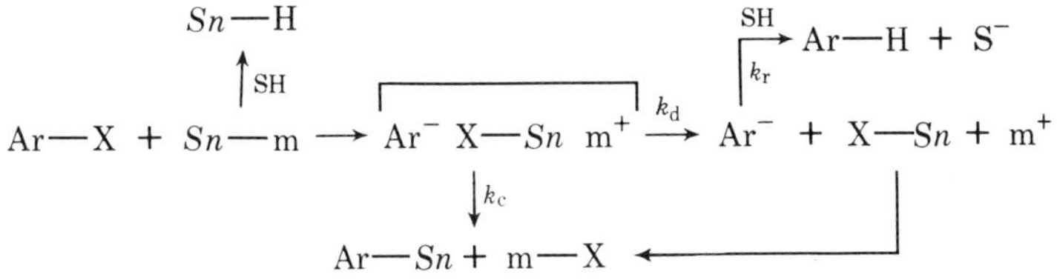

duction product to substitution should vary inversely with the square root of the viscosity if the same model holds for our system.   That this indeed is the case is shown in Figure 2.   The fact that the intercept of the line is zero within the experimental error indicates that the only reaction occurring in the cage is coupling to form substitution product, but reversal of the forward step cannot be excluded.

Another mechanism which is not excluded by the data presented thus far is one in which the species in the cage are formed by an electron transfer from stannylanionoid to aryl halide to form the haloaryl radical anion.   This dissociates,

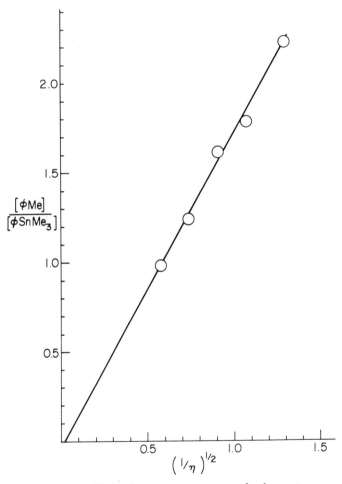

Figure 2.   Effect of viscosity on ratio of toluene to p-tolytrimethylstannane in tetraglyme–dimethoxyethane mixtures.   (Lowest viscosity with 20/80 v/v DME/TGl; highest viscosity with pure TG)

and the resultant species in the cage are aryl radical, halide ion, organostannyl radical, and positive counterion. These could react within the cage and after diffusion to yield the products found. Although chemically induced dynamic polarization experiments which have been attempted gave negative results, this mechanism cannot be rigorously excluded. Fortunately, further experiments can be carried out to distinguish between the polar and radical mechanisms.

We prefer the polar mechanism as the simplest one because of the results of experiments conducted with *o*-dibromobenzene. As shown in Table IV the products of reaction with trimethylstannylsodium are *o*-bistrimethylstannyl-benzene and the mono-reduction product, trimethylphenylstannane. If the mechanism shown in Scheme I is correct, then one would expect that benzyne should be formed rapidly from the initially formed *o*-bromophenylanion as shown in Reaction 22. That this does in fact happen was demonstrated by the obser-

$$\text{(22)}$$

vation that the presence of furan in the initial reaction mixture resulted in the isolation of 1,4-endoxodihydronaphthalene, Reaction 23c. The formation of the other products obtained in the normal reaction are depicted in Reactions 23a and 23b. Reaction 23a constitutes a novel reaction of benzyne in which identical substituents become attached to the two ends of the yne unit.

$$\text{(23a)}$$

$$\text{(23b)}$$

$$\text{(23c)}$$

Returning to Scheme I, note that in the experiments in which the ratio of *tert*-butyl alcohol to trimethylstannylsodium was 20/1 about 85% of the *p*-bromotoluene was consumed. This means that the stannylanionoid reacts by nucleophilic substitution on bromine at least 100 times as fast as it abstracts a proton from the hydroxyl group of the alcohol.

The Course and Mechanism of the Reaction of Trimethylstannylsodium with Carbon Tetrachloride

As shown in Reactions 5, 6, 7, and 8, the reaction of carbon tetrachloride with Group IV anionoids appears to be capricious with the lead derivatives reacting via replacement of each of the four chlorines with lead atoms and the germanium derivatives giving none of the analogous products. Thus we decided to examine the behavior of the intermediate element in Group IV using trimethylstannyl-sodium as the nucleophile (19). When an excess was allowed to react at the lowest feasible temperature (ca. $-40°C$) colorless, crystalline tetrakistrimethyl-stannylmethane precipitated. This substance showed the correct elemental analysis, does not melt but sublimes at 360°C and shows two types of carbon atom in the ^{13}C magnetic resonance spectrum at 3.58 and 35.7 ppm upfield from internal tetramethylsilane. The former signal is assigned to the methyl carbons which usually appear nearer to 10 ppm upfield; the latter signal is assigned to the highly shielded central carbon and represents the most highly shielded tetracoordinate carbon atom known to us.

$$\delta = -35.7\,ppm(CH_2Cl_2)$$

$$CCl_4 + Me_3SnNa \xrightarrow{TG} [Me_3Sn]_4C + [Me_3Sn]CH + [Me_3Sn]_2 \qquad (24)$$
$$(to\ 41\%) \qquad <5\%$$

The mechanism by which the four trimethylstannyl groups become attached to the central carbon was of particular interest to us because it is well known that nucleophiles react with carbon tetrachloride at the chlorine atoms rather than at carbon. In the present instance the first step would be as shown in Reaction 25 yielding the trichloromethyl anionoid and trimethyltin chloride. This is the

$$Me_3SnCCl_3 \text{ (nil)}$$
$$\uparrow$$
$$Me_3Sn^- + CCl_4 \longrightarrow Me_3SnCl + {}^-CCl_3 \xrightarrow{EtOH} HCCl_3 \text{ (to 86\%)} \qquad (25)$$
$$\downarrow -Cl^-$$
$$Me_2C{-}CMe_2 \xleftarrow{} Me_2C{=}CMe_2 \quad :CCl_2$$
$$\diagdown\diagup$$
$$CCl_2$$

case as verified by conducting the reaction in the presence of ethanol which gave up to 86% chloroform by trapping of the anion. Coupling of trimethylchloros-tannane and the trichloromethyl anion is either a slow reaction or the product is highly reactive in the reaction mixture because no trimethyltrichloromethyl-stannane was detected (by NMR) in reaction mixtures in the absence of ethanol. The trichloromethyl anion does have time to dissociate to chloride ion and di-chlorocarbene as demonstrated by the isolation of tetramethyldichlorocyclo-propane when the reaction was conducted in the presence of tetramethylethylene, Reaction 25. How can the dichlorocarbene serve as an intermediate in the formation of the ultimate product? Two possible initial steps are shown in Reaction 26. Reaction with trimethylstannylanion can lead to the dichlorotrimethylstannyl anion which in turn can either react with trimethylchlorostannane to form bis-

$$:CCl_2 + Me_3Sn^- \longrightarrow Me_3Sn\overset{-}{C}Cl_2 \xrightarrow{Me_3SnCl} Cl^- + [Me_3Sn]_2CCl_2 \quad (26)$$

$$\begin{array}{c} Me_3Sn \\ \diagdown \\ C: + Cl^- \\ \diagup \\ Cl \end{array}$$

trimethylstannyldichlorostannane (vide infra) or dissociate to form a new carbene and chloride ion. The new chlorotrimethylstannyl carbene can in turn react with another trimethylstannyl anion as in Reactions 27 to form bistrimethylstannyl-chloromethyl anion. This can couple with trimethylchlorostannane to form the tristrimethylstannylchloromethane shown, or it can dissociate to form chloride

$$\begin{array}{c} Me_3Sn \\ \diagdown \\ C: + Me_3Sn^- \longrightarrow [Me_3Sn]_2\overset{-}{C}Cl \xrightarrow{Me_3SnCl} [Me_3Sn]_3CCl + Cl^- \quad (27) \\ \diagup \\ Cl \end{array}$$

$$[Me_3Sn]_2C: + Cl^-$$

and bistrimethylstannylcarbene. Further reaction of this carbene with tri-methylstannylsodium, followed by reaction of the resulting tristrimethyls-tannylanionoid with trimethylchlorostannane can yield tetrakistrimethyls-tannylmethane, Reaction 28.

$$[Me_3Sn]_2C: + Me_3Sn^- \longrightarrow [Me_3Sn]_3C^- \xrightarrow{Me_3SnCl} [Me_3Sn]_4C + Cl^- \quad (28)$$

To test the viability of the carbene depicted in Reaction 28 as an intermediate an attempt was made to prepare it by the sequence of reactions shown in Reaction 29. When bistrimethylstannyldichloromethane (*20*) was treated with trimeth-ylstannylsodium, tetrakistrimethylstannylmethane was formed in 85% yield. To

$$(Me_3Sn)_2CCl_2 + Me_3Sn^- \longrightarrow (Me_3Sn)_4C \ (85\%)$$
$$\downarrow \qquad\qquad \uparrow Me_3SnCl$$
$$Me_3SnCl + (Me_3Sn)_2\overset{-}{C}Cl \qquad (Me_3Sn)_3C^- \qquad (29)$$
$$\diagdown \qquad \diagup Me_3Sn^-$$
$$Cl^- + (Me_3Sn)_2C:$$

ascertain whether or not this product might be formed by two direct nucleophilic sbustitutions on carbon the reaction was conducted in ethanol. In this case none of the product was obtained. This is clear evidence that anionic intermediates which are trapped by the alcohol are involved on the pathway to the product, and the circuitous one shown is the simplest and most direct.

Applications in Conformational and Structural Studies via ^{13}C NMR

Our ability to synthesize rigid bicyclic compounds in the adamantyl, nor-bornyl, norbornenyl, and nortricyclyl series enabled us to gather ^{13}C NMR spectral parameters which can be used to establish configurations and structures of organotin compounds. The key is the existence of the ^{119}Sn and ^{117}Sn nuclei with spin of one-half which comprise 8.6% and 7.6%, respectively, of the tin isotopes. In addition to the usual chemical shift data which can be used for these purposes the $^3J(^{119}$Sn$—^{13}$C) are of considerable value. Studies with 3-nortri-cyclyltrimethyltin, for example, provide considerable data as shown below (21,

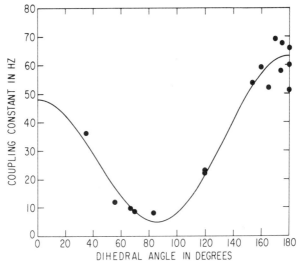

22). The ^{13}C NMR spectrum showed discrete signals from each of the seven ring carbons and from the methyls on tin. The coupling constants through three bonds are shown in parentheses with the dihedral angles for carbons 6, 1, 7, and 5 being observed from models to be 154°, 83°, 67°, and 174°, respectively. Data such

Figure 3. Karplus plot for $^3J(^{119}$Sn$—^{13}C)$

as these could then be used to construct the Karplus plot shown in Figure 3. Although more data at the lower dihedral angles are desirable, it is evident that the angular dependence on ^3J values can be used in conformational and structural studies.

An example in which the structures of two isomeric organotins could be unambiguously assigned by application of these data is shown in the products of the Reactions 20 (23). Because of the bulk of the trimethylstannyl groups,

$$(20)$$

their assignment to exo configurations is reasonably sound. The relative locations of the trimethylstannyl and carbonyl groups in the 2,5- and 2,6-positions could be assigned by examining the coupling patterns around the very low field carbonyl carbons. In one $^3J(^{119}Sn-^{13}C)$ was 61 Hz, and in the other it was 5 Hz. In the exo-2-trimethylstannyl-6-norbornanone the tin and carbonyl carbon atoms display a dehedral angle near 180° in models. This is obviously the isomer with the 62 Hz coupling, while the other with a ^3J coupling of only 5 Hz to the carbonyl carbon is the 2,5-isomer. To have established these structures by chemical means would have been extremely time consuming, and other spectral methods would be of little value.

Acknowledgments

The author thanks those of his colleagues mentioned in this report for having done all of the experimental work discussed, and he is grateful to them and his other collaborators who have made and continue to make research in his laboratory a stimulating and pleasant experience. Thanks are also due to the National Science Foundation for support of much of this research.

Literature Cited

1. Johnson, W. K., Pollatt, K. A., *J. Org. Chem.* (1961) **26**, 4092.
2. Wittenberg, D., Gilman, H., *J. Am. Chem. Soc.* (1958) **80**, 2677.
3. Smith, F. B., Kraus, C. A., *J. Am. Chem. Soc.* (1952) **74**, 1418.

4. Willemsens, L. C., van der Kerk, G. J. M., *J. Organometal. Chem.* (1970) **23**, 471.
5. Bulten, E. J., Noltes, J. G., *J. Organometal. Chem.* (1971) **29**, 409.
6. Sakurai, H., Okada, A., Kira, M., Yonezawa, K., *Tetrahedron Lett.*, (1971) 1511.
7. Reeves, W. G., Ph.D. Thesis, State University of New York at Albany, 1976.
8. Koermer, G. S., Hall, M. L., Traylor, T. G., *J. Am. Chem. Soc.* (1972) **94**, 7205.
9. Considine, J. L., unpublished data.
10. Sisido, K., Kozima, S., Takizawa, K., *Tetrahedron Lett.* (1967) 33.
11. Kuivila, H. G., Considine, J. L., Kennedy, J. D., *J. Am. Chem. Soc.* (1972) **94**, 7206.
12. Jensen, F. R., Davis, D. D., *J. Am. Chem. Soc.* (1971) **93**, 4047.
13. Scarpa, N., Ph.D. Thesis, State University of New York at Albany, 1970.
14. Davis, D. D., Gray, C. E., *Organometal. Chem. Rev. A.* (1970) **6**, 283.
15. Kraus, C. A., Greer, W. N., *J. Am. Chem. Soc.* (1925) **47**, 2568.
16. Kuivila, H. G., Wursthorn, K. R., *J. Organometal. Chem.* (1976) **105**, C6.
17. Kuivila, H. G., Wursthorn, K. R., *Tetrahedron Lett.* (1975) 4357.
18. Koenig, T., Deinzer, M., *J. Am. Chem. Soc.* (1968) **90**, 7014.
19. DiStefano, F. V., unpublished data.
20. Seyferth, D., Armbrecht, F. M., Jr., Schneider, B., *J. Am. Chem. Soc.* (1969) **91**, 1954.
21. Doddrell, D. D., Burfitt, I., Kitching, W., Bulliptt, M., Lee, C-H., Mynott, R. J., considine, J. L., Kuivila, H. G., Sarma, R. H., *J. Am. Chem. Soc.* (1974) **96**, 1640.
22. Kuivila, H. G., Considine, J. L., Sarma, R. H., Mynott, R. J., *J. Organometal. Chem.*, in press.
23. Kuivila, H. G., Maxfield, P. L., Tsia, K-H., Dixon, J. E., *J. Am. Chem. Soc.* (1976) **98**, 104.

RECEIVED May 3, 1976.

Organotin Phosphines, Arsines, Stibines, and Bismuthines: Starting Materials for New Catalysts

HERBERT SCHUMANN, JOACHIM HELD, WOLF-W. DU MONT, GISBERT RODEWALD, and BERND WÖBKE

Institut für Anorganische und Analytische Chemie der Technischen Universität Berlin, D-1000 Berlin, Germany

The preparation, properties, and reactions of organotin phosphines, arsines, stibines, and bismuthines are reviewed, and some newer results are described in detail. The title compounds react with some transition metal carbonyls to form organotin phosphine, arsine, stibine, and bismuthine substituted transition metal carbonyl complexes, which show unusual stability and may be useful homogeneous catalysts. The preparation and properties of some new systems for catalytical purposes are described: (1) organotin phosphines like (tert-$C_4H_9)_2[(CH_3)_2SnCl]P$ react with metal carbonyls to form the corresponding complexes $LM(CO)_{n-1}$, which can be fixed on the surface of e.g., an aerosol by a strong Sn-O-Si bond; (2) p-styryl-substituted organotin phosphines can be synthesized. Transition metal complexes with these compounds as ligands can be copolymerized with olefins, yielding systematically synthesized polymeric organotin phosphine substituted transition metal carbonyl complexes.

In contrast to the organotin compounds in which tin is bound to a main-group element, like a halogen or chalcogen, organotin compounds with bonds between tin and a heavy element of the fifth main group of the periodic table are comparatively new. A patent (1) in 1936 as well as several papers by Arbuzov et al., (2, 3, 4) and Malatesta (5) from 1947 to 1950 describe the synthesis of organotin phosphines, arsines, stibines, and bismuthines. However, it was shown later that these compounds did not contain direct bonds between tin and the group V element (6). The first real organotin phosphine was prepared in

1959 by Kuchen and Buchwald (7) and one year later by Bruker and co-workers (8):

$$(C_2H_5)_3SnBr + NaP(C_6H_5)_2 \longrightarrow (C_2H_5)_3SnP(C_6H_5)_2 + NaBr \qquad (1)$$

$$3(CH_3)_3SnBr + 3NaPH_2 \longrightarrow [(CH_3)_3Sn]_3P + 3NaBr + 2PH_3 \qquad (2)$$

In 1964 Jones and Lappert (9) prepared trimethyltin diphenylarsine, and Schumann and Schmidt (10) isolated tris(triphenyltin)arsine as the first organotin arsenic compounds. In the same year Campbell, Fowles, and Nixon (11) and Schumann and Schmidt (10) isolated the first antimony derivatives as well as the first bismuth compounds. Razuvaev et al. (12) then showed that the organotin bismuth derivatives could be prepared by the reaction of organotin hydride with triethylbismuth.

$$(CH_3)_3SnN(CH_3)_2 + HAs(C_6H_5)_2 \longrightarrow (CH_3)_3SnAs(C_6H_5)_2 + (CH_3)_2NH \qquad (3)$$

$$3(C_6H_5)_3SnLi + ECl_3 \longrightarrow [(C_6H_5)_3Sn]_3E + LiCl \qquad (4)$$

$$E = As, Sb, Bi$$

$$(CH_3)_3SnCl + NaSb(C_6H_5)_2 \longrightarrow (CH_3)_3SnSb(C_6H_5)_2 + NaCl \qquad (5)$$

$$3(C_2H_5)_3SnH + Bi(C_2H_5)_3 \longrightarrow [(C_2H_5)_3Sn]_3Bi + 3C_2H_6 \qquad (6)$$

Analogous to the reaction of sulfur with tetraorganotin compounds (13), we began in 1960 to prepare organotinphosphines by the reaction of tetraphenyltin with phosphorus in a sealed tube above 220°C. Triphenylphosphine and alloy-like tin phosphides were also formed (14). Although this reaction was not suitable for synthesizing specific organotin phosphines, we could isolate at 230° to 250°C, for example, cyclic trimeric diphenyltin phenylphosphine as well as small quantities of a tetrameric phenyltin phosphine having a cubane structure.

One method for preparing organotin phosphines is the reaction of organotin chlorides with lithium diphenylphosphide in tetrahydrofuran (THF) to form organotin phosphines (15). The reaction of sodium diorganophosphide with organotin halides in liquid ammonia (15) is a better method. This reaction makes possible not only the synthesis of trialkyltin diorganophosphines, but also the isolation of diphenylbis(diphenylphosphino)tin. In addition this reaction proved its worth because it was possible to remove unreacted organotin halides from the reaction mixture as their solid ammonia complexes. It is also possible to obtain organotin phosphines by removing alkali halides from alkali organotin compounds and organochlorophosphines in TMF but only in low yields (*16*). The reaction of organotin halides with phosphine or organophosphines in the presence of a base like triethylamine as a hydrogen halide acceptor is quite specific (*17*). The organotin amines are equally suitable as starting materials for synthesizing organotin phosphines. Trimethylbis(dimethylamino)tin, e.g., reacts not only with many other proton-active substances but also with diphenylphosphine or di-(*tert*-butyl)phosphine with the liberation of dimethylamine (*18*); trimethylsilylphosphines react with organotin chlorides and eliminate trimethylchlorosilane to form organotin phosphines (*19*). Trimethyltin chloride reacts with PCl_3 or organochlorophosphines and magnesium in hexamethylphosphoramide to form organotin phosphines in high yields (*20*). The same methods could be used to synthesize organotin arsines, stibines, or bismuthines (*21, 22*).

$$\begin{aligned}
&\text{>Sn—Cl} + \text{Li—P<} \\
&\text{>Sn—Li} + \text{Cl—P<} \\
&\text{>Sn—Cl} + \text{H—P<} \\
&\text{>Sn—NR}_2 + \text{H—P<}
\end{aligned}
\quad\longrightarrow\quad \text{>Sn—P<} \quad
\begin{aligned}
&\text{>Sn—Cl} + (CH_3)_3Si—P< \\
&\longleftarrow \text{Sn(C}_6\text{H}_5)_4 + P_4 \\
&\text{>Sn—Cl} + \text{Cl—P<} + \text{Mg}
\end{aligned}$$

$$(7)$$

The tin–phosphorus bond is very reactive, and the organotin phosphines give many interesting reactions. The sensitivity of organotin phosphines toward oxygen (*23*) depends mainly on the number and nature of the organic groups used to block the undesirable valences of tin and phosphorus. Alkyl compounds are more sensitive than the corresponding phenyl derivatives toward oxygen. The oxidation, either on simple exposure to air or with hydrogen peroxide in ethanol, forms organotin phosphinates. The possible intermediates, phosphine oxide or phosphinite, have never been isolated in this reaction. Organotin phosphines cannot be hydrolyzed in the absence of air, but hydrogen halides split the compounds forming organotin halides and organophosphines. Triphenyltin diphenylphosphine reacts with elemental sulfur to form a triphenyltin substituted thiophosphinate (*23*). In the same way, organic azides such as phenylazide break the bonds between phosphorus (*24*) or arsenic (*25*) and tin to form novel iminophosphoranes or iminoarsanes. The reaction, e.g., for the tin phosphine may proceed by initial formation of an iminophosphorane $\text{>Sn-PR}_2\text{=NC}_6\text{H}_5$, which is unstable, and rearranges into an aminophosphine $\text{>Sn-NC}_6\text{H}_5\text{-}\overline{\text{P}}\text{HR}_2$ which takes up a further phenylimine residue to form the stable end product.

The tin–phosphorus bond can be cleaved by 1,2-dipolar reagents (*26*) like CS_2 forming organotin phosphinodithioformates. The reaction presumably proceeds analogous to the insertion into Sn–N compounds (*27*) by a polar four-center mechanism. Triphenyltin diphenylphosphine reacts with olefinic compounds like allyl chloride or styrene to form chloromethyltriphenylstannyl ethyldiphenylphosphine and phenyl triphenylstannyl ethyldiphenylphosphine, respectively. In both cases the products are two isomeric forms which have not been separated (28).

Organotin phosphines do not form stable phosphonium salts. They react with Lewis acids only at very low temperatures to give addition products which decompose on warming by cleavage of the tin–phosphorus bond. NMR spectroscopic data show an exchange reaction between tris(trimethyltin)phosphine and trimethylmetal halides. In the same way, tris(trimethyltin)phosphine reacts with tris(trimethylsilicon)phosphine and tris(trimethylgermanium)phosphine with redistribution of the trimethyl (group IV elements). All 10 possible compounds are detected by ^{31}P-NMR (29). The reactions of organotin phosphines and arsines with phenylphosphorus or phenylarsenic chlorides lead to stable products containing two, three, or four directly linked phosphorus or arsenic atoms and compounds in which a phosphorus or arsenic atom is linked to one, two, or three arsenic or phosphorus atoms, respectively (30).

The chemical and physical properties of organotin phosphines indicate that the lone pair of electrons on the phosphorus in these compounds makes a certain contribution to the phosphorus–tin bond through a $(p{\rightarrow}d)_\pi$ interaction. The organotin phosphines therefore should not be capable of forming quarternary salts. This theory could be proved by the reaction between organotin phosphines and methyl iodide and by the oxidation of organotin phosphines.

In the first case it was not possible to isolate a quarternized organotin phosphine and in the second, no organotin phosphine oxide was formed. On the other hand, there are some compounds which contradict this theory, e.g.,

$$\text{>Sn—}\overline{\text{P}}\text{<} + CH_3I \longrightarrow \left\{ \text{>Sn—}\overset{CH_3}{\underset{\oplus}{\overset{|}{P}}}\text{<} I^{\ominus} \right\} \xrightarrow{+CH_3I} \text{>SnI} + (CH_3)_2\overset{\oplus}{P}\text{<} I^{\ominus} \qquad (7)$$

$$\text{>Sn—}\overline{\text{P}}\text{<} + 1/2\ O_2 \longrightarrow \left\{ \text{>Sn—}\overset{O}{\overset{\|}{P}}\text{<} \right\} \xrightarrow{+1/2\ O_2} \text{>Sn—O—}\overset{O}{\overset{\|}{P}}\text{<} \qquad (8)$$

trimethyltin di-*tert*-butylphosphine trimethylsilylimine (*31*) and trimethylsilyl trimethylphosphonium tetracarbonylcobaltate (*32*):

$$(CH_3)_3SnCl + (t\text{-}C_4H_9)_2P\text{—}N\overset{Li}{\underset{Si(CH_3)_3}{\overset{\diagup}{\diagdown}}} \longrightarrow (t\text{-}C_4H_9)_2P\overset{NSi(CH_3)_3}{\underset{Sn(CH_3)_3}{\overset{\diagup}{\diagdown}}} + LiCl \qquad (9)$$

$$(CH_3)_3Si\text{—}Co(CO)_4 + P(CH_3)_3 \longrightarrow [(CH_3)_3Si\overset{\oplus}{P}(CH_3)_3][Co(CO)_4]^{\ominus} \qquad (10)$$

These considerations suggested a study of the reactions of organotin phosphines with transition metal carbonyl complexes. Hexacarbonylchromium, hexacarbonylmolybdenum and hexacarbonyltungsten react under UV irradiation with organotin phosphines (*33*), arsines (*34*), stibines (*35*), and bismuthines (*36*) with displacement of one CO ligand to form the corresponding pentacarbonyl organotinphosphine-, arsine, stibine, and bismuthine–chromium, –molybdenum, and –tungsten complexes. Organotin phosphine tetracarbonyl iron complexes could be prepared from organotin phosphines and $Fe_2(CO)_9$ by elimination of $Fe(CO)_5$. The same compounds are formed by UV irradiation of solutions of organotin phosphines and $Fe(CO)_5$ in THF (*37, 38*). Tetracarbonyl nickel reacts with organotin phosphines (*39, 40*), arsines (*34*), stibines (*41*), and bismuthines (*36*) in THF at room temperature to form nickel(0) complexes in which these new ligands with tetracoordinated phosphorus, arsenic, antimony, or bismuth are bonded to the transition metal. All these complexes are formed in high yields as colorless crystals which, unlike the starting ligands, are surprisingly stable to atmospheric oxygen but are often thermally very unstable. This stability supports

$$Cr(CO)_6 + P[Sn(CH_3)_3]_3 \xrightarrow{h\nu} (CO)_5Cr\text{—}P[Sn(CH_3)_3]_3 + CO \qquad (11)$$

$$Cr(CO)_6 + As[Sn(CH_3)_3]_3 \xrightarrow{h\nu} (CO)_5Cr\text{—}As[Sn(CH_3)_3]_3 + CO \qquad (12)$$

$$Cr(CO)_6 + Sb[Sn(CH_3)_3]_3 \xrightarrow{h\nu} (CO)_5Cr\text{—}Sb[Sn(CH_3)_3]_3 + CO \qquad (13)$$

$$Cr(CO)_6 + Bi[Sn(CH_3)_3]_3 \xrightarrow{h\nu} (CO)_5Cr\text{—}Bi[Sn(CH_3)_3]_3 + CO \qquad (14)$$

$$Mo(CO)_6 + P[Sn(CH_3)_3]_3 \xrightarrow{h\nu} (CO)_5Mo\text{—}P[Sn(CH_3)_3]_3 + CO \qquad (15)$$

$$W(CO)_6 + P[Sn(CH_3)_3]_3 \xrightarrow{h\nu} (CO)_5W\text{—}P[Sn(CH_3)_3]_3 + CO \qquad (16)$$

$$Fe(CO)_5 + P[Sn(CH_3)_3]_3 \xrightarrow{h\nu} (CO)_4Fe\text{—}P[Sn(CH_3)_3]_3 + CO \qquad (17)$$

$$Ni(CO)_4 + P[Sn(CH_3)_3]_3 \longrightarrow (CO)_3Ni\text{—}P[Sn(CH_3)_3]_3 + CO \qquad (18)$$

the hypothesis that on oxidation of organotin-substituted phosphines, oxygen attack takes place on the lone pair of electrons on the phosphorus, which is blocked in the complexes by coordination with the transition metal and is thus no longer available for electrophilic attack.

Pentacarbonylmanganese and rhenium halogenides react with trimethylsilyl diphenylphosphine with cleavage of the silicon–phosphorus bond to yield binuclear complexes containing diphenylphosphino groups as bridging ligands (42). We prepared analogous complexes with bis(trimethyltin)phosphine groups as bridging ligands between manganese and rhenium by the reaction of tris(trimethyltin)phosphine with pentacarbonylmanganese and rheniumbromide in diglyme as orange–red crystals (43). The reaction of tris(trimethyltin)phosphine with octacarbonyldicobalt is somewhat more complicated. Reaction of both components in benzene, pentane, or cyclohexane between $-15°$ and $60°C$ gives a black mixture from which two products can be isolated. One of them, trimethyltin tetracarbonylcobaltate itself reacts with tris(trimethyltin)phosphine but only under UV irradiation to form trimethyltin tricarbonyl tris(trimethyltin)phosphine cobaltate in addition to carbon monoxide.

$$2(CO)_5MnBr + 2P[Sn(CH_3)_3]_3 \longrightarrow (CO)_4Mn \underset{\underset{(CH_3)_3Sn \quad Sn(CH_3)_3}{P}}{\overset{\overset{(CH_3)_3Sn \quad Sn(CH_3)_3}{P}}{\diagup}} Mn(CO)_4 + 2(CH_3)_3SnBr \quad (19)$$

$$Co_2(CO)_8 + P[Sn(CH_3)_3]_3 \longrightarrow (CH_3)_3SnCo(CO)_4 + Co_2(CO)_6\{P[Sn(CH_3)_3]_3\}_2 + CO \quad (20)$$

$$(CH_3)_3SnCo(CO)_4 + P[Sn(CH_3)_3]_3 \longrightarrow (CH_3)_3SnCo(CO)_3P[Sn(CH_3)_3]_3 + CO \quad (21)$$

Dimethylbis(di-*tert*-butylphosphino)tin reacts with two equivalents of tetracarbonylnickel to form binuclear μ[dimethylbis(di-*tert*-butylphosphino)tin]bis[tricarbonylnickel(0)] which is a yellow, soluble powder of normal stability. The same phosphine may react with one equivalent of tetracarbonylnickel, to form selectively tricarbonyldimethylbis(di-*tert*-butylphosphine)tin nickel. The ¹H-NMR spectrum shows a doublet signal for the uncomplexed di-*tert*-butylphosphine groups at -1.4 ppm with a coupling constant $J(^1HCC^{31}P) = 11.5$ Hz and a second doublet for the coordinated group at -1.5 ppm with $J(^1HCC^{31}P) = 13.3$. Hz. The two different ^{31}P-NMR signals at -49.6 and -61.8 ppm are also split into doublets with a coupling constant $J(^{31}PSn^{31}P) = 14.0$ Hz. This benzene soluble compound decomposes on standing, losing carbon monoxide and yielding a benzene-insoluble complex dicarbonyl[dimethylbis(di-*tert*-butylphosphino)tin]nickel which may be a four-membered chelate like that described by Abel et al. (44) or a polymeric complex. Analogously we believe a polymeric complex is formed by the reaction between dimethylbis(di-*tert*-butylphosphino)tin and hexacarbonylchromium under UV irradiation as well as with tetracarbonyl–norbornadienechromium in pentane (45).

$$(CH_3)_2Sn\begin{array}{c}PR_2\\PR_2\end{array} + 2Ni(CO)_4 \longrightarrow (CH_3)_2Sn\begin{array}{c}PR_2{-}Ni(CO)_3\\PR_2{-}Ni(CO)_3\end{array} + 2CO \quad (22)$$

$$(CH_3)_2Sn\begin{array}{c}PR_2\\PR_2\end{array} + Ni(CO)_4 \longrightarrow (CH_3)_2Sn\begin{array}{c}PR_2\\PR_2{-}Ni(CO)_3\end{array} + CO \quad (23)$$

$$(CH_3)_2Sn\begin{array}{c}PR_2\\PR_2{-}Ni(CO)_3\end{array} \longrightarrow (CH_3)_2Sn\begin{array}{c}PR_2\\PR_2\end{array}Ni(CO)_2 + CO \quad (24)$$

$$(CH_3)_2Sn\begin{array}{c}PR_2\\PR_2\end{array} + \begin{array}{c}Cr(CO)_6 \xrightarrow{h\nu} 2CO\\C_7H_8Cr(CO)_4 \longrightarrow C_7H_8\end{array} + (CH_3)_2Sn\begin{array}{c}PR_2\\PR_2\end{array}Cr(CO)_4 \quad (25)$$

$$R = t\text{-}C_4H_9$$

Synthesizing the bidentate ligand dimethylbis(di-*tert*-butylphosphino)tin can be revealing.   Trimethyl(di-*tert*-butylphosphino)silane reacts with dimethyltin dichloride in benzene at room temperature in a short time to form chlorodimethyl(di-*tert*-butylphosphino)tin in nearly quantitative yield. Methyltin trichloride forms, accordingly, dichloromethyl(di-*tert*-butylphosphino)tin.   An excess of trimethyl(di-*tert*-butylphosphino)silane reacts with chlorodimethyl(di-*tert*-butylphosphino)tin only after 4 days at 55°C/12 torr yielding dimethylbis(di-*tert*-butylphosphino)tin.   With methyltin trichloride, however, the bidentate chloromethylbis(di-*tert*-butylphosphino)tin may form. In each case, trimethylchlorosilane can be removed by distillation.   The substitution of all three chlorine atoms is not possible, even using a high excess of trimethyl(di-*tert*-butylphosphino)silane at 60°C.   Chlorodimethyl(di-*tert*-butylphosphino)tin is also formed in the reaction of trimethyl(di-*tert*-butylphosphino)tin with dimethyltin dichloride in high yields.

$$(CH_3)_2SnCl_2 + (CH_3)_3SiP(t\text{-}C_4H_9)_2 \longrightarrow \underset{\underset{Cl}{|}}{(CH_3)_2SnP(t\text{-}C_4H_9)_2} + (CH_3)_3SiCl \quad (26)$$

$$CH_3SnCl_3 + (CH_3)_3SiP(t\text{-}C_4H_9)_2 \longrightarrow \underset{\underset{CH_3}{|}}{Cl_2SnP(t\text{-}C_4H_9)_2} + (CH_3)_3SiCl \quad (27)$$

$$(CH_3)_2SnCl_2 + 2(CH_3)_3SiP(t\text{-}C_4H_9)_2 \longrightarrow (CH_3)_2Sn[P(t\text{-}C_4H_9)_2]_2 + 2(CH_3)_3SiCl \quad (28)$$

$$CH_3SnCl_3 + 2(CH_3)_3SiP(t\text{-}C_4H_9)_2 \longrightarrow \underset{\underset{Cl}{|}}{CH_3Sn[P(t\text{-}C_4H_9)_2]_2} + 2(CH_3)_3SiCl \quad (29)$$

$$(CH_3)_2SnCl_2 + (CH_3)_3SnP(t\text{-}C_4H_9)_2 \longrightarrow \underset{\underset{Cl}{|}}{(CH_3)_2SnP(t\text{-}C_4H_9)_2} + (CH_3)_3SnCl \quad (30)$$

^{1}H- and ^{31}P-NMR spectra of these compounds show that the signals of the *tert*-butyl protons are doublets because of the coupling $J(^1HCC^{31}P)$.   Chloromethー

Table I. ^{1}H-NMR and ^{31}P-NMR Spectroscopic

Compound	δCH_3C	$J(^1HCC^{31}P)$	δCH_3Sn
$(CH_3)_3SnP(t\text{-}C_4H_9)_2$	-1.36 (d)	11.4	-0.34 (d)
$(CH_3)_2ClSnP(t\text{-}C_4H_9)_2$	-1.41 (d)	12.1	-0.76
$CH_3(Cl_2)SnP(t\text{-}C_4H_9)_2$	-1.43 (d)	12.2	-1.07
$(CH_3)_2Sn[P(t\text{-}C_4H_9)_2]_2$	-1.47 (d)	11.3	-0.76 (t)
$CH_3(Cl)Sn[P(t\text{-}C_4H_9)_2]_2$	-1.67 (d)	11.8	-1.24 (t)
	-1.48 (d)	11.3	
$Cl_2Sn[P(t\text{-}C_4H_9)_2]_2$	-1.50 (d)	12.7	—

$^a\delta$ in ppm against TMS, or 85% H_3PO_4 as external standard, J in Hz, Varian

ylbis(di-*tert*-butylphosphino)tin shows two doublets with different coupling constants. This situation may be caused by steric hindrance of the rotation of the di-*tert*-butylphosphine groups around the P–Sn bond. This means that one of each of the two *tert*-butyl groups of the phosphine residues is orientated toward the chlorine whereas the others point at the methyl group bonded to tin. Both phosphorus atoms remain equivalent, as shown by the ^{31}P-NMR spectrum, which shows that there should be only one signal (Table I).

The reaction of trimethyl(di-*tert*-butylphosphino)silane with SnCl₄ is much more complicated. Whereas the reaction of the silicon phosphine with SiCl₄ or GeCl₄ under certain conditions leads to trichloro(di-*tert*-butylphosphino)silane and trichloro(di-*tert*-butylphosphino)germane, the reaction with SnCl₄ proceeds exclusively via oxidative cleavage of the Si–P bond. On the other hand, both substitution and redox reactions take place concurrently if the silicon phosphine is in excess on reaction with SnCl₄ in benzene. A complete separation of the reaction products of Equations 32a and 32b has not yet been achieved. The reason could be that α-elimination reactions take place, in which di-*tert*-butyl-chlorophosphine is also formed along with chloro(di-*tert*-butylphosphino)tin from dichlorobis(di-*tert*-butylphosphino)tin (*46*). The latter compound could however be identified by NMR, and the novel tin(II) phosphine could be syn-

$$ECl_4 + (CH_3)_3SiP(t\text{-}C_4H_9)_2 \xrightarrow{E = Si, Ge} Cl_3EP(t\text{-}C_4H_9)_2 + (CH_3)_3SiCl \quad (31a)$$

$$\xrightarrow{E = Sn} (t\text{-}C_4H_9)_2PCl + ECl_2 + (CH_3)_3SiCl \quad (31b)$$

$$SnCl_4 + 2(CH_3)_3SiP(t\text{-}C_4H_9)_2 \nearrow Cl_2Sn[P(t\text{-}C_4H_9)_2]_2 + 2(CH_3)_3SiCl \quad (32a)$$

$$\searrow (t\text{-}C_4H_9)_2PSnCl + (t\text{-}C_4H_9)_2PCl + 2(CH_3)_3SiCl \quad (32b)$$

$$SnCl_2 + (CH_3)_3SiP(t\text{-}C_4H_9)_2 \longrightarrow (t\text{-}C_4H_9)_2PSnCl + (CH_3)_3SiCl \quad (33)$$

Data of Chlorostannyl Phosphines[a]

$J(^1HCSn^{31}P)$	$J(^1HC^{117/119}Sn)$	$\delta^{31}P$	$J(^{31}P^{117/119}Sn)$
1.4	45.0/47.2	−20.7	760.6/795.4
—	49 /51	−40.5	—
—	62 /64	−72.3	
0.9	40 /42	−31.3	92 / 970
0.9	48 /50	−49.3	1281/1343
—	—	−84.5	1720/1800

XL-100-15, 60/40.5 MHz.

thesized by an alternative route and isolated in a pure form.   Thus chloro-(di-*tert*-butylphosphino)tin is formed in the reaction of trimethyl(di-*tert*-butylphosphino)silane with $SnCl_4$ as a yellow powder which decomposes above 175°C.   The Mössbauer data (δ = 2.96 ± 0.06 mm/s (SnO_2); Δ = 1.68 ± 0.12 mm/s) prove the presence of Sn(II).

Functionally substituted organotin phosphines are interesting starting materials for new catalysts.   On the one hand, it is possible to prepare catalytically active transition metal complexes with organotin phosphines as ligands which show a high stability on the transition metal phosphorus tin linkage; on the other hand, there should be a way via their reactivity at tin to fix these compounds to the surface of a carrier like aerosol by a strong Si–O–Sn bond.   We found that halogen-substituted organotin phosphines do react with tetracarbonyl nickel, as with hexacarbonylchromium, molybdenum, or tungsten to form the corresponding tricarbonylnickel–, pentacarbonylchromium–, pentacarbonylmolybdenum–, and pentacarbonyl–tungsten complexes, respectively (47, 48).

$$Ni(CO)_4 + (CH_3)_2SnP(t\text{-}C_4H_9)_2 \longrightarrow (CO)_3Ni\text{—}P(t\text{-}C_4H_9)_2Sn(CH_3)_2 + CO \quad (34)$$
$$\underset{X}{|} \qquad\qquad\qquad\qquad\qquad \underset{X}{|}$$

$$X = Cl, Br$$

$$M(CO)_6 + (CH_3)_2SnP(t\text{-}C_4H_9)_2 \longrightarrow (CO)_5M\text{—}P(t\text{-}C_4H_9)_2Sn(CH_3)_2 + CO \quad (35)$$
$$\underset{Cl}{|} \qquad\qquad\qquad\qquad\qquad \underset{Cl}{|}$$

$$M = Cr, Mo, W$$

The new complexes are pale yellow substances which decompose above 100°C to yield polymeric products of unknown structure.   On the other hand, they show a high stability toward atmospheric oxygen.   The reaction of chlorodimethyl(di-*tert*-butylphosphino)tin with sodium trimethylsilanolate gives dimethyltrimethylsiloxy(di-*tert*-butylphosphino)tin which is the first model substance for an organotin phosphine fixed on the surface of aerosol (49).   The product reacts with tetracarbonylnickel to form the corresponding tricarbonylnickel complex. The homologous germanium compound (*tert*-C$_4$H$_9$)$_2$[(CH$_3$)$_2$GeOSi(CH$_3$)$_3$]P-Ni(CO)$_3$ is a liquid with a melting point of 3°C. The NMR and IR data for this stable compound with the structural unit

$$-\overset{|}{\underset{|}{Si}}-O-\overset{|}{\underset{|}{Ge}}-\overset{|}{\underset{|}{P}}-\overset{|}{\underset{|}{Ni}}-C\equiv O$$

show, that the Ni(CO)$_3$ group is not perceptibly influenced by the siloxy group bound to germanium: this means that such complexes should show similar catalytic activity like their siloxy-free analogs.

$$(CH_3)_2\underset{\underset{Cl}{|}}{Sn}P(t\text{-}C_4H_9)_2 + NaOSi(CH_3)_3 \longrightarrow (CH_3)_2\underset{\underset{O-Si(CH_3)_3}{|}}{Sn}-P(t\text{-}C_4H_9)_2 + NaCl \quad (36)$$

$$Ni(CO)_4 + (CH_3)_2\underset{\underset{O-Si(CH_3)_3}{|}}{Sn}P(t\text{-}C_4H_9)_2 \longrightarrow (CO)_3Ni-P[\underset{\underset{O-Si(CH_3)_3}{|}}{Sn}(CH_3)_2](t\text{-}C_4H_9)_2 + CO \quad (37)$$

First attempts have been made to react either the chlorine substituted starting phosphine (CH$_3$)$_2$Sn(Cl)P(*tert*-C$_4$H$_9$)$_2$ or its tricarbonylnickel complex with aerosol which was converted to an aerosol with Si–O–Li groups on its surface by reacting with butyllithium. IR, Raman, and ESCA measurements should show if there are some of the new molecules fixed on the surface of aerosol. Definite conclusions cannot be drawn as yet from the preliminary results. This work is in progress.

$$(p\text{-}CH_3C_6H_4)_2SnCl_2 + p\text{-}CH_2\text{=}CHC_6H_4MgCl \longrightarrow$$

$$MgCl_2 + (p\text{-}CH_3C_6H_4)_2Sn\overset{C_6H_4\text{-}p\text{-}CH\text{=}CH_2}{\underset{Cl}{<}} \quad (38)$$

$$(p\text{-}CH_3C_6H_4)_2Sn\overset{C_6H_4\text{-}p\text{-}CH\text{=}CH_2}{\underset{Cl}{<}} + (CH_3)_3Si-PR_2 \longrightarrow$$

$$(p\text{-}CH_3C_6H_4)_2Sn\overset{C_6H_4\text{-}p\text{-}CH\text{=}CH_2}{\underset{PR_2}{<}} + (CH_3)_3SiCl \quad (39)$$

$$(p\text{-}CH_3C_6H_4)_2Sn\overset{C_6H_4\text{-}p\text{-}CH\text{=}CH_2}{\underset{PR_2}{<}} + Ni(CO)_4 \longrightarrow$$

$$(p\text{-}CH_3C_6H_4)_2Sn\overset{C_6H_4\text{-}p\text{-}CH\text{=}CH_2}{\underset{\underset{\downarrow}{PR_2}}{<}} + CO \quad (40)$$

$$\underset{Ni(CO)_3}{}$$

$$R = C_6H_5, \ t\text{-}C_4H_9$$

We looked for another way to synthesize new catalysts on the basis of polymers. Organotin compounds with two very reactive functions like both an olefinic double bond and a tin–phosphorus bond are very unstable. Thus all attempts to prepare alkyltin organophosphines with a vinyl group bonded to tin have not been successful. However now we have been able to synthesize some compounds stable at room temperature starting from di(p-tolyl)p-styryltin chloride and trimethyl(di-*tert*-butylphosphino)silane or trimethyl(diphenyl-phosphino)silane, respectively. They react with tetracarbonylnickel in pentane at room temperature yielding the corresponding tricarbonylnickel complexes in high yields as yellow to pink crystals, which polymerize above 90 °C (Reactions 38–40).

Table II. Trimerization Reactions Catalyzed by Organotinphosphine-Substituted Tricarbonylnickel Complexes

Monomer	Product	Conditions	Yield with I^a	Yield with II^a
$HC{\equiv}CC_6H_5$	(1,2,4-trisubstituted benzene: C_6H_5 / C_6H_5 / C_6H_5)	benzene 80° C, 1 hr	70	95
$HC{\equiv}CCO_2C_2H_5$	(1,2,4-trisubstituted benzene: $CO_2C_2H_5$ / $CO_2C_2H_5$ / $CO_2C_2H_5$)	benzene 80° C, 2 hr	51	70
$HC{\equiv}CCH_2OH$	(1,2,4-trisubstituted benzene: CH_2OH / CH_2OH / CH_2OH)	benzene 80° C, 12 hr	55	47

[a] Catalyst I: $(CH_3C_6H_4)_2(CH_2 = CHC_6H_4)SnP(t\text{-}C_4H_9)_2Ni(CO)_3$; catalyst II: $(CH_3C_6H_4)_2(CH_2 = CHC_6H_4)SnP(C_6H_5)_2Ni(CO)_3$

These monomers show catalytic activities in some trimerization reactions. Initial investigations (Table II) showed that some substituted acetylenes are trimerized yielding mainly 1,2,4-trisubstituted benzene derivatives in high yields under mild conditions. Further experiments including polymer catalysts as well as organotin phosphine substituted π-allyl nickel, rhodium, and ruthenium complexes are in progress.

Literature Cited

1. Standard Oil Development Co., British Patent **445 813** (1936).
2. Arbuzov, B. A., Grechkin, N. P., *Zh. Obshch. Khim.* (1947) **17**, 2166.
3. *Ibid.* (1950) **20**, 107.

4. Arbuzov, B. A., Pudovik, A. N., *Zh. Obshch. Khim.* (1947) **17**, 2158.
5. Malatesta, L., Sacco, A., *Gazz. Chim. Ital.* (1950) **80**, 658.
6. Arbuzov, B. A., Grechkin, N. P., *Izv. Akad. Nauk SSSR, Otdel. Khim. Nauk* (1956) 440.
7. Kuchen, W., Buchwald, H., *Chem. Ber.* (1959) **92**, 227.
8. Bruker, A. B., Balashova, L. D., Soborovskii, L. Z., *Dokl. Akad. Nauk SSSR* (1960), **135**, 843.
9. Jones, K., Lappert, M. F., *Proc. Chem. Soc. London* (1964) 22.
10. Schumann, H. Schmidt, M., *Angew. Chem.* (1964) **76**, 344.
11. Campbell, I. G. M., Fowles, G. W. A., Nixon, L. A., *J. Chem. Soc.* (1964) 3026.
12. Kruglaya, O. A. Vyazankin, N. S., Razuvaev, G. A., *Zh. Obshch. Khim.* (1965) **35**, 394.
13. Schmidt, M., Dersin, H. J., Schumann, H., *Chem. Ber.* (1962) **95**, 1428.
14. Schumann, H., Köpf, H., Schmidt, M., *Chem. Ber.* (1964) **97**, 1458.
15. Campbell, I. G. M., Fowles, G. W. A., Nixon, L. A., *J. Chem. Soc.* (1964) 1389.
16. Schumann, H., Köpf, H., Schmidt, M., *Chem. Ber.* (1964) **97**, 2395.
17. Schumann, H., Köpf, H., Schmidt, M., *J. Organometal. Chem.* (1964) **2**, 159.
18. Jones, K., Lappert, M. F., *J. Organometal. Chem.* (1965) **3**, 295.
19. Schumann, H., Rösch, L., *Chem. Ber.* (1974) **107**, 854.
20. Schumann, H., Rösch, L., *J. Organometal.Cem.* (1973) **55**, 257.
21. Schumann, H., Schumann-Ruidisch, I., Schmidt, L. M. in "Organotin Compounds," Vol. II, p. 581 , A. K. Sawyer, Ed., M. Dekker, New York, 1971.
23. Schumann, H., Jutzi, P., Roth, A., Schwabe, P., Schauer, E., *J. Organometal. Chem.* (1967) **10**, 71.
24. Schumann, H., Roth, A., *J. Organometal. Chem.*, (1968) **11**, 125.
25. Schumann, H., Roth, A., *Chem. Ber.*, (1969) **102**, 3731.
26. Schumann, H., Jutzi, P., *Chem. Ber.*, (1968) **101**, 24.
27. George, T. A., Jones, K., Lappert, M. F., *J. Chem. Soc.* (1965), 2157.
28. Schumann, H., Jutzi, P., Schmidt, M., *Angew. Chem.*, (1965) **77**, 912.
29. Schumann, H., Kroth, H. J., Rösch, L., *Z. Naturforsch.*, (1974) **29b** 608.
30. Schumann, H., Roth, A., Stelzer, O., *J. Organometal. Chem.* (1970) **24** 183.
31. Scherer, O. J., Schieder, G., *Chem. Ber.*, (1968) **101**, 4184.
32. Bald, J. F., MacDiarmid, A. G., *J. Organometal. Chem.* (1970) **22**, C22.
33. Schumann, H., Stelzer, O., Kuhlmey, J., Niederreuther, U., *Chem. Ber.* (1971) **104**, 993.
34. Schumann, H., Pfeifer, G., Röser, H., *J. Organometal. Chem.* (1972) **44**, C 10.
35. Schumann, H., Breunig, H. J., Frank, U., *J. Organometal. Chem.* (1973) **60**, 279.
36. Schumann, H., Breunig, H. J., *J. Organometal. Chem.* (1975) **87**, 83.
37. Schumann, H., Stelzer, O., Niederreuther, U., Rösch, L., *Chem. Ber.* (1970) **103**, 2350.
38. Schumann, H., Rösch, L., Kroth, H. J., Neumann, H., Neudert, B., *Chem. Ber.* (1975) **108**, 2487.
39. Schumann, H., Stelzer, O., Niederreuther, U., Rösch, L., *Chem. Ber.* (1970) **103**, 1383.
40. Schumann, H., Rösch, L., Neumann, H., Kroth, H. J., *Chem. Ber.* (1975) **108**, 1630.
41. Schumann, H., Breunig, H. J., *J. Organometal. Chem.* (1974) **76**, 225.
42. Abel, E. W., Sabherwal, I. H., *J. Organometal. Chem.* (1967) **10**, 491.
43. Schumann, H., Kroth, H. J., *J. Organometal. Chem.* (1971) **32**, C47.
44. Abel, E. W., Crow, J. P., Illingworth, J. M., *J. Chem. Soc. A* (1969) 1631.
45. Schumann, H., du Mont, W. W., *Z. Naturforschg.* (1976) **31b**, 90.
46. du Mont, W. W., Schumann, H., *J. Organometal. Chem.* (1975) **85**, C 45.
47. Schumann, H., Held, J., Kroth, H. J., du Mont, W. W., Wöbke, B., *J. Organometal. Chem.* (1976) **105**, 393.

48. Schumann, H., Held, J., Kroth, H. J., du Mont, W. W., *Chem. Ber.* (1976) **109**, 246.
49. Schumann, H., du Mont, W. W., Wöbke, B., *Chem. Ber.*, (1976) **109**, 1017.

RECEIVED May 3, 1976. Work supported by Deutsche Forschungsgemeinschaft, Bonn-Bad Godesberg and Senator für Wirtschaft des Landes Berlin in connection with the ERP program.

5

Di- and Trivalent Trimethylsilyl-Substituted Tin Amides and Related Compounds Such as $Sn[N(SiMe_3)_2]_2$ or $_3$

MICHAEL F. LAPPERT and PHILIP P. POWER

School of Molecular Sciences, University of Sussex, Brighton BN1 9QJ, England

New kinetically stable homoleptic tin(II) and tin(III) amides [Sn(NX$_2$)$_2$ (X = Me$_3$Si, Me$_3$Ge, Et$_3$Si, Et$_3$Ge, or Ph$_3$Ge) and Sn(NX'$_2$)$_3$ (X' = Me$_3$Si, Me$_3$Ge, or Et$_3$Ge)] were prepared: Sn(NX$_2$)$_2$, from LiNX$_2$ and SnCl$_2$ in diethyl ether at 0°C; and Sn(NX'$_2$)$_3$ by photolysis of Sn(NX'$_2$)$_2$ in n-hexane at 20°C. The diamides are yellow-orange at 20°C but become red in the vapor state. ^{1}H NMR spectra indicate a singlet monomer structure, cryoscopic molecular weights are those of the monomers, and mass spectra reveal parent molecular cations. Six types of reactions were observed for the diamides: Lewis base properties, ligand exchange, metathesis with protic reagents HA, insertion into SnN by PhNCO, oxidative addition, and disproportionation by irradiation. Tin(III) amides have long half-lives at 20°C and show the expected ^{119}Sn, ^{117}Sn, and ^{14}N hyperfine couplings appropriate for the pyramidal metal-centered radicals.

lthough the di- and tetravalent states for tin are represented by ligands such as Cl$^-$ or O^{2-}, the dialkylamides have only been represented by quadrivalent tin. We therefore wished to synthesize stable divalent amides and, if possible, trivalent compounds. The corresponding di- and trialkyl tins Sn[CH(SiMe$_3$)$_2$]$_2$ or $_3$ have been successfully prepared (1, 2, 3, 4). Accordingly we reported the synthesis and physical properties of the isoelectronic Sn[N(SiMe$_3$)$_2$]$_2$ or $_3$ (5, 6) and described the reactions of the diamide, summarized in Figure 1 (5); additionally its photolysis yielded Ṡn[N(SiMe$_3$)$_2$]$_3$ (6). The diamide is an orange, low melting (37°–38°C) crystalline solid, subliming at ca. 60°C/10^{-3} mm Hg to produce a red vapor; cryoscopy showed it to be a monomer in benzene, in which it is diamagnetic by ^{1}H NMR. Zuckerman and Schaeffer prepared this compound independently and report it as a dimer in solution or

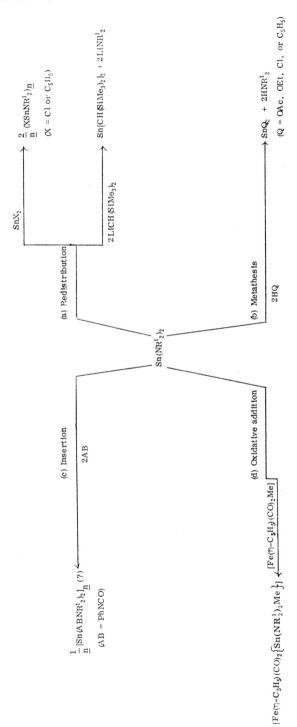

Figure 1. Some reactions of $Sn[N(SiMe_3)_2]_2$ from Ref. 5 ($R^1 = Me_3Si$, $Cp = \eta\text{-}C_5H_5$)

vapor. They proposed alternative possibilities of μ-amido bridging (7). Since then, other interesting and relevant compounds have been described: $Sn[N(CMe_3)SiMe_3]_2$ (5), $\dot{S}n[N(CMe_3)SiMe_3]_3$ (8), $Sn(THF)N(SiMe_3)$-$C_6H_4.NSiMe_3$-o (7), $Sn(THF)N(SiMe_3)C_{10}H_6.NSiMe_3$-8 (7), and the cyclic compounds I and II.

I and II were obtained from the appropriate N,N'-dilithioamine and $SnCl_2$ and I, a red liquid, was monomeric in benzene (9); $[Sn(NMe_2)_2]_2$ was μ-bis(di-methylamido) bridged (10). ESCA studies of $Sn[N(SiMe_3)_2]_2$ vapor show core-binding energies (eV) as follows: Sn $(3d_{5/2})$, 491.93 ± 0.05; N (1s), 402.01 ± 0.07; Si $(2p_{3/2})$, 105.97 ± 0.07; and C (1s), 289.36 ± 0.07 (11); these were interpreted on the basis of a monomer model with SiN $d \leftarrow p$ π-bonding. The He(I) photoelectron spectrum shows bands at low IP assigned to combinations of nitrogen lone-pair and Sn lone-pair orbitals, respectively (12). ^{119}Sn Mössbauer spectra are available (7).

Here we report extensions of earlier studies (5, 6) into (a) a wider range of amido ligands, (b) a more detailed investigation of the chemical properties of $Sn[N(SiMe_3)_2]_2$, (c) a study of the photolysis of the new amides, and (d) the ESR characterization of trivalent tin amides which result from (c) in all but the sterically most hindered situations.

Table I.   Yields and Some

Compound	Yield (%)	Appearance
$Sn[N(SiMe_3)_2]_2$[d]	79	orange crystals
$Sn[N(GeMe_3)_2]_2$	82	orange crystals
$Sn[N(SiEt_3)_2]_2$	63	orange liquid
$Sn[N(GeEt_3)_2]_2$	69	orange liquid
$Sn[N(GePh_3)_2]_2$	73	orange liquid
$Sn(Cl)N(SiMe_3)_2$[d]	51	colorless crystals
$Sn(Cl)N(GeMe_3)_2$	40	colorless crystals

[a] In benzene relative to solvent benzene.
[b] In CH_2Cl_2 relative to solvent CH_2Cl_2

The use of the bulky bis(trimethylsilyl)amido ligand $\overline{N}(SiMe_3)_2$ was pioneered initially by Bürger and Wannagat and subsequently over a broad range of first row transition elements and lanthanides by Bradley and co-workers. This field has been comprehensively reviewed (*13*). The choice of this ligand means that: (a) the parent amine hexamethyldisilazane, $HN(SiMe_3)_2$, is readily available and is easily converted into its N-lithio derivative, and (b) the size of the ligand stabilizes metal low coordination numbers, for the d and f-block metals the rare coordination number of three being particularly common. There is probably significant Si–N $(d \leftarrow p)$ π-bonding. The absence of β-hydrogen atoms and the presence of β-silicon atoms makes metal amide decomposition by a β-elimination energetically unfavorable. Metal amides are often useful synthetic precursors of other metal derivatives (*14, 15, 16*), and this is true also of the bis(silyl) amides. A striking recent example is in the synthesis of $Ge[CH(SiMe_3)_2]_2$ from $Ge[N-(SiMe_3)_2]_2$ (*17*), particularly because germanium(II) halides are not suitable precursors.

Results

Tin(II) Amides: Preparation, Characterization, Structure, and Bonding. The compounds were obtained by the conventional procedure shown in Reactions 1 and 2 (X = Me_3Si, Me_3Ge, Et_3Si, Et_3Ge, or Ph_3Ge). [HNX_2 compounds were kindly provided by P. Rivière and M. Rivière-Baudet.]

$$n\text{-BuLi} + HNX_2 \xrightarrow[0\ ^{\circ}C]{Et_2O/n\text{-}C_6H_{14}} LiNX_2 + C_4H_{10} \qquad (1)$$

$$2LiNX_2 + SnCl_2 \xrightarrow[0\ ^{\circ}C]{Et_2O} Sn(NX_2)_2 + 2LiCl \qquad (2)$$

Yields and physical data are summarized in Tables I and II, and more details on cryoscopic molecular weight data are in Table III; we note that with cyclohexane

Properties of Tin(II) Amides

Mp ($^{\circ}C$)	Bp ($^{\circ}C/mm\ Hg$)	^1H NMR[a] (τ)	$\nu_{(M-N)}$[c] (cm^{-1})
37–38	84/0.04	9.6	400w
35–37	70/10⁻³	9.61	406w
	150/10⁻³	8.8 complex	411w, br
	150/10⁻³	8.79 complex	412w, br
		2.69 br[b]	
150 (dec)	90/10⁻³	9.53	388w
150 (dec)	100/10⁻³	9.55	380w

[c] As neat liquid.
[d] In part from Ref. 5.

Table II.   Analytical and Molecular Weight Data for Tin(II) Diamides

	% C		% H		% N		Molecular Weight	
Compound	Calc.	Found	Calc.	Found	Calc.	Found	Calc.	Found[a]
$Sn[N(SiMe_3)_2]_2$	32.8	32.8	8.3	8.1	6.4	6.2	438.5	440[b]
$Sn[N(GeMe_3)_2]_2$	23.3	23.3	5.88	5.75	4.54	4.6	617.4	630
$Sn[N(SiEt_3)_2]_2$	47.4	47.5	9.96	10.1	4.7	4.8	607.6	650
$Sn[N(GeEt_3)_2]_2$	36.7	37.1	7.7	7.8	3.56	3.62	785.8	850
$Sn[N(GePh_3)_2]_2$	63.5	63.3	4.44	4.6	2.06	1.96		
$Sn(Cl)N(SiMe_3)_2$	22.9	22.7	5.8	5.5	4.4	4.4	dimer	
$Sn(Cl)N(GeMe_3)_2$	17.85	17.9	4.49	4.6	3.47	3.4		

[a] Cryoscopy in cyclohexane.
[b] Cryoscopy in benzene.

as solvent very substantial freezing point depressions are observed and the margins for error are therefore quite small.

The compounds are monomers in vapor and solution and, as deduced from (a) the low boiling points [e.g., compare $Sn\{N(SiMe_3)_2\}_2$, b.p. 84°C/0.04 mm Hg and $Sn(NEt_2)_4$, b.p. 90°C/0.05 mm Hg, which have $M = 440$ and 406, respectively], and (b) less definitively, by mass spectrometry when the parent molecular mono cation is the ion at highest m/e.   Attempts are in progress to obtain x-ray (18) and, for the vapor, electron diffraction (19) data.   By analogy with the solid state structure for $[Sn\{CH(SiMe_3)_2\}_2]_2$ (17), it is likely that the diamagnetic crystalline diamide will have a similar structure; schematic representations for bonding and structure in the monomer III and dimer IV are therefore (X = Me_3Si, Me_3Ge, Et_3Si, Et_3Ge, or Ph_3Ge):

sp_xp_y hybrid orbitals
p_z orbital vacant
III, Monomer (solution or vapor)

Bent Sn–Sn double bond
IV, Dimer (crystal)

Table III.   Molecular Weight Data for Two Divalent Tin Amides by Cryoscopy in Cyclohexane

Compound	ΔT^a	Molality by Weight	Cryoscopic Molality[b,c]
$Sn[N(SiEt_3)_2]_2$	0.54	0.0282	0.027
$Sn[N(GeEt_3)_2]_2$	0.68	0.036	0.034

[a] ΔT = freezing point depression in degrees.
[b] In cyclohexane.
[c] Cryoscopic constant, k for cyclohexane = 20.

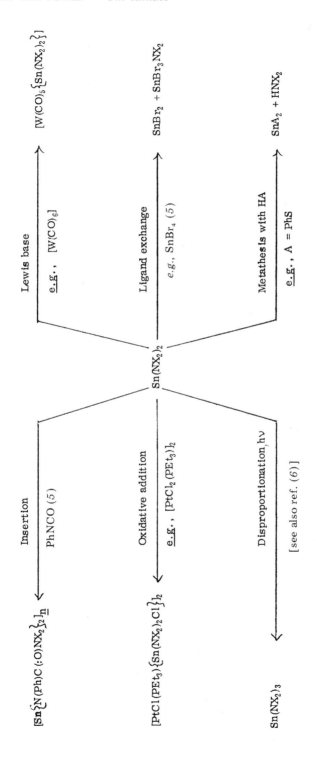

Figure 2.   Typical reactions of Sn(NX₂)₂ (X = Me₃Si)

Tin(II) Amides: Chemical Properties. These are classified in Figure 2 for $Sn[N(SiMe_3)_2]_2$. This compound's electron-rich character, as indicated by He(I) photoelectron spectra (12) and monomer structure III, is well illustrated by its tendency to act as a Lewis base or substrate for oxidative addition. On the other hand, unlike its carbon isoelectronic analog $Sn[CH(SiMe_3)_2]_2$ (4), we have been unable to detect significant Lewis acid character; for example there is no evidence for coordination with pyridine whereas the dialkyl forms a white crystalline 1,1-adduct although even this dissociates upon warming in vacuo. Further the tin(II) amide does not form (but see Ref. 20) a complex with α,α'-bipyridyl (bipy) at $0°-20°C$ (although there is a reaction with o-phenanthroline). The metathetical exchange reactions with protic compounds, insertion of heterocumulenes, and ligand exchange reactions are found with a wide range of metal dialkylamides [as first shown for a tin(IV) compound Me_3SnNMe_2 (13)] because the nitrogen centers act as nucleophiles; hence, apart from the last reaction type (to a limited extent only) the others are not found for the dialkyl (4). However, the photo-induced disproportionation is common to the two systems (6, 8).

Qualitative differences in the chemical behavior of $Sn[N(SiMe_3)_2]_2$ and $Sn[CH(SiMe_3)_2]_2$ (21) are attributable either to the relative ease of fission of the Sn–N or Sn–C bond (N > C) or (cf., reactions with N-bases) the electrophilicity of the tin center (C > N). Thus, whereas the Sn–N bond in the amide is readily cleaved by a wide range of C-, P-, O-, S-, Hal, or metal-centered protic reagents (Figures 1 and 2), under similar conditions the alkyl (21) only reacts with HHal or some metal hydrides—e.g., $[Mo(\eta-C_5H_5)(CO)_3H]$—and then affords the tin(IV) products of oxidative addition. Quantitative differences show themselves in (a) the greater nucleophilicity of the alkyl [cf., the trend in first ionization potential, $SnR_2 > Sn(NR'_2)_2$ (12)] and (b) the higher propensity of the alkyl to undergo oxidative addition, especially with an alkyl or aryl halide (22). As an illustration of (a) we cite the different reactions of amide or alkyl with $[PtCl_2(PEt_3)]_2$ (vide infra) and the fact that R_2Sn complexes of Rh(I) and Fe(I) have been obtained (21), $[RhCl(PPh_3)_2(SnR_2)]$ and $[Fe_2(\eta-C_5H_5)_2(CO)_3(SnR_2)]$ whereas there was no reaction under similar conditions between the amide and $[RhCl(PPh_3)_3]$, $[RhCl(C_2H_4)(PPh_3)_2]$, or $[Fe_2(\eta-C_5H_5)_2(CO)_4]$. Factors responsible for these differences between $Sn[CH(SiMe_3)_2]$ and $Sn[N(SiMe_3)_2]_2$ include (a) electronegativity (N > C), (b) Sn–X bond polarity (N > C), and (c) π-bonding effects,

$$\left[\begin{array}{c} CH_2 \quad Cl \quad CH_2 \\ CH \text{---} Pd \quad Pd \text{---} CH \\ CH_2 \quad Cl \quad CH_2 \end{array} \right] + 2Sn[N(SiMe_3)_2]_2 \xrightarrow[25\ °C]{n\text{-}C_6H_{14}}$$

$$2 \left[\begin{array}{c} CH_2 \quad Cl \\ CH \text{---} Pd \\ CH_2 \quad Sn[N(SiMe_3)_2]_2 \end{array} \right] \quad (3)$$

especially Ṅ—Si $(p \rightarrow d)$. The properties of the tin(II) amide may also now be compared with those of the few other tin(II) organic compounds known, namely $Sn(C_5H_5)_2$ or $Sn(acac)_2$ or related derivatives (23).

Lewis bases are well known to cleave many di-μ-chloride bridged binuclear complexes, and this feature is also noted in Reaction 3.

Photolysis of $[W(CO)_6]$ and the tin(II) amide in n-hexane or benzene for 40 hr at 20°C yields the monostannylene derivative $[W(CO)_5Sn\{N(SiMe_3)_2\}_2]$.

$Sn[N(SiMe_3)_2]_2$ behaves as a substrate for oxidative addition as shown in Reaction 4, where insertion into the Pt–Cl bonds is favored over bridge cleavage.

Under similarly mild conditions the tin(II) alkyl reacts with the same diplatinum reagent (V) to give the product of insertion as well as bridge cleavage $[PtCl(PEt_3)(SnR_2)(SnR_2Cl)]$ (21).

Reaction of an alkyl or aryl halide, R'X, (1 mol) in n-hexane with the tin(II) amide (1 mol) gave the 1:1-tin(IV) adduct, VI, shown for R'X = MeI or PhBr. ^{1}H NMR spectra of VI showed a singlet Me$_3$Si signal, indicating either a planar heavy atom environment about each N (NSi$_2$Sn) or an energetically facile inversion at N. Thus, the corresponding carbon compounds, $Sn[CH(SiMe_3)_2]_2(R)X$, showed two diastereotopically distinct Me$_3$Si ^{1}H signals (22). The bromobenzene reaction was catalyzed by the addition of $\leqslant 0.05M$ bromoethane.

The products of the reaction with benzenethiol in benzene were identified as hexamethyldisilazane and the tin(II) benzenethiolate $[Sn(SPh)_2]_n$ (25).

Photolysis experiments in n-hexane at 20°C led to the formation and ESR characterization of two new tin-centered radicals, VII, which on the basis of their hyperfine splitting constants (Table IV) are found to be pyramidal. [These experiments were carried out by M. J. S. Gynane.]   When R_3M' was relatively bulky (Et_3Si or Ph_3Ge), the radical was not formed.

$$(R_3M')_2N\underset{\underset{VII}{\overset{\cdot}{Sn}}}{\overset{N(M'R_3)_2}{\diagdown\ |\ \diagup N(M'R_3)_2}}$$

$$[R_3M' = Me_3Si\ (6),\ Me_3Ge,$$
$$\text{or } Et_3Ge]$$

Analytical and spectroscopic data and yields on the new compounds are shown in Tables I–V. Data are also provided in the experimental section on attempts to prepare and characterize $[Sn(NR_2)_2]_x$ (R = i-Pr or Ph) and $[Sn(N$-i-$Pr_2)\{N(SiMe_3)_2\}]_y$.

Table IV.   Parameters of the Tris(amido) Radicals (ESR)[a]

Radical	g	a (N)[b]	a (M)[b]
$\dot{S}n[N(SiMe_3)_2]_3$[c]	1.9912	1.09	$\begin{cases}317.6\ (^{117}Sn)\\342.6\ (^{119}Sn)\end{cases}$
$\dot{S}n[N(CMe_3)(SiMe_3)]_3$[d]	1.9928	1.27	—
$\dot{S}n[N(GeMe_3)_2]_3$	1.9924	1.07	—
$\dot{S}n[N(GeEt_3)_2]_3$	1.9939	1.19	—

[a] These have $t_{1/2} >$ ca. 1 month at 20°C except for $\dot{S}n[N(CMe_3)(SiMe_3)]_3$ (8) (ca. 5 min).
[b] In mT = milliteslas.
[c] From Ref. 6.
[d] From Ref. 8.

Table V.   Analytical and Spectroscopic Data on Derivatives of $Sn(NR_2)_2$ (R = Me_3Si)[a]

Compound	% C		% H		% N	
	Calc.	Found	Calc.	Found	Calc.	Found
$[W(CO)_5Sn(NR_2)_2]$[b]	27.6	26.6	4.83	4.3	3.51	3.73
$[Pd(Cl)(\pi\text{-allyl})Sn(NR_2)_2]$[c]	28.9	27.4	6.17	6.64	4.2	4.5
$Sn(Me)I(NR_2)_2$[d]	26.8	26.9	6.7	6.88	4.83	4.85
$Sn(Br)Ph(NR_2)_2$[e]	36.2	36.4	6.93	6.85	4.69	4.74
$[Pt(Cl)(PEt_3)\{Sn(NR_2)_2Cl\}]_2$[f]	26.2	26.1	6.24	6.27	3.39	3.27

[a] ^1H NMR (τ) in C_6H_6 or $CDCl_3$ for compound c; IR as Nujol mulls, except for compound b (C_6H_{14}); all crystalline; yellow for compounds b, c, and f or colorless.
[b] $\nu_{(CO)}$ (cm^{-1}) 2073m, 1969s, 1958vs, 1931s, sh; ^1H NMR (τ) (Me_3Si) 9.59.
[c] ^1H NMR (τ) (π-allyl) 4.96m, 6.88d of d; (Me_3Si) 9.72.
[d] $\nu_{(Sn-N)}$ (cm^{-1}) 411s, 371s; $\nu_{(Sn-C)}$ (cm^{-1}) 535s; ^1H NMR (τ) (Me_3Si) 9.54; (CH_3) 8.83.
[e] $\nu_{(Sn-N)}$(cm^{-1}) 412s, 372s; $\nu_{(Sn-C)}$ (cm^{-1}) 465w; ^1H NMR (τ) (Me_3Si) 9.6.
[f] $\nu_{(Sn-N)}$ (cm^{-1}) 405s, 370s; $\nu_{(Sn-Cl)}$ (cm^{-1}) 305s; ^1H NMR (τ) (Me_3Si) 9.3.

Experimental

Since all the tin(II) and tin(III) compounds are exceedingly air and moisture sensitive, Schlenk tube and vacuum line manipulative techniques were used throughout. Compounds were thus handled in vacuo or under dry and anaerobic conditions at ambient N_2 or Ar pressures.

Preparation of the Tin(II) Amides. Bis(trimethylsilyl)amidolithium monoetherate was added slowly to a stirred suspension of anhydrous tin(II) chloride in a 2:1 molar ratio and gave an immediate reaction with precipitation of white lithium chloride and formation of a yellow solution. For R = Me_3Si or Me_3Ge, the yellow color disappeared with further stirring; only after half the lithium amide was added did the yellow color persist. After stirring at room temperature for 2 hr, the ether was removed in vacuo. The tin amide was extracted with n-hexane, and filtration gave an orange solution. Removal of volatiles under vacuum and distillation of the residue in a molecular still gave the tin(II) amide.

Bis[trimethylsilyl (or germyl)]amidochlorotin(II), $SnCl(NR''_2)(R'' = Me_3Si$ or Me_3Ge). Bis[trimethylsilyl (or germyl)]amidolithium monoetherate and anhydrous tin(II) chloride (equimolar proportions) reacted similarly in diethyl ether at 20°C. At first the reaction mixture turned yellow, but after several minutes the color was discharged. After 3 hr stirring at 20°C, the solution was filtered, and volatiles were removed. Extraction with toluene and repeated recrystallization from toluene/n-hexane yielded the product as colorless crystals.

Preparation of the Stannylene Complexes. Photolysis of an equimolar mixture (0.5 g) of hexacarbonyltungsten(0) and bis[bis(trimethylsilyl)amido]tin(II) in n-hexane (20 cc) for 40 hr using a medium pressure mercury lamp led, after removal of excess $[W(CO)_6]$ by filtration of a cooled (−78°C) solution and reduction in volume of the filtrate to ca. 2 cc, to yellow crystals of penta(carbonyl){bis[bis(trimethylsilyl)amido]stannylene}tungsten(0); these were filtered at −78°C, and were recrystallized from n-pentane.

1:1-Addition, on a 1-mmolar scale, of bis[bis(trimethylsilyl)]amidotin(II) in n-hexane to a suspension of η-diallyl-μ-dichlorodipalladium(II) in n-hexane gave an immediate reaction; the yellow crystalline η-allyl(chloro){bis[bis(trimethylsilyl)amido]stannylene}palladium(II) was obtained after filtration and concentration of the filtrate. It was purified by recrystallization (n-C_6H_{14}/C_6H_5Me).

Oxidative Addition Reactions. Using the procedure described above for the palladium(II) experiment, reaction of the diplatinum(II) compound (1 mol) and the tin(II) amide (2 or 4 mol, in separate experiments) on a ca. 1-mmolar scale in n-hexane gave pale yellow crystals of bis[(triethyphosphine){chlorobis[bis-(trimethylsilyl)amido]stannyl}]di-μ-dichlorodiplatinum(II).

Interaction of methyl iodide or bromobenzene (1 mol) and the tin(II) amide (1 mol) in n-hexane (ca. 1 M) at 20°C gave a reaction whose rate depended upon the nature of the halide, requiring ca. 0.2 hr or ca. 5 days for MeI and PhBr, respectively. For the latter, the presence of ca. $0.05M$ EtBr resulted in the reaction's being essentially complete after 18 hr. There was no catalysis by $[n\text{-}Bu_4N]^+I^-$ or $SnBr[N(SiMe_3)_2]_2Et$. The reaction rate was followed initially by the disappearance of the color of the tin(II) amide but later also by ^1H NMR. Removal of volatiles at the end of each of these reactions at ca. $20°C/10^{-3}$ mm Hg afforded principally the 1:1-adduct, purified by distillation or sublimation.

Reaction with Benzenethiol. Reaction of the thiol (2 mol) and the tin(II) amide (1 mol) in benzene (ca. $1M$) instantly produced a yellow solution. Removal of solvent in vacuo gave the tin(II) dithiolate (found: C, 43.1; H, 2.97. calc. for $C_{24}H_{20}S_4Sn_2$: C, 42.7; H, 2.99%).

Photolysis Experiments. Each tin(II) amide $Sn[N(M'R_3)_2]_2$ (R_3M' = Me_3Ge, Et_3Si, Et_3Ge, or Ph_3Ge) in benzene (ca. $1M$) was irradiated in the cavity of a Varian E3 ESR spectrometer, using high microwave power (e.g., 50 mW) and high modulation amplitude (e.g., 0.5 mT). A strong ESR signal characteristic of $\dot{S}n[N(M'R_3)_2]_3$ (R_3M' = Me_3Ge or Et_3Ge) (Table V) was noted in all except two experiments (R_3M' = Et_3Si or Ph_3Ge), being a septet showing high field ^{117}Sn and ^{119}Sn satellites, which overlapped to give a 10-line pattern; the relative intensities were consistent with the overlapping of the fourth line of one septet with the first line of the other. A computer simulation of the two binomial septets (splitting a, separation $3a$) was in agreement with the spectrum obtained; septet structure was not found on the low field ^{117}Sn line. A line of 47.7 mT was assigned to the low field satellite of ^{117}Sn.

Attempts to Prepare and Characterize $[Sn(NR_2)_2]_x$ (R = i-Pr or Ph) and $[Sn(N\text{-}i\text{-}Pr_2)\{N(SiMe_3)_2\}]$. Addition of two equivalents of $LiNR_2$ (R = i-Pr or Ph) to one equivalent of $SnCl_2$ in diethyl ether at $-78°C$ and subsequent warming to $20°C$ gave a colored solution. Removal of solvent yielded an oil (brown for R = i-Pr or red for R = Ph), which was soluble in aliphatic and aromatic hydrocarbons for R = i-Pr, but for R = Ph soluble only in aromatic solvents. IR showed a medium intensity band assigned to $\nu_{asym}(SnN_2)$ at 370 (R = i-Pr) or 390 (R = Ph) cm^{-1}, and ^1H NMR showed resonances for R = i-Pr (in C_6H_6) at τ 8.60/8.71 (doublet) and 6.64 (septet), and for R = Ph (in cyclo-C_6H_{12}) at τ 2.84 (singlet) and 3.69 (multiplet).

Similarly, from equimolar amounts of $SnCl[N(SiMe_3)_2]$ and $LiN\text{-}i\text{-}Pr_2$ a brown oil was obtained, with $\nu_{(SnN_2)}$ at 400 (weak) and 369 (medium) cm^{-1} and ^1H NMR (in benzene) at τ 8.6/8.7 (doublet) and 6.62 (septet) and ($SiMe_3$) at τ 9.63.

Acknowledgment

We thank D. H. Harris, M. J. S. Gynane, P. Rivière, and M. Rivière-Baudet for useful discussions and other contributions and J. J. Zuckerman for an invitation to M. F. L. to present the material of this paper at the Organotin symposium.

Literature Cited

1. Davidson, P. J., Lappert, M. F., Pearce, R., *Acc. Chem. Res.* (1974) **7**, 209.
2. Lappert, M. F., *Adv. Chem. Ser.* (1976) **150**, 256.
3. Davidson, P. J., Lappert, M. F., *J. Chem. Soc., Chem. Commun.* (1973) 317.
4. Davidson, P. J., Hudson, A. Lappert, M. F., Lednor, P. W., *J. Chem. Soc., Chem. Commun.* (1973) 829.
5. Harris, D. H., Lappert, M. F., *J. Chem. Soc., Chem. Commun.* (1974) 895.
6. Cotton, J. D., Cundy, C. S., Harris, D. H., Hudson, A., Lappert, M. F., Lednor, P. W., *J. Chem. Soc., Chem. Commun.* (1974) 651.
7. Schaeffer, C. D., Zuckerman, J. J., *J. Amer. Chem. Soc.* (1974) **96**, 7160.
8. Lappert, M. F., Lednor, P. W., *Adv. Organomet. Chem.* (1976) **14**, 345.
9. Veith, M., *Angew. Chem., Internat. Edn.* (1975) **14**, 263.
10. Foley, P., Zeldin, M., *Inorg. Chem.* (1975) **14**, 2264.
11. Jolly, W. L. Avanzino, S., personal communication (1975).
12. Harris, D. H., Lappert, M. F., Pedley, J. B., Sharp, G. J., *J. Chem. Soc., Dalton* (1976) 945.
13. Harris, D. H., Lappert, J. F., *Organomet. Chem. Library* (1976) **2**, 13.
14. Jones, K., Lappert, M. F., *Proc. Chem. Soc.* (1964) 22.
15. Jones, K., Lappert, M. F., *Proc. Chem. Soc.* (1962) 358.
16. George, T. A., Lappert, M. F., *Chem. Commun.* (1966) 463.
17. Goldberg, D. E., Harris, D. H., Lappert, M. F., Thomas, K. M., *J. Chem. Soc., Chem. Commun.* (1976) 261.
18. Hursthouse, M. B., unpublished data.
19. Hedberg, K., Robbiette, A. K., unpublished data.
20. Zuckerman, J. J., "Abstracts of Papers," Centennial ACS Meeting, New York, April, 1976, INOR 008.
21. Cotton, J. D., Davidson, P. J., Lappert, M. F., *J. Chem. Soc., Dalton*, in press.
22. Gynane, M. J. S., Lappert, M. F., Miles, S. J., Power, P. P., *J. Chem. Soc., Chem. Commun.* (1976) 256.
23. Bos, K. D., Bulten, E. J., Noltes, J. G., *J. Organometal. Chem.* (1975) **99**, 397.
24. Cornwell, A. B., Harrison, P. G., Richards, J. A., *J. Organometal. Chem.* (1976) **108**, 47.
25. Harrison, P. G., Stobart, R. S., *Inorg. Chim. Acta* (1973) 306.

RECEIVED May 3, 1976.

6

Organotin Alkoxides and Amines: New Chemistry and Applications

JEAN-CLAUDE POMMIER and MICHEL PEREYRE

Laboratoire de Chimie Organique, Laboratoire des Composés Organiques du Silicium et de l'Etain, associé au C.N.R.S., Université de Bordeaux I, 33405–Talence, France

This paper deals with some aspects of organotin alkoxide and amine chemistry, mainly centered on the work done in Bordeaux. The first part includes some new preparations of organotin alkoxides and some of their properties: exchange reactions in organometallic chemistry, nucleophilic substitution either internal (with halides or oxiranes) or intermolecular (α-haloketones, alkylation of enolates), and finally oxidation to carbonyl derivatives. The behavior of organotin amines in nucleophilic displacement of α-haloketones and organic dihalides is investigated. They undergo either substitution or addition with carbonyl derivatives, depending on electronic and steric factors. An extension of these reactions to organotin enolates affords a route to organic enamines or, more interestingly, to organotin enamines. Some new preparations of this last class of compounds are reported.

T his review discusses some aspects of organotin (alkoxide and amine) chemistry mainly centered on the work of the Bordeaux group during the past five years and does not claim to be complete, the references cited being only examples. In particular, work done before 1970 has been omitted since it was been the subject of other papers or books (*1, 2, 3, 4, 5, 6, 7*).

Introduction

Organotin derivatives of alcohol and amines exhibit special properties owing to two conflicting factors:

(1)   Formally they are metal alkoxides or amides, the nucleophilic character of the oxygen or nitrogen being enhanced, as in the corresponding compounds derived from Group I or II metals.

(2)   The electropositive nature of tin is quite low among metals, and, consequently, bonds involving oxygen or nitrogen are rather covalent and exhibit

special reactivity. They are covalent in physical properties, they are generally liquids or low melting solids, and they may often react neat.

The reactions reported here relate to both these properties. A striking and general observation is that although they behave like their more electropositive analogues, they are less reactive and, therefore, often much more selective.

Organotin Alkoxides

Preparations. Depending on the final use of alkoxide, different methods of preparation may be used. To take advantage of the alkoxide properties (which is generally the case), the best and easiest route is that described by Davies et al., (*8, 9, 10, 11*) involving a fast and rapid reaction between the commercially available organotin oxides and alkyl carbonates.

$$R_6SnO + CO_3R_2' \longrightarrow 2R_3SnOR' + CO_2$$

$$R_2SnO + CO_3R_2' \longrightarrow R_2Sn(OR')_2 + CO_2$$

From these tin alkoxides, other compounds containing tin–oxygen bonds can be prepared using a transalkoxylation or a transesterification procedure (*see*, for example, Refs. *12, 13, 15, 16, 17, 18, 19, 21, 22, 23*):

$$\equiv SnOR' + ROH \longrightarrow \equiv SnOR + R'OH$$

$$\equiv SnOR' + AcOR \longrightarrow \equiv SnOR + AcOR'$$

In the past few years, other methods have been reported but mainly with interest in the organotin compound itself: (1) the metathesis between a metal alkoxide and an organotin halide (*24, 25, 26*), (2) reaction between tin oxides and alcohols (*10, 27, 28*) (particularly efficient for high boiling alcohols), and (3) alcoholysis of organotin amides (*14, 20, 29, 30, 31*).

Other miscellaneous methods have been reported in special cases using an interfacial technique (*32*), disproportionation (*33, 34*), the ring opening of oxiranes (*35, 36*), and reactions of magnesium alcoholates with dialkyltin oxides (*37*). A new and very convenient method was found here (*38*) for preparing trimethyl- and dimethylphenyltin alkoxides based on an exchange between the easily available tributyltin alkoxides and the corresponding organotin bromides. For example:

$$Bu_3SnOR + Me_3SnBr \longrightarrow Me_3SnOR + Bu_3SnBr$$
$$(R = Me\ 75\%;\ Et:80\%)$$

$$Bu_3SnOMe + \phi Me_2SnBr \longrightarrow \phi Me_2SnOMe + Bu_3SnBr\ (80\%)$$

It seems probable that the more efficient coordinating ability of the final organotin alkoxide provides the driving force for this type of reaction. Although tributylin methoxide is involved in the exchange of the methoxide groups observed by NMR

(10, 39), no evidence of coordination appears in its IR (41), whereas a strong association was found in the case of the trimethyltin analogue (40). This brief overview of the different preparations of organotin alkoxides shows that such compounds are easily prepared either to study their own chemistry or for synthetic purposes.

Exchange Reactions: Use in Organometallic Chemistry. According to the Pearson theory, tin is a softer acid than silicon or germanium and thus links preferentially to soft bases. Accordingly, many exchanges involving the final formation of a silicon or germanium–oxygen bonds have been reported such as with M–X (X = halogen), M–S, M–N, or even M–H (M = Si or Ge).

Since our initial discovery (42) of the exchange between tin alkoxides and chlorosilanes,

$$\equiv\!SnOR \;+\; \equiv\!SiCl \;\longrightarrow\; \equiv\!SnCl \;+\; \equiv\!SiOR$$

many applications of this reaction have been applied to the preparation of a variety of organosilicon or germanium derivatives (43–55). Many of these reports deal with the preparations of organosilicon or germanium enolates from the corresponding organotin derivative. An extension of these reactions to other heteroelements like sulfur (49, 56, 57) or nitrogen (58) may also be found in the literature.

During the past five years, our interest in this aspect of organometallic chemistry has been concentrated on the mechanism of the exchange between silanes and organotin alkoxides, a basic reaction for the preparation of organotin hydrides (44, 59, 60, 61, 62).

$$R_3SnOR' \;+\; R''_3SiH \;\longrightarrow\; R_3SnH \;+\; R''_3SiOR'$$

Optically active organosilicon hydrides were used to study isotopic and substituent effects on the rate of reaction. The results (63, 64, 65) show a net retention of configuration at the silicon center, a positive primary kinetic isotopic effect, and a high activation entropy. Furthermore, the rate enhancement by either electron withdrawing groups on the silane or electron releasing groups on the organotin compound is in accordance with a rate determining step in which

$$
\begin{array}{c}
\equiv\!Sn\text{----}O\!\!\diagup^{\displaystyle R} \\
\vdots \qquad | \\
\vdots \;\text{----}\,Si\!\equiv \\
H^{\diagdown} \\
\mathbf{1}
\end{array}
$$

$$
\equiv\!Si\!-\!H \;+\; \equiv\!Sn\!-\!OR \;\underset{k_{-1}}{\overset{k_1}{\rightleftharpoons}}\; \equiv\!Sn\!-\!\overset{+}{O}\!-\!R \;\overset{k_2}{\rightarrow}\; \equiv\!Sn\!-\!H \;+\; \equiv\!Si\!-\!OR
$$
$$
\underset{\equiv Si\,-\,H}{\overset{|\;\_}{}} \qquad\qquad\qquad\quad \mathbf{2}
$$

the incipient O–Si bond is more developed than the Sn–H bond.   These facts agree with either an Sn_i–Si mechanism (1) or with a two-step mechanism via an unstable pentacoordinated silicon intermediate occuring between two transition states of comparable energy (2).

Nucleophilic Substitution at a Carbon Center.   Organotin alkoxides act as nucleophiles toward organic and organometallic halides.   However, with alkyl halides their reactivity is quite low (66) since good yields are only obtained with the more reactive halides (allylic or benzylic derivatives).   However, in some cases this reaction is synthetically useful such as with chloromethylethers (55), tosyl chloride (23) sulfamoyl chloride (96), acid chlorides (69, 70, 67), cyanogen chloride (71), thyonyl chloride (72), phosgene (73), etc.

Intramolecular Nucleophilic Substitution

Organotin Halogenoalkoxides.   The general scheme of this reaction is an intramolecular substitution leading to an oxygenated heterocycle:

$$R_3Sn—O\text{~~~}X \longrightarrow R_3SnX + O\Big\rangle$$

The starting organotin alkoxide can be readily obtained from tributyltin methoxide and the corresponding halohydrin:

$$R_3SnOMe + HO\text{~~~}X \longrightarrow R_3Sn—O\text{~~~}X + MeOH$$

These alkoxides are generally thermally unstable and decompose at various temperatures depending on ring size of the product heterocycle, the nature of the halogen, the substitution of the carbon bearing the oxygen and, in some cases, the configuration of the molecule (74).

For oxiranes (75), some interesting features have been observed: the reaction is completely a stereospecific process:

$$
\begin{array}{c}
\text{Me—CH—CH—Me} \quad \xrightarrow[30\min]{140\,°C} \quad \text{Me—CH—CH—Me} \\
|| \diagdown\diagup \\
\text{Bu}_3\text{SnO} \quad \text{Br} O
\end{array}
$$

Bu$_3$SnBr
+

erythro	trans-90% (+10% of ketone, see below)
meso	cis-90% (+10% of ketone, see below)

The necessity of this condition is demonstrated by the following examples taken from the cyclohexyl series (76):

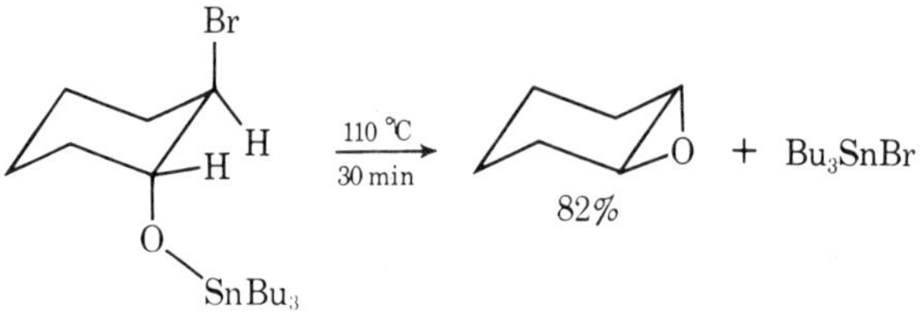

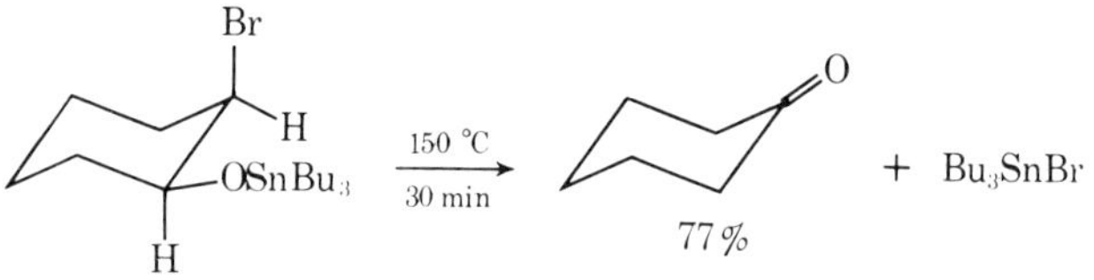

$$82\%$$

$$77\%$$

The former reaction proceeds via a nucleophilic intramolecular displacement by the oxygen:

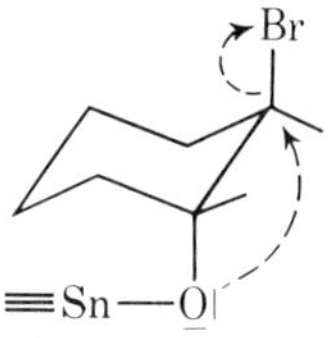

The latter by a hydride migration:

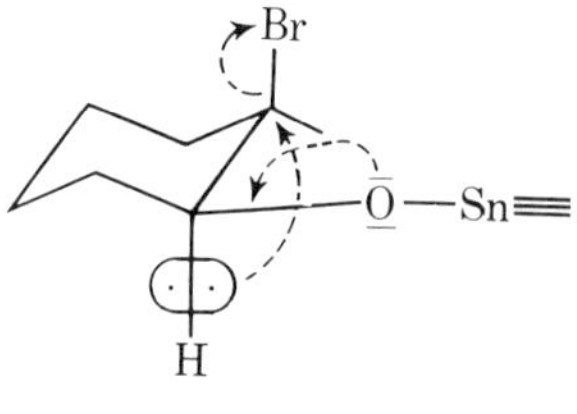

When the antiposition is substituted as in the example shown below, a transposition occurs leading to a reduction in ring size.

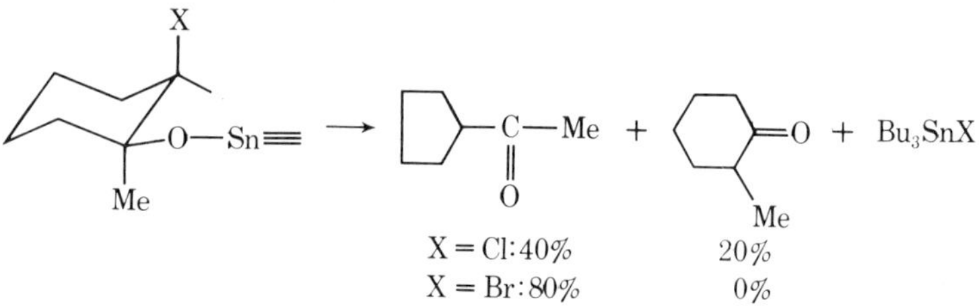

X = Cl: 40%	20%	
X = Br: 80%	0%	

The hydride migration is a much more energetic process than the formation of oxiranes, and thus the decomposition of the halogenalkoxides at high temperatures leads to a mixture of oxiranes and ketones while formation of the latter can be avoided by using milder conditions:

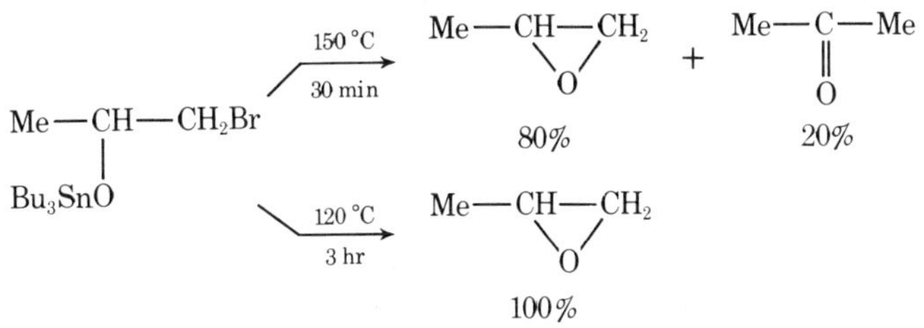

$$
\text{Me}-\text{CH}-\text{CH}_2\text{Br}
\begin{array}{l}
\quad \xrightarrow[\text{30 min}]{150\,°C} \quad \text{Me}-\overset{\diagup\text{CH}-\text{CH}_2}{\underset{O}{\diagdown}} \; + \; \text{Me}-\overset{\text{C}}{\underset{\parallel\,O}{}}-\text{Me} \\
\qquad\qquad\qquad\qquad\qquad 80\% \qquad\qquad 20\% \\
\text{Bu}_3\text{SnO} \\
\quad \xrightarrow[\text{3 hr}]{120\,°C} \quad \text{Me}-\overset{\diagup\text{CH}-\text{CH}_2}{\underset{O}{\diagdown}} \\
\qquad\qquad\qquad\qquad 100\%
\end{array}
$$

Obviously, the latter conditions are more favorable for synthetic purposes.

The study of this type of compound also shows clearly the influence of the substitution of the carbon bearing the oxygen; the more extensive the substitution, the easier the decomposition of the corresponding alkoxide. For example, for total reaction in 30 min, the organotin derivative of 1-bromo-2-propanol needs to be heated to 150 °C, whereas the derivative of 1-bromo-2-methyl-2-propanol is decomposed at temperatures as low as 70 °C. This fact parallels the nucleophilic character of the oxygen and corresponds to the situation observed in all the cases which have been studied.

A further important factor on the relative ease of the reactions is the nature of the halogen. Two striking examples taken in the cyclohexyl series are given below:

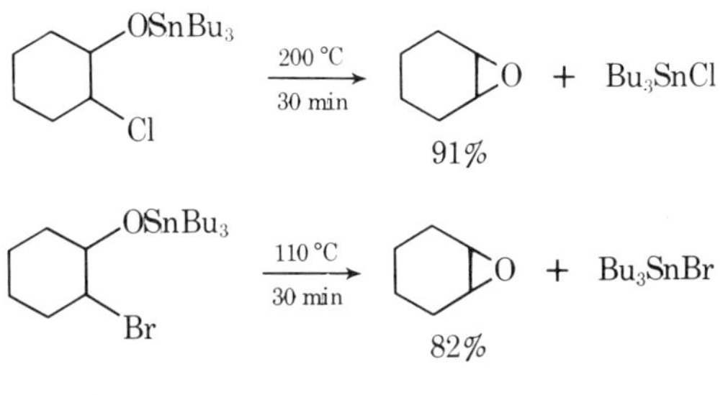

Considering these observations, the hydride migration and the effect of the halogen on the rate of the reaction, it is expected that side reactions will occur more readily with the chlorine derivatives. Thus, for synthetic purposes, it is better to start with a bromoalkoxide.

The case of the 3-haloalkoxide is of particular interest since it provides a very useful and sometimes unique route to oxetanes (77, 78):

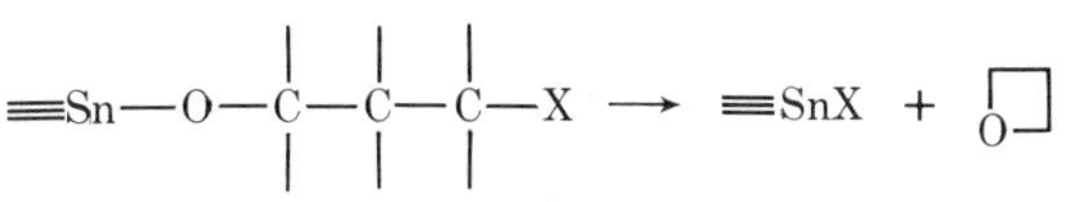

$$
\equiv\text{Sn}-\text{O}-\overset{|}{\underset{|}{\text{C}}}-\overset{|}{\underset{|}{\text{C}}}-\overset{|}{\underset{|}{\text{C}}}-\text{X} \longrightarrow \equiv\text{SnX} \; + \; \overset{\square}{O}
$$

The following examples are typical:

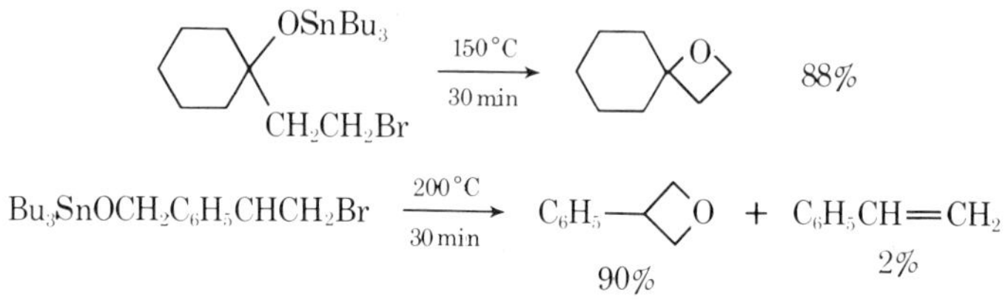

The latter reaction is significant since this oxetane is generally difficult to obtain because large amounts of styrene form readily. 3-Halogenoalkoxides can also be prepared from organotin oxides by a transesterification procedure analogous to that published by Matsuda et al. (79); their decomposition normally gives the corresponding oxetane (80).

An identical decomposition occurs with 4- and 5-haloalkoxides leading to tetrahydrofuranes or tetrahydropyranes (81), demonstrated by the following examples:

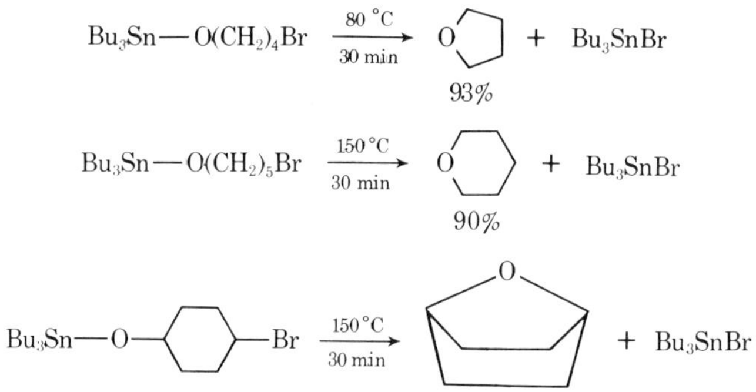

Other reactions based on this idea were studied with polyhaloalkoxides which lead to α-halogencarbonyl compounds via a halogenoxirane (isolated in the case of the chloride) (82):

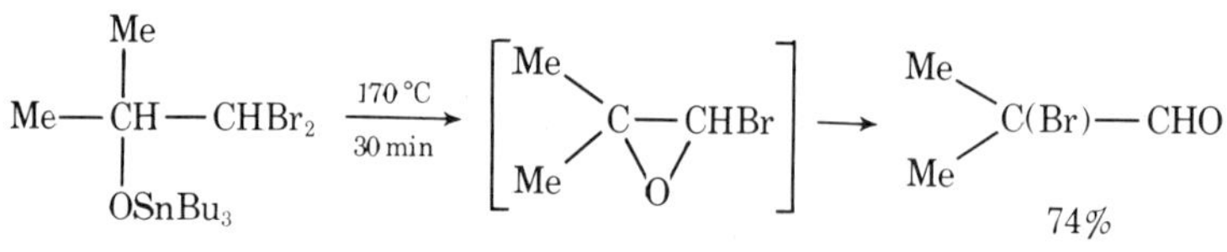

Other heterocycles may also be formed by either geminal or vicinal dibromides reacting with organotin dialkoxides (83). This route is less easy than the preceding ones although it provides a sometimes useful access to certain dioxanes or dioxolanes:

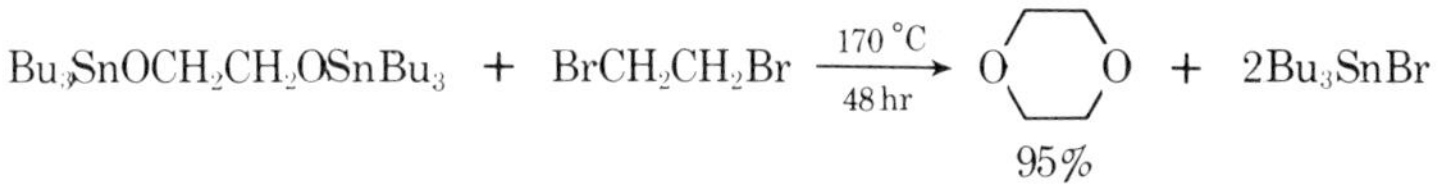

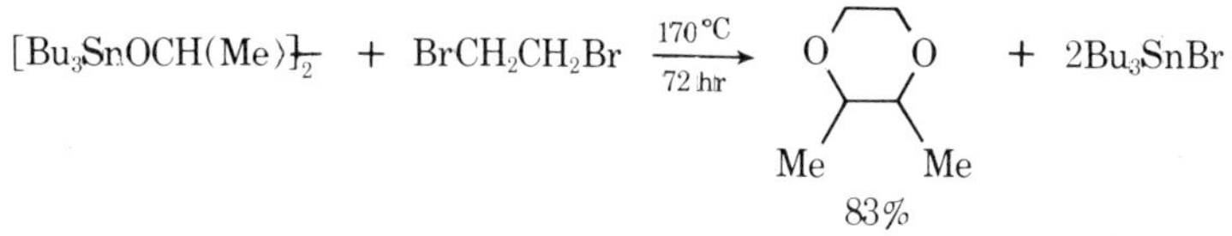

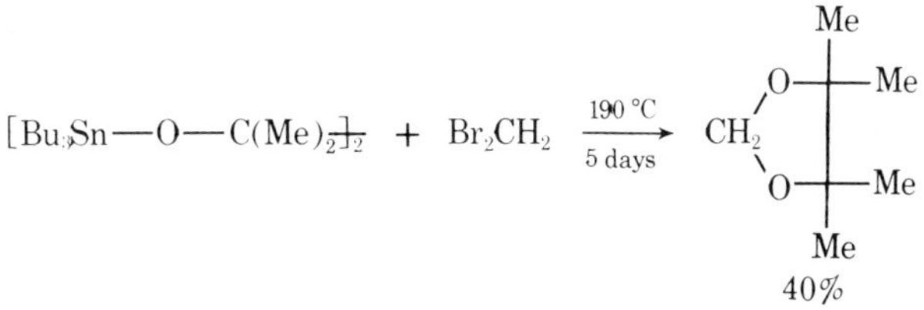

$$\text{Bu}_3\text{SnOCH}_2\text{CH}_2\text{OSnBu}_3 \; + \; \text{BrCH}_2\text{CH}_2\text{Br} \; \xrightarrow[48\,\text{hr}]{170\,^\circ\text{C}} \; \underset{95\%}{\text{O}\diagdown\diagup\text{O}} \; + \; 2\text{Bu}_3\text{SnBr}$$

$$[\text{Bu}_3\text{SnOCH(Me)}]_2 \; + \; \text{BrCH}_2\text{CH}_2\text{Br} \; \xrightarrow[72\,\text{hr}]{170\,^\circ\text{C}} \; \underset{\underset{83\%}{\text{Me} \qquad \text{Me}}}{\text{O}\diagdown\diagup\text{O}} \; + \; 2\text{Bu}_3\text{SnBr}$$

$$[\text{Bu}_3\text{Sn}\!-\!\text{O}\!-\!\text{C(Me)}_2]_2 \; + \; \text{Br}_2\text{CH}_2 \; \xrightarrow[5\,\text{days}]{190\,^\circ\text{C}} \; \text{CH}_2 \quad \underset{\underset{\underset{40\%}{\text{Me}}}{\text{Me}}}{\overset{\text{Me}}{\underset{}{\langle\substack{\text{O}\!-\!\!\!+\!\!-\text{Me} \\ \text{O}\!-\!\!\!+\!\!-\text{Me}}}}}$$

These reactions are believed to proceed by a two-step mechanism via a halogenalkoxide intermediate.

Organotin Alkoxyoxiranes. We have found (*84*) another interesting and clean intramolecular nucleophilic substitution involving the decomposition or organotin β-alkoxyoxiranes. This reaction led to two different heterocycles depending on the site of attack by the oxygen:

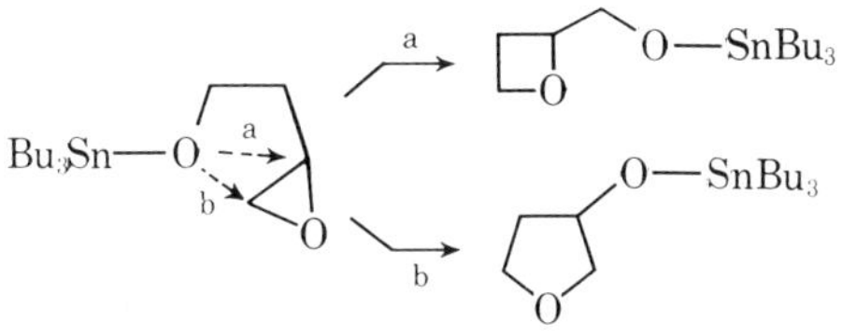

The new organotin alkoxide formed can be readily transformed into the corresponding alcohol by treating it with phthallic acid which precipitates the insoluble organotin phthallate. The orientation of the reaction is mainly a function of the substitution of the oxirane; if the terminal carbon atom of the oxirane is less substituted than the internal one, the formation of the oxetane is largely favored, but if it is equally or more substituted, a tetrahydrofurane is formed.

These facts support a mechanism in which a positive charge is developed on the carbons of the oxirane ring in the transition state. A representative selection of results is given at the top of p. 90.

$(Me)_2C-CH_2-CH-CH_2$ (epoxide), with $OSnBu_3$ substituent $\xrightarrow[\text{2) }C_6H_4(COOH)_2]{\text{1) }200\ °C,\ 3hr}$ oxetane product 84%

$(Me)_2C-C(Me)_2-CH-CH_2$ (epoxide), with $OSnBu_3$ substituent $\xrightarrow[\text{2) }C_6H_4(COOH)_2]{\text{1) }180\ °C,\ 3hr}$ oxetane product 75%

cyclohexyl$-CH_2-\underset{OSnBu_3}{\overset{Me}{C}}-CH_2$ (epoxide) $\xrightarrow[\text{2) }C_6H_4(COOH)_2]{\text{1) }190\ °C,\ 3hr}$ spiro product 80%

$(Me)_2C-C(CH_3)_2-\underset{OSnBu_3}{C}-CH(Me)$ (epoxide) $\xrightarrow[\text{2) }C_6H_4(COOH)_2]{\text{1) }190\ °C,\ 3hr}$ oxetane product $+$ tetrahydrofuran product
75%   25%
75%

$(Me)_2C-C(Me)_2-CH-CH(CH_3)$ (epoxide), with $OSnBu_3$ $\xrightarrow[\text{2) }C_6H_4(COOH)_2]{\text{1) }170\ °C,\ 3hr}$ tetrahydrofuran product 80%

These reactions present a convenient access to functionally substituted oxetanes and tetrahydrofuranes and promise some important applications.

Intermolecular Nucleophilic Substitution

Synthesis of α-Alkoxyketones.   The reactions of α-haloketones with the alkoxides of electropositive metals lead generally to a Favorski rearrangement according to various mechanisms.   In organotin alkoxides, the nucleophilic properties of the oxygen are more pronounced than its basic ones, and when they are allowed to react with α-haloketones, they undergo a normal substitution leading to α-alkoxyketones (68).   The cleanest results are obtained with the chlorine derivatives, whereas the bromoketones side reactions yield mainly a 2-hydroxy ketal.   Some typical examples are shown at the top of p. 91.

These reactions provide a convenient preparation of α-alkoxyketones in cases where a Favorski rearrangement is likely when using the alkoxide of a more electropositive metal.

Alkylation of Carbonyl Compounds. The direct nucleophilic substitution between organotin alkoxides and alkylhalides has been applied successfully to organotin enolates. These compounds, which are easily obtained by various methods (hydrostannation of α-enones, transesterification from enol esters), undergo substitution with various alkyl halides giving monoalkylated products in good yield (*85, 86, 87*). Organotin enolates exist as two metallotropic (C and O) forms in equilibrium. However, no evidence has been found for a difference in reactivity between the two isomers, and the form used in the text is that of the more stable one, and thus even the C isomer will be considered as an alkoxide.

As with other bimolecular nucleophilic substitution processes, the steric hindrance of the halide plays an important role, and the iodides tend to react more readily than the bromides.

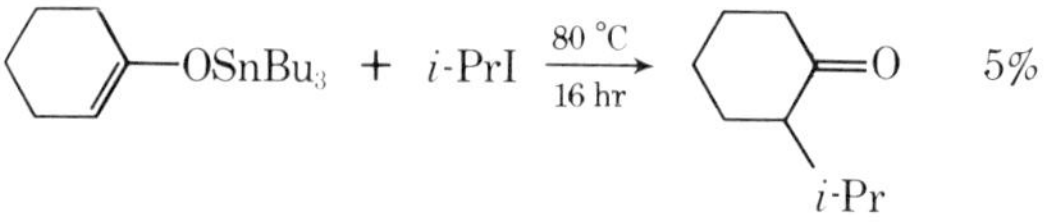

$$\text{(cyclohexenyl)—OSnBu}_3 \; + \; i\text{-PrI} \; \xrightarrow[16\,\text{hr}]{80\,^\circ\text{C}} \; \text{(cyclohexanone, } i\text{-Pr)} \quad 5\%$$

$$\text{Bu}_3\text{SnCH}_2\text{COMe} \; + \; \text{BuI} \; \xrightarrow[16\,\text{hr}]{80\,^\circ\text{C}} \; \text{BuCH}_2\text{COMe} \quad 43\%$$

$$\text{Bu}_3\text{SnCH}_2\text{COMe} \; + \; \text{BuBr} \; \xrightarrow[16\,\text{hr}]{80\,^\circ\text{C}} \; \text{BuCH}_2\text{COMe} \quad \text{traces}$$

In no case does O-alkylation occur, and no polyalkylation is observed. In fact, the problem of polyalkylation appears to be related to an exchange between the initial enolate and the already alkylated compound leading to a new enolate which is available for further alkylation:

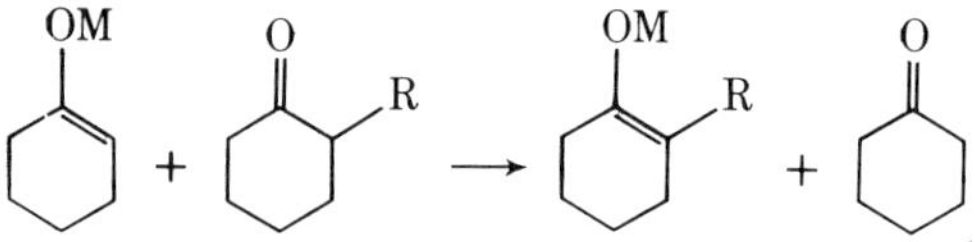

For tin (M = Bu₃Sn—), at least at 80 °C, the above reaction proceeded very slowly, and the polyalkylation was almost completely avoided. A further interesting aspect of this reaction is that it is completely regiospecific:

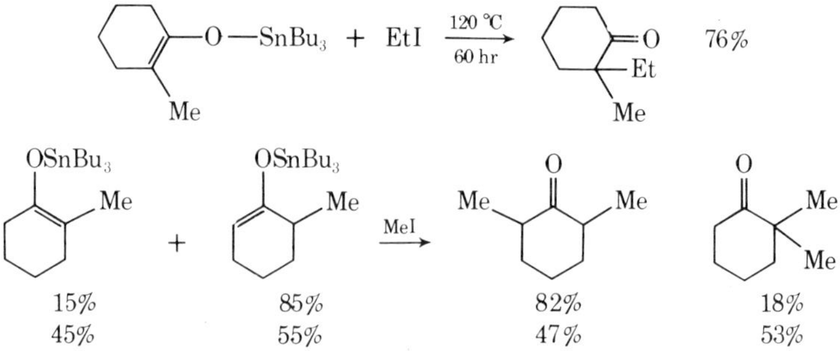

$$\text{(enol ether)—O—SnBu}_3 \; + \; \text{EtI} \; \xrightarrow[60\,\text{hr}]{120\,^\circ\text{C}} \; \text{(ketone, Et, Me)} \quad 76\%$$

OSnBu₃ / Me	OSnBu₃ / Me		O Me Me	O Me Me
15%	85%	MeI →	82%	18%
45%	55%		47%	53%

In the formation of aldehydes from enolates, good alkylation yields can be obtained if the steric hindrance of the alkylated aldehyde is substantial enough to prevent a side reaction involving its addition to the organotin enolate:

$$(\text{Me})_2\text{C}{=}\text{CHOSnBu}_3 \; + \; \text{IMe} \; \xrightarrow[14\,\text{hr}]{90\,^\circ\text{C}} \; t\text{-BuCHO} \quad 86\%$$

$$\text{Et}_2\text{C}{=}\text{CHOSnBu}_3 \; + \; \text{IMe} \; \xrightarrow[14\,\text{hr}]{90\,^\circ\text{C}} \; \text{Et}_2\text{MeCHO} \quad 82\%$$

$$\text{BuCH}{=}\text{CHOSnBu}_3 \; + \; \text{IMe} \; \xrightarrow[14\,\text{hr}]{90\,^\circ\text{C}} \; \text{Bu—CHCHO} \quad 36\%$$
$$\overset{|}{\underset{\text{Me}}{}}$$

The alkylation of organotin enolates with various functionally substituted organic halides has also been performed:

$$Bu_3SnCH_2COMe \ + \ BrCH_2COOEt \ \longrightarrow \ MeCO[CH_2]_2COOEt \quad 50\%$$

$$Bu_3SnCH_2COMe \ + \ ClCH_2OMe \ \longrightarrow \ \underset{\underset{OCH_2OMe}{|}}{Me-C}{=}CH_2 \quad 60\%$$

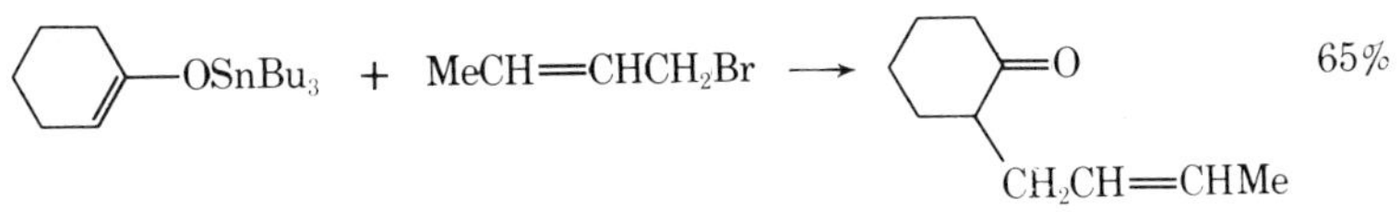

In summary, these reactions are very convenient synthetic procedures for monoalkylation, and since the starting enolates are easily prepared, the method can compete successfully with the more classical alkylation routes. Good alkylation yields can be obtained again via the organotin enolates by reaction either in the presence of magnesium or lithium salts or after a metal exchange reaction with a Grignard or organolithium reagent. In these cases, however, some polyalkylation occurs and, whenever possible, the direct alkylation route is prefered:

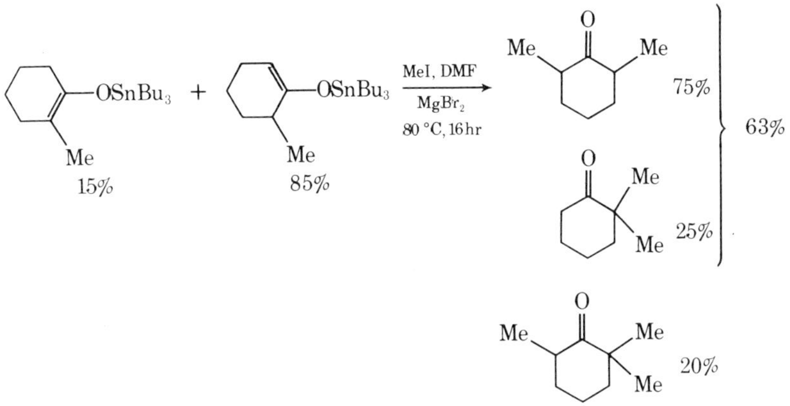

Oxidation of Organotin Alkoxides to Carbonyl Derivatives

Our first investigations concerning this type of reaction (88, 89) involved organotin alkoxides and polyhalomethanes which reacted in a free radical process (initiated by either UV light or A.I.B.N.) to give the oxidation product in good yields. Some examples are:

$$C_6H_5CH_2OSnBu_3 \ + \ CHBr_3 \ \xrightarrow{\text{UV, 80 °C}} \ \underset{61\%}{C_6H_5CHO} \ + \ Br_2CH_2 \ + \ Bu_3SnBr$$

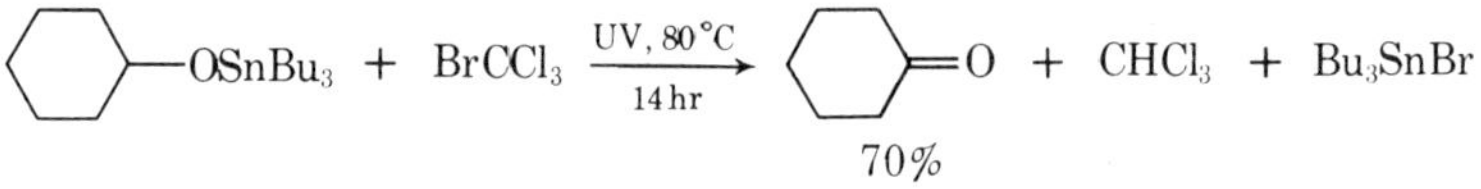

$$C_6H_5MeCHOSnBu_3 \ + \ BrCCl_3 \ \xrightarrow[\text{14 hr}]{\text{UV, 80 °C}} \ \underset{83\%}{C_6H_5COMe} \ + \ CHCl_3 \ + \ Bu_3SnBr$$

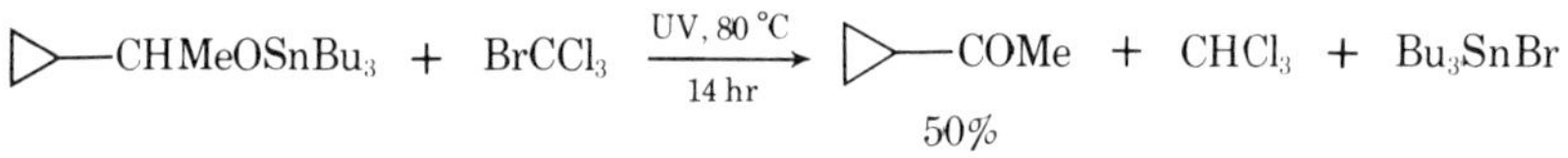

The mechanism of this reaction appeared initially to involve abstraction of the hydrogen on the functional carbon atom as a first step followed by the elimination of an organotin free radical:

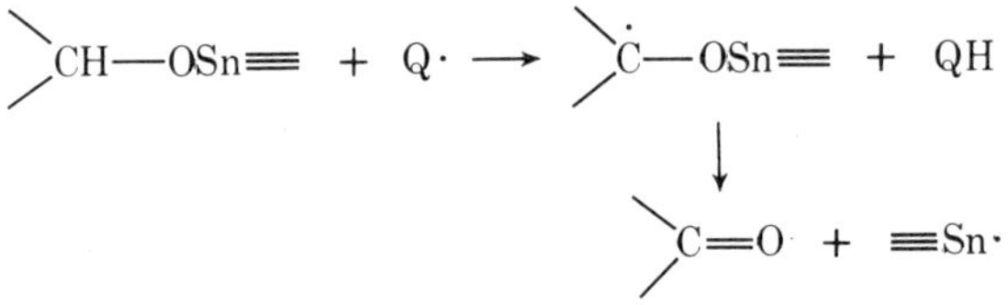

However, results obtained with the cyclopropyl derivatives (for example to give a cyclic ketone) apparently contradicted this hypothesis. Indeed, our results from the free radical reaction of tin hydrides with cyclopropyl ketones (91) show that the open chain ketones obtained are formed through an intermediate stannyloxy radical:

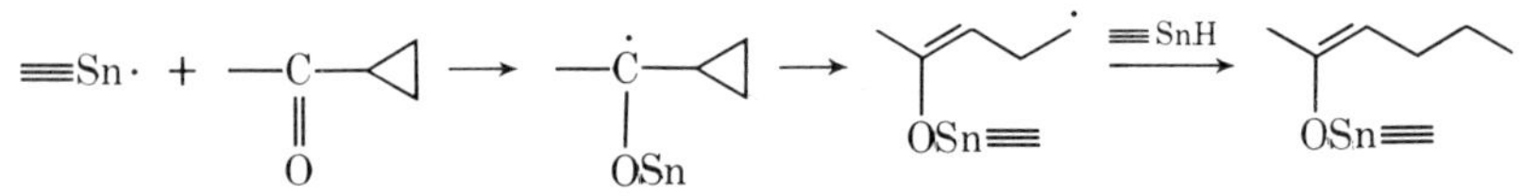

The postulated intermediates were the same in the two processes

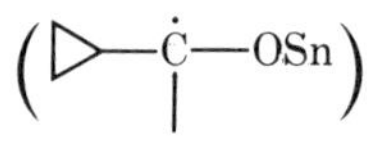

and, to gain insight into this point, we investigated the oxidation of the cyclopropyl derivatives with di-*tert*-butyl peroxide. The results (90) show that the ketone obtained is mainly the cyclic one but is generally accompanied by a small amount of the open chain compound:

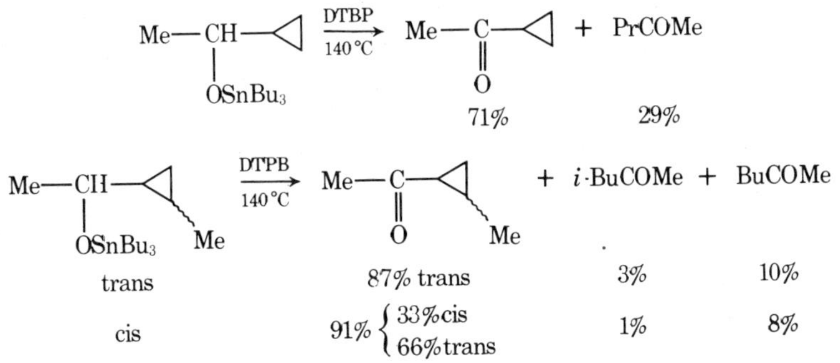

$$\text{Me}-\underset{\underset{\text{OSnBu}_3}{|}}{\text{CH}}-\triangleleft \quad \xrightarrow[140\,°C]{\text{DTBP}} \quad \text{Me}-\underset{\underset{O}{\|}}{C}-\triangleleft \quad + \quad \text{PrCOMe}$$

71%   29%

$$\underset{\text{trans}}{\underset{\underset{\text{OSnBu}_3}{|}}{\text{Me}-\text{CH}}-\triangleleft\backslash_{\text{Me}}} \quad \xrightarrow[140\,°C]{\text{DTPB}} \quad \underset{\text{87\% trans}}{\text{Me}-\underset{\underset{O}{\|}}{C}-\triangleleft\backslash_{\text{Me}}} \quad + \quad i\text{-BuCOMe} \quad + \quad \text{BuCOMe}$$

| | | | 3% | 10% |
| cis | | $91\% \begin{cases} 33\%\,\text{cis} \\ 66\%\,\text{trans} \end{cases}$ | 1% | 8% |

These observations indicate that the cyclic ketone is the main oxidation product, and, in the case of the *cis*-2-methylcyclopropyl derivative, an isomerization occurs during the reaction while the open chain ketone is mainly the linear one. The following scheme agrees with all the observations:

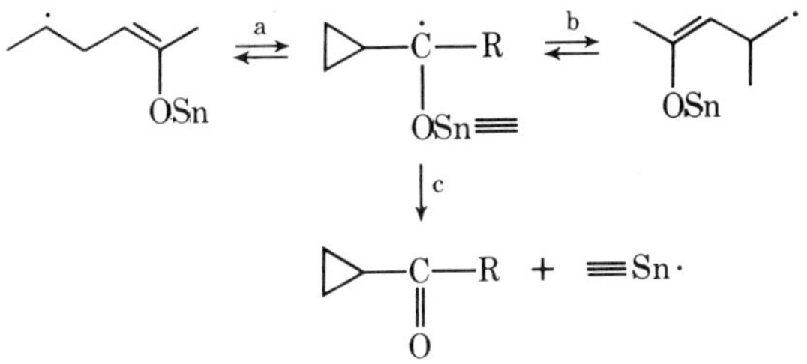

In the oxidation case, Reaction c should be favored since no good hydrogen donors are present in the reaction mixture (the only C–H able to act as a donor seems to be the starting alkoxide itself). On the other hand the relative proportions of the open chain ketones reflect the thermodynamic stabiility of the two free radicals at equilibrium.

The situation is completely different in the tin hydride reduction since the trapping of the radicals once formed is very rapid. The difference between the two types of behavior is very clear if one considers the relative proportions of the open chain ketone obtained:

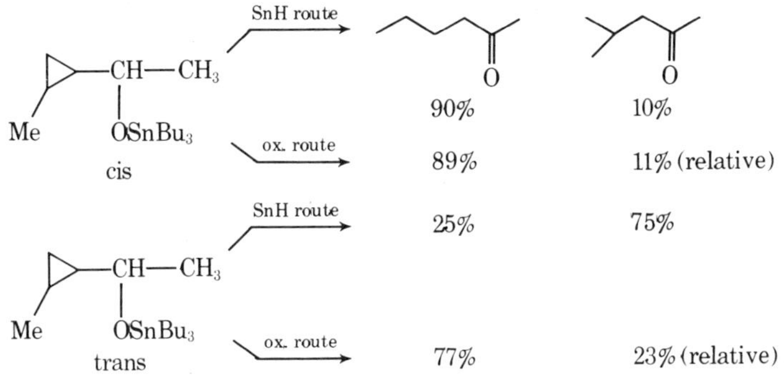

	SnH route	90%	10%
cis	ox. route	89%	11% (relative)
	SnH route	25%	75%
trans	ox. route	77%	23% (relative)

In the tin hydride case, it was found that polar (*trans*-cyclopropylketone) and stereoelectronic effects (*cis*-cyclopropylketone) direct the ring opening (*92*). An ESR study was then designed in collaboration with A. G. Davies (University College, London) to detect the first formed radical at low temperature. The results (*93*) show that at −60 °C the photolysis of the organotin cyclopropylmethyl alkoxide in the presence of (*tert*-BuO)$_2$, leads to a ring opening, the cis compound giving principally the secondary alkyl radical, and the trans only the primary one.

These ESR results concur with the idea of the fast trapping of the initially formed radical by the tin hydride and with an equilibration of the open chain radicals in the absence of a good hydrogen donor via an unstable cyclopropyl-stannyloxy radical, which in this case, decomposes into the cyclic ketone and a tributylstannyl radical.

Two excellent reports have appeared recently in which the organotin al-koxides are very useful intermediates. The first concerns the oxidation of cyclic organotin dialkoxides with bromine which provides a very convenient route to the acyloins (*94*):

$$\begin{array}{c}\text{CH}\!-\!\text{O} \\ | \qquad\quad \text{SnBu}_2 \ + \ \text{Br}_2 \\ \text{CH}\!-\!\text{O}\end{array} \ \xrightarrow{\text{CH}_2\text{Cl}_2} \ \begin{array}{c}\text{C}\!=\!\text{O} \\ | \\ \text{CHOH}\end{array} \ + \ \text{Bu}_2\text{SnBr}_2$$

The second procedure involves the oxidation by bromine of various triethyltin alkoxides in the presence of triethyltin methoxide (*95*). For example:

$$\text{C}_6\text{H}_5\text{CH}_2\text{OSnEt}_3 \ + \ \text{Br}_2 \ \xrightarrow{\text{Et}_3\text{SnOMe}} \ \text{C}_6\text{H}_5\text{CHO} \ + \ 2\text{Et}_3\text{SnBr} \ + \ \text{MeOH}$$
$$81\%$$

$$\text{C}_7\text{H}_{15}\text{CH}_2\text{OSnEt}_3 \ + \ \text{Br}_2 \ \xrightarrow{\text{Et}_3\text{SnOMe}} \ \text{C}_7\text{H}_{15}\text{CHO} \ + \ 2\text{Et}_3\text{SnBr} \ + \ \text{MeOH}$$
$$85\%$$

The latter method, however, requires one mole of triethyltin methoxide just to trap the hydrogen bromide. It is probable that this aspect of the reaction can be improved, and that this process will then become a very popular oxidation procedure.

Organotin Amines

Like the organotin alkoxides, organotin amines exhibit some quite interesting properties. However, where it is possible to propose several easy preparations of the alkoxides, this is not so for compounds having a tin–nitrogen covalent bond, especially in the case of amino groups R$_2$N where R = alkyl. The only universal method consists of a transmetallation between a lithium or magnesium amine and an organotin halide or oxide, and except in special cases, this method is the most commonly used (*see* for example *97, 98, 99, 100, 101, 102, 103, 104, 105*):

$$R_2NM + R'_3SnX \longrightarrow R'_3SnNR_2 + MX$$
$$(M = Li, MgX)$$

Several other specific routes to the tin amines have been reported although none is generally applicable. Among them, we would like to emphasize the metatheses of organosilicon amines and organotin alkoxides (*106*), the direct amination of tin halides by amines (*107, 108, 109*), and an interfacial technique described by Carraher (*110, 111, 112*).

The first example of a primary organotin amine formed from sterically crowded organotin compounds is also worthy of note (*113, 114*). For example:

$$t\text{-Bu}_3SnC_6H_5 \xrightarrow{\text{KNH}_2} t\text{-Bu}_3SnNH_2$$

In spite of the efforts made by several groups of workers (taking account of the number of private communications of unsuccessful attempts) no significant improvement has been found during the past five years. Thus the usefulness of these compounds in synthesis is limited by their difficulty of preparation. However, some properties of the organotin amines are sufficiently interesting to be further studied, and others should be kept in mind until simple preparations make these compounds more readily available. Our group has been involved mainly in substitution and addition reactions, and we report here some of our major results.

Nucleophilic Displacement. The reaction of organotin amines with organic halides has been investigated several times and reported as giving either an elimination (*115*) or, in the more general case, a substitution (*116, 117, 118, 119, 120, 121, 122*). During the past five years we have applied this reaction in two special cases: (1) reactions with α-haloketones and (2) reactions of stannazanes with organic dihalides.

Reactions of Organotin Amines with α-Haloketones. The overall reaction consists of a substitution of the halogen to form an α-amino ketone (*123*). For example

$$\text{Me}-\text{CO}-\text{CH}_2\text{Cl} + \text{Bu}_3\text{SnNEt}_2 \longrightarrow \text{Me}-\text{CO}-\text{CH}_2-\text{NEt}_2 + \text{Bu}_3\text{SnCl}$$
$$53\%$$

$$\text{Me}-\text{CO}-\text{CHCl}-\text{Me} + \text{Bu}_3\text{SnNEt}_2 \longrightarrow \text{Me}-\text{CO}-\underset{\underset{\text{NEt}_2}{|}}{\text{CH}}-\text{Me} + \text{Bu}_3\text{SnCl}$$
$$60\%$$

Two special features have to be mentioned: (1) contrary to the case of substitution with alkyl halides, all these reactions are exothermic: the experimental procedure consists only of mixing the two components and distilling immediately the product amino derivative, and (2) the initial step of the reaction is an addition to the carbonyl group as shown by the disappearance of the C=O vibration in the IR.

From this a mechanism in which the adduct decomposes into an unstable oxirane giving the observed amino ketone by rearrangement can be drawn (124, 125, 126):

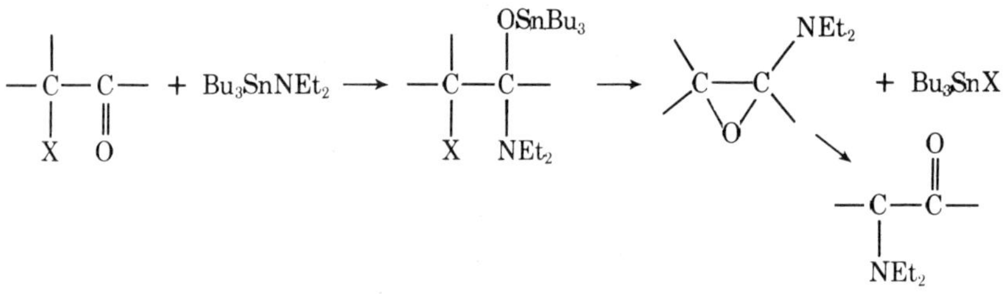

However, the observation of the adduct does not indicate that it lies on the reaction coordinate since it is conceivable that it may be the kinetic product of the reaction:

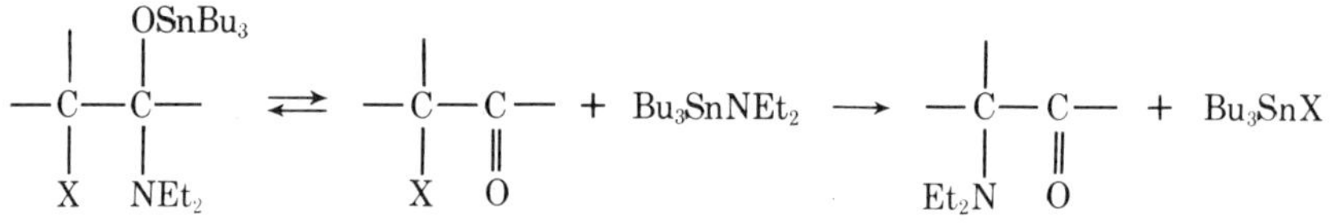

In fact, the pathway by direct substitution must be taken into account in the case of 1-chloro-3-pentanone which gives 1-diethylamino-3-pentanone in high yield; the formation of an intermediate oxetane is unlikely:

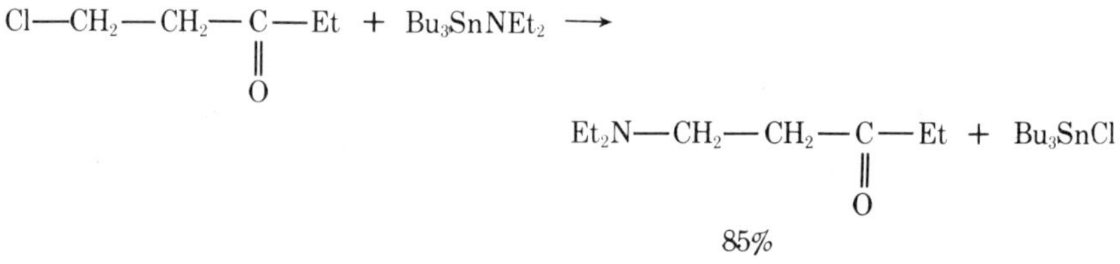

$$Cl-CH_2-CH_2-C-Et + Bu_3SnNEt_2 \longrightarrow$$
$$\underset{O}{\overset{\|}{}}$$

$$Et_2N-CH_2-CH_2-C-Et + Bu_3SnCl$$
$$\underset{O}{\overset{\|}{}}$$
$$85\%$$

It is difficult to choose between these two mechanisms since they would both give the same product (if we admit that during the rearrangement the only migrating group is the amino). For example, 2-chloro-3-pentanone gives 2-diethylamino-3-pentanone as the only product.

However, these reactions seem to be very convenient for preparing amino ketones since the above yields correspond to reaction from a stoichiometric mixture of the reactants, and this can be improved by using a slight excess (0.3 *M*) of organotin amine:

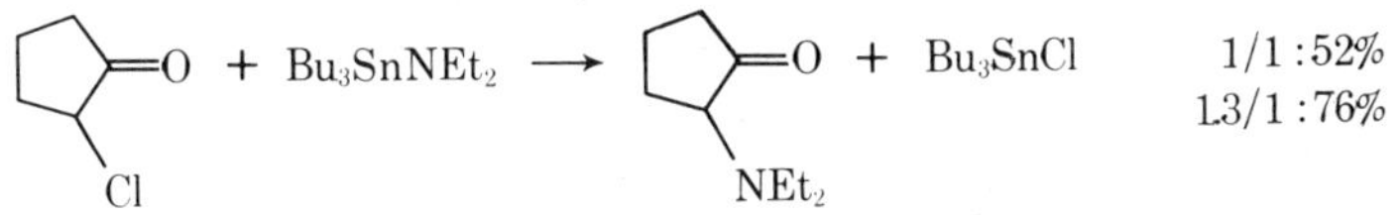

Reactions of Stannazanes with Alkyl Dihalides.   We found that the two tin–nitrogen bonds of the stannazanes can be substituted separately:

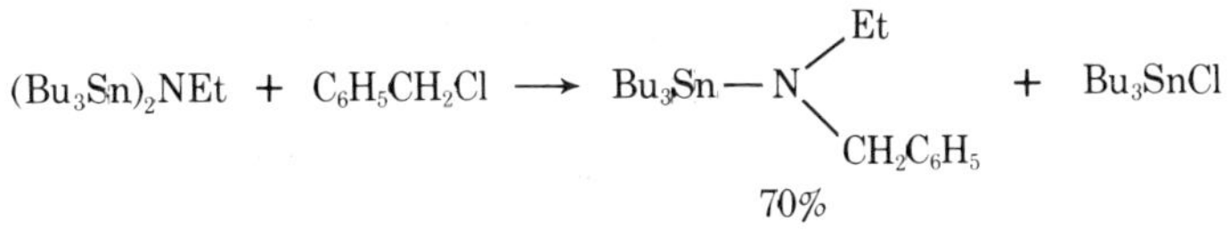

Thus, with organic dihalides, it is possible to imagine first a substitution leading to an haloalkylaminotin compound which by cyclization, as in the case of organotin haloalkoxides, would give a nitrogen heterocycle:

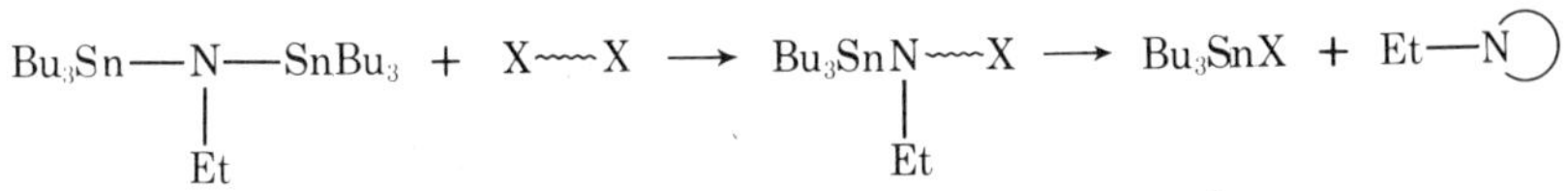

This reaction proceeds as expected except in the case of the 1,2-dibromoethane where elimination was largely favored. Some problems arose in the case of the 1,3-dibromopropane; it proved impossible to extract the corresponding azetidine from the reaction mixture because of its strong complexation with the organotin bromide.

In all the other cases (*127*), the expected heterocycle was obtained by a continuous distillation of a stoichiometric mixture of dihalide and stannazane. Some characteristic examples are shown below:

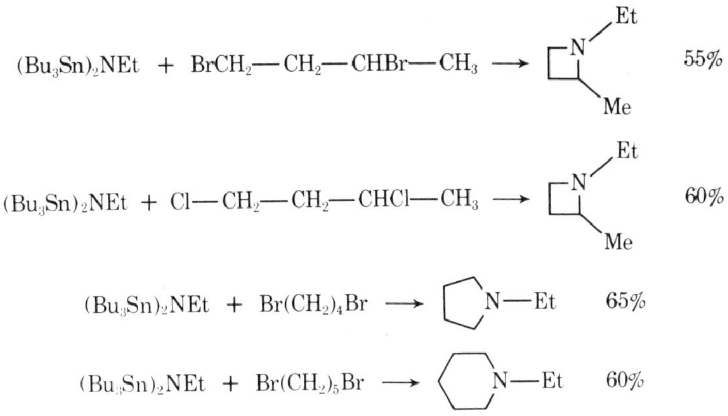

Reactions with Carbonyl Derivatives. ORGANOTIN AMINES AND AL-
DEHYDES OR KETONES. Organotin amines are now well known to react readily
with carbonyl derivatives to give 1,2 dipolar addition (*128, 129, 130, 131*); this
property has been well studied in the case of isocyanates (*132, 133, 134, 135, 136,
137, 138, 139, 140*). For our part, we have investigated the behavior of organotin
amines with various carbonyl derivatives and found that the reactions are not
generally simple. An initial paper reports the reaction between organotin amines
and acetone as an aldolization owing to the basic properties to the nitrogen atom
(*141*). However, contrary to this, our preliminary experiments have shown that
the reactions with cyclohexanone and cyclopentanone led to the formation of
enamines, and with methyl *tert*-butyl ketone to an organotin enolate (*142*).

$$\text{\Large$\bigcirc$}=O \ + \ Bu_3SnNEt_2 \ \longrightarrow \ \text{\Large$\bigcirc$}-NEt_2 \ + \ 1/2\,Bu_6Sn_2O \ + \ 1/2\,H_2O$$
60%

$$\text{\Large$\bigcirc$}=O \ + \ Bu_3SnNEt_2 \ \longrightarrow \ \text{\Large$\bigcirc$}-NEt_2 \ + \ 1/2\,Bu_2Sn_2O \ + \ 1/2\,H_2O$$
50%

$$Me-CO-t\text{-}Bu \ + \ Bu_3SnNEt_2 \ \longrightarrow \ Bu_3SnCH_2-CO-t\text{-}Bu$$
66%

$$Bu_3Sn-O-C\begin{matrix}\nearrow CH_2 \\ \searrow t\text{-}Bu\end{matrix}$$
33%

$$\Bigg\} \quad 60\% \ + \ Et_2NH$$

Lorberth (*116*) and Manoussakis (*143*) reported similar findings. To obtain a
greater insight into these reactions, we reinvestigated the reaction with acetone
itself and found a variety of very different results according to the nature of the
amino group involved (*144*).

$$Bu_3SnNMe_2 \ \xrightarrow{\ MeCOMe\ }$$

$$CH_2=\underset{\underset{NMe_2}{|}}{C}-Me_3 \ + \ Me_2C=CH-\underset{\underset{O}{\|}}{C}-Me \ + \ Me_2C=CH-\underset{\underset{NMe_2}{|}}{C}=CH_2$$
18%   32%   46%

$$Bu_3SnNEt_2 \ \xrightarrow{\ MeCOMe\ } \ CH_2=\underset{\underset{NEt_2}{|}}{C}-Me_3 \ + \ Bu_3SnCH_2COMe$$
62%   8%

$$Bu_3SnNBu_2 \ \xrightarrow{\ MeCOMe\ } \ CH_2=\underset{\underset{NBu_2}{|}}{C}-Me$$
94%

$$\text{Bu}_3\text{SnN(C}_6\text{H}_5)_2 \xrightarrow{\text{MeCOMe}} \text{Bu}_3\text{SnCH}_2\text{COMe}$$
$$10\%$$

These different results were rationalized by complementary experiments (*145*) which led us to conclude that beside the aldolization (observed only when a di-methylamino group is used), two other mechanisms are available:

(1)   When both the organotin amine and the carbonyl group are not very crowded, an addition followed by elimination occurs leading to an enamine:

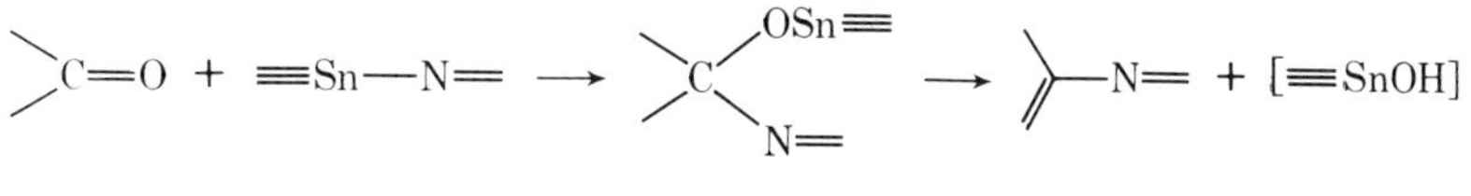

$$\text{>C=O} + \text{≡Sn—N=} \longrightarrow \text{>C}\overset{\text{OSn≡}}{\underset{\text{N=}}{<}} \longrightarrow \text{>}{=}\text{—N=} + [\text{≡SnOH}]$$

(2)   In the other cases, the abstraction of an α-hydrogen occurs to give an organotin enolate:

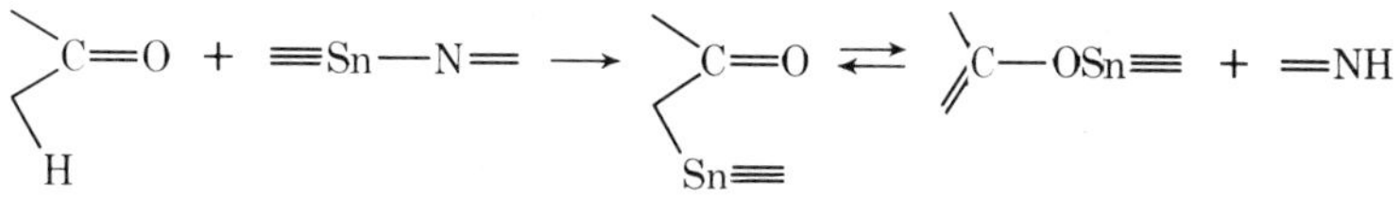

$$\underset{\text{H}}{\overset{\text{>}}{\text{C=O}}} + \text{≡Sn—N=} \longrightarrow \underset{\text{Sn≡}}{\overset{\text{>}}{\text{C=O}}} \rightleftarrows \text{>C—OSn≡} + \text{=NH}$$

The cases of cyclohexanone and 2-methylcyclohexanone illustrate this point of view; the former gives an enamine, and the latter gives a tin enolate:

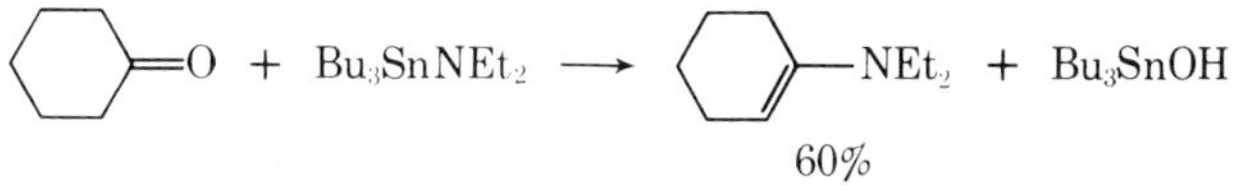

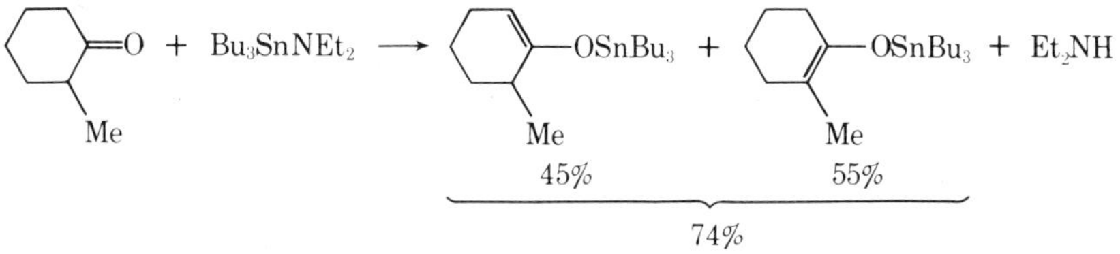

(cyclohexanone)=O + Bu₃SnNEt₂ ⟶ (cyclohexene)—NEt₂ + Bu₃SnOH
60%

(2-methylcyclohexanone)=O + Bu₃SnNEt₂ ⟶ —OSnBu₃ (Me) + —OSnBu₃ (Me) + Et₂NH
45%   55%
74%

During this last reaction, the amounts of different isomers obtained were 45% and 55%, the major product corresponding to the more thermodynamically stable isomer.   The relative yields of the isomers do not agree with the predicted orientation of the reaction by abstraction of the more acidic hydrogen.   In fact, under the experimental conditions used (140 °C, 15 hr), a mixture of the isomers underwent isomerization in the presence of the starting ketone, as shown by the following experiment:

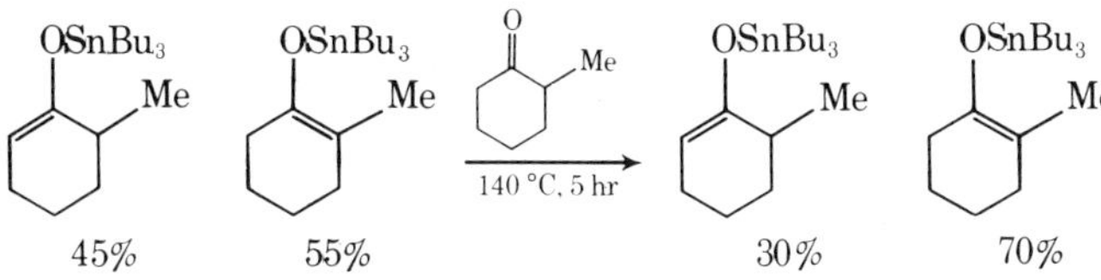

OSnBu₃ / Me (45%)   OSnBu₃ / Me (55%)   (ketone with Me) $\xrightarrow{140\,°C,\,5\,hr}$ OSnBu₃ / Me (30%)   OSnBu₃ / Me (70%)

The possibility that the minor product enolate was the major kinetic product arose and was verified by performing the reaction following a modified experimental procedure: the ketone was added slowly in vapor phase (argon stream) to the organotin amine at 140°C, to avoid a large excess of the starting ketone. At the end of the reaction, the organotin enolate isomers had the following relative concentrations:

70%   30%

This experiment shows clearly that the kinetic product is the less substituted enolate. The fact that this kind of abstraction is never observed for dibutyldiaminotin compounds, but only enamine formation is interesting (*147*).

STANNAZANES AND ALDEHYDES OR KETONES. Some addition reactions of stannazanes with chloral or acetaldehyde had been reported by Davies et al. (*129*) as leading to imines. We extended this process to various carbonyl compounds and found (*146*) that the formation of imines is general and that, contrary to the preceding case, no abstraction of hydrogen occurs:

We found also that this reaction procedes via an intermediate adduct which could be trapped by reaction with acetyl chloride:

$$(Bu_3Sn)_2NEt \ + \ \langle \ \rangle = O \ \longrightarrow$$

In these cases also, the steric requirements are important since hindered ketones only give the imines in low yield (less than 20% for the methyl *tert*-butyl ketone).

ORGANOTIN AMINES AND α-ENONE COMPOUNDS. Some preliminary results on the addition of organotin amines to conjugated carbonyl derivatives were reported as a 3,4-addition pathway (*148*). For example:

$$C_6H_5-CH=CH-CHO \ + \ Et_3SnNMe_3 \ \longrightarrow \ C_6H_5-\overset{\overset{\displaystyle NEt_2}{|}}{CH}-\underset{\underset{\displaystyle SnEt_3}{|}}{CH}-CH-CHO$$

Following our studies of the reactivity or organotin amines with carbonyl derivatives, we investigated more thoroughly these reactions, and found a quite puzzling set of processes depending on the nature of the starting carbonyl compound investigated (*149*). One representative example of each type of behavior is outlined below.

Cinnamaldehyde: Formation of Enediamine. I. The reaction proceeded with the quantitative formation of a 1,2 adduct (observed by NMR) which transforms slowly at room temperature to an enediamine with regeneration of half of the initial cinnamaldehyde and formation of stannoxane:

$$C_6H_5-CH=CH-CHO \ + \ Bu_3SnNEt_2 \ \longrightarrow \ C_6H_5-CH=CH-CH\overset{\displaystyle OSnBu_3}{\underset{\displaystyle NEt_2}{\diagdown}} \ \longrightarrow$$

$$C_6H_5-\underset{\underset{\displaystyle NEt_2}{|}}{CH}-CH=CHNEt_2 \ + \ 1/2Bu_6Sn_2O \ + \ 1/2C_6H_5CH=CH-CHO$$

$$I$$

With an excess of tributyltin amine, the enediamine is obtained in 75% yield.

α-*Methyl Cinnamaldehyde: Formation of Enediamine and Aminal. II.*

$$C_6H_5-CH=\underset{\underset{Me}{|}}{C}-CHO \ + \ 2Bu_3SnNEt_2 \ \longrightarrow$$

$$C_6H_5-CH=\underset{\underset{\underset{II\ 85\%}{Me}}{|}}{C}-CH(NEt_2)_2 \ + \ C_6H_5-\underset{\underset{NEt_2}{|}}{CH}-\overset{\overset{Me}{|}}{C}=CHNEt_2$$

$$\underbrace{}_{45\%}$$

β-*Methyl Cinnamaldehyde: Formation of Dieneamine. III.*

$$C_6H_5-\underset{\underset{Me}{|}}{C}=CH-CHO \ + \ 2Bu_3SnNEt_2 \ \longrightarrow \ C_6H_5-\underset{\underset{CH_2}{||}}{C}-CH=CH-NEt_2$$

$$70\%$$
$$III$$

Benzalacetone: Formation of Organotin Dienolate. IV.

$$C_6H_5-CH=CH-\underset{\underset{O}{||}}{C}-CH_3 \ + \ Bu_3SnNEt_2 \ \longrightarrow$$

$$C_6H_5-CH=CH-\underset{\underset{O}{||}}{C}-CH_2SnBu_3 \ + \ C_6H_5-CH=CH-\underset{\underset{Bu_3SnO}{|}}{C}=CH_2$$

$$\underbrace{\qquad\qquad 40\% \qquad\qquad\qquad\qquad 60\% \qquad\qquad}_{60\%}$$
$$IV$$

Benzalacetophenone: Formation of a Reversible 1,4 Adduct. V. This reaction was exothermic and a 1,4 adduct was formed which was identified by NMR. Any attempt at distillation decomposed the product into the starting materials, which distilled together and reacted again in the receiving flask.

$$C_6H_5-CH=CH-CO-C_6H_5 \ + \ Bu_3SnNEt_2 \ \rightleftharpoons \ C_6H_5-\underset{\underset{NEt_2}{|}}{CH}-CH=\overset{\overset{OSnBu_3}{|}}{C}-C_6H_5$$

$$V$$

Dypnone: Formation of Organotin Dienolate by Elimination. VI.

$$C_6H_5-\overset{\overset{Me}{|}}{C}=CH-\underset{\underset{O}{\|}}{C}-C_6H_5 \ + \ Bu_3SnNEt_2 \ \longrightarrow \ C_6H_5-\underset{\underset{CH_2}{\|}}{C}-CH=\underset{\underset{OSnBu_3}{|}}{C}-C_6H_5$$

78%

VI

Except in the case of abstraction of a hydrogen IV, most of the results can be rationalized in terms of a 1,4 adduct which is formed either by elimination of amine (III, VI) or via the 3,4 metallotropic form. Indeed, this last compound can either give a β elimination V or add a new molecule of the organotin amine which is followed by elimination I. For this last reaction, we proposed the following mechanism:

$$-\underset{\underset{NEt_2}{|}}{C}-CH=C-OSn\equiv \ \rightleftharpoons \ -\underset{\underset{NEt_2}{|}}{C}-\overset{\overset{SnBu_3}{|}}{C}H-CHO \xrightarrow{\equiv SnNEt_2}$$

$$-\underset{\underset{Et_2N}{|}}{C}-\overset{\overset{Sn\equiv}{|}}{C}H-\overset{\overset{OSn\equiv}{|}}{C}H \ \longrightarrow \ -\underset{\underset{Et_2N}{|}}{C}-CH=CH-NEt_2 \ + \ Bu_6Sn_2O$$

To verify this last hypothesis, we investigated the reactions of various organotin enolates to see whether the carbonyl group of the C isomer was able to give normal addition reactions.

ORGANOTIN AMINES AND ORGANOTIN ENOLATES. We have already seen that organotin enolates exist in two metallotropic forms in equilibrium, the C form having a carbonyl group which may be capable of reacting with organotin amines. Reaction is effectively restricted to those cases where no severe steric hindrance occurs in the C form of the enolates (derivatives of aldehydes and some ketones) (*151*). This fact is indicative of the formation of a transient 1,2 adduct which on decomposition gives an enamine:

$$\overset{\diagup}{\underset{\diagdown}{C}}=\overset{\diagup}{\underset{\underset{OSn\equiv}{C}}{}} \ \rightleftharpoons \ -\underset{\underset{\equiv Sn}{|}}{C}-\underset{\underset{O}{\|}}{C}- \xrightarrow[105\ °C,\ 2hr]{\equiv SnNEt_2}$$

$$-\underset{\underset{\equiv Sn}{|}}{C}-\underset{\underset{OSn\equiv}{|}}{C}-NEt_2 \ \longrightarrow \ \overset{\diagup}{\underset{\diagdown}{C}}=\overset{\diagup}{\underset{\underset{NEt_2}{}}{C}} \ + \ \equiv Sn_2O$$

Table I. Some Enamine Preparations

Starting Carbonyl Derivative	CSn/OSn, %	Enamine	Yield, %
PrCHO	5/95	$EtCH$=$CHNEt_2$	94%
$C_5H_{11}CHO$	5/95	$BuCH$=CH—NEt_2	65%
$\begin{matrix} Me \\ \quad\diagdown \\ \qquad CH—CHO \\ \quad\diagup \\ Me \end{matrix}$	5/95	Me_2C=$CHNEt_2$	16%
MeCOMe	95/5	CH_2=$\underset{\underset{NEt_2}{\vert}}{C}$—$CH_3$	53%
EtCOEt	30/70	—	0%

The enamine is formed even in the cases where no detectable C form is present at equilibrium (by NMR). Some results are given in Table I. These results also lend support to the mechanism we proposed in the reactions of organotin amines with α-enones.

REACTIONS OF STANNAZANES WITH ORGANOTIN ENOLATES. When the organotin amine is replaced by a stannazane in the preceding experiment, we obtained the first example of an organotin eneamine instead of an organic enamine.

Under our conditions (150°C, 2 hr), the reaction occurs with the same restrictions as the preceeding one regarding the steric hindrance of the organotin enolate (150). However, by this method, we were able to prepare the organotin eneamines derived from aldehydes and some ketones. Some typical examples are given below:

$$
Bu\!-\!CH\!=\!CH\!-\!OSnBu_3 \;+\; (Bu_3Sn)_2NEt \;\longrightarrow\;
\left\{
\begin{array}{l}
\overset{\displaystyle Et}{\underset{\displaystyle |}{Bu\!-\!CH\!=\!CH\!-\!N\!-\!SnBu_3}} \\[4pt]
86\% \\[10pt]
Bu\!-\!\underset{\displaystyle Bu_3Sn}{\underset{|}{CH}}\!-\!CH\!=\!N\!-\!Et \\[4pt]
14\%
\end{array}
\right\} \; 58\%
$$

$$
Me\!-\!CO\!-\!CH_2SnBu_3 \;+\; (Bu_3Sn)_2NEt \;\longrightarrow\;
\left\{
\begin{array}{l}
CH_3\!-\!\underset{\underset{Et \quad SnBu_3}{N}}{C}\!=\!CH_2 \\[6pt]
79\% \\[10pt]
CH_3\!-\!\underset{\underset{N\!-\!Et}{\parallel}}{C}\!-\!CH_2SnBu_3 \\[4pt]
21\%
\end{array}
\right\} \; 40\%
$$

Since this preparative route was rather limited, we investigated several other possible routes. We found particularly that the metallo derivatives of imines react with tributyltin chloride to give the expected derivatives; for example, the reaction of tributyltin chloride with a lithium derivative of imine prepared either from a lithium alkyl (*150*) or, more interestingly from a lithium amide (*152*).

$$
BuCH_2CH\!=\!N\!-\!i\text{-}Bu \;\xrightarrow[\;2)\;Bu_3SnCl\;]{\;1)\;BuLi\;}\;
\left\{
\begin{array}{l}
Bu\!-\!\underset{\displaystyle Bu_3Sn}{\underset{|}{CH}}\!-\!CH\!=\!N\!-\!i\text{-}Bu \\[4pt]
33\% \\[10pt]
Bu\!-\!CH\!=\!CH\!-\!\underset{\displaystyle Bu_3Sn}{\underset{|}{N}}\!-\!i\text{-}Bu \\[4pt]
66\%
\end{array}
\right\} \; 38\%
$$

$$
Me\!-\!CH_2\!-\!CH\!=\!N\!-\!i\text{-}Bu \;\xrightarrow[\;2)\;Bu_3SnCl\;]{\;1)\;i\text{-}Pr_2NLi\;}\;
\left\{
\begin{array}{l}
Me\!-\!\underset{\displaystyle Bu_3Sn}{\underset{|}{CH}}\!-\!CH\!=\!N\!-\!i\text{-}Bu \\[4pt]
60\% \\[10pt]
Me\!-\!CH\!=\!CH\!-\!\underset{\displaystyle Bu_3Sn}{\underset{|}{N}}\!-\!i\text{-}Bu \\[4pt]
40\%
\end{array}
\right\} \; 70\%
$$

$$\text{Et—CH}_2\text{—CH}=\text{N—}i\text{-Bu} \xrightarrow[\text{2) Bu}_3\text{SnCl}]{\text{1) }i\text{-Pr}_2\text{NLi}}
\left.
\begin{array}{c}
\text{Et—CH—CH}=\text{N—}i\text{-Bu} \\
\big| \\
\text{Bu}_3\text{Sn} \\
53\% \\
\\
\text{Et—CH}=\text{CH—N—}i\text{-Bu} \\
\big| \\
\text{Bu}_3\text{Sn} \\
47\%
\end{array}
\right\} 78\%$$

These are the reactions of tributyltin chloride with the magnesium derivatives of imines under more special experimental conditions:

$$i\text{-PrCH}=\text{N—}i\text{-Bu} \xrightarrow[\text{2) Bu}_3\text{SnCl}]{\text{1) }i\text{-PrMgCl}}
\left.
\begin{array}{c}
\underset{\text{Me}}{\overset{\text{Me}}{>}}\text{C}=\text{CH—N}\underset{i\text{-Bu}}{\overset{\text{SnBu}_3}{<}} \\
72\% \\
\\
\underset{\underset{\text{SnBu}_3}{|}}{\overset{\text{Me}}{\underset{\text{Me}}{>}}}\text{C—CH}=\text{N—}i\text{-Bu} \\
28\%
\end{array}
\right\} 40\%$$

$$i\text{-PrCH}=\text{N—Et} \xrightarrow[\text{2) Bu}_3\text{SnCl}]{\text{1) }i\text{-PrMgCl}}
\left.
\begin{array}{c}
\underset{\text{Me}}{\overset{\text{Me}}{>}}\text{C}=\text{CH—N}\underset{\text{Et}}{\overset{\text{SnBu}_3}{<}} \\
80\% \\
\\
\underset{\underset{\text{SnBu}_3}{|}}{\overset{\text{Me}}{\underset{\text{Me}}{>}}}\text{C—CH}=\text{N—Et} \\
20\%
\end{array}
\right\} 28\%$$

$$\text{Et}_2\text{—CH}=\text{N—}i\text{-Bu} \xrightarrow[\text{2) Bu}_3\text{SnCl}]{\text{1) }i\text{-PrMgCl}}
\left.
\begin{array}{c}
\underset{\text{Et}}{\overset{\text{Et}}{>}}\text{C}=\text{CH—N}\underset{i\text{-Bu}}{\overset{\text{SnBu}_3}{<}} \\
85\% \\
\\
\underset{\underset{\text{SnBu}_3}{|}}{\overset{\text{Et}}{\underset{\text{Et}}{>}}}\text{C—CH}=\text{N—}i\text{-Bu} \\
15\%
\end{array}
\right\} 54\%$$

Although these two latter pathways are known to work more readily with hindered imines while the reverse is true for the method via the organotin enolates, we think we have developed a sufficient set of preparative routes to obtain most organotin eneamines. At this time, studies of the metallotropic equilibrium between the C and N organotin isomers and also of the reactivity of these compounds are in progress.

Acknowledgments

The authors thank J. Valade who initiated organotin chemistry in Bordeaux for his stimulating interest; to J. A. Richards for checking the manuscript and to M. F. Penna for accurate and patient typing.

Literature Cited

1. Neumann, W. P., "The Organic Chemistry of Tin," Wiley, New York, 1970.
2. Poller, R. C., "The Chemistry of Organotin Compounds," Logos, London, 1970.
3. Sawyer, A. K., "Organotin Compounds," Vols. I, II, III, Dekker, New York, 1971.
4. Smith, J. D., Walton, D. R. M., in "Advances in Organometallic Chemistry," Vol. 13, F. G. A. Stone and R. West Eds., p. 453 Academic, New York, 1975.
5. Bloodworth, A. J., Davies, A. G., *Chem. Ind., London* (1972) **12**, 490.
6. Davies, A. G., *Synthesis* (1969) 56.
7. Jones, K., Lappert, M. F., *Organomet. Chem. Rev.* (1966) **1**, 67.
8. Davies, A. G., Palan, P. R., Vasishta, S. C., *Chem. Ind.* (1967) 229.
9. Davies, A. G., U.S. Patent, **3**, 492,327; *Chem. Abs.* (1970) **72**, 90638.
10. Davies, A. G., Kleinschmidt, D. C., Palan, P. R., Vasishta, S. C., *J. Chem. Soc., C,* (1971) 3972.
11. Sakai, S., Fujimura, Y., Ishii, Y., *J. Org. Chem.* (1970) 2344.
12. David, S., Thieffry, A., *C. R. Acad. Sci., Ser., C* (1974) **279**, 1045.
13. Gaur, D. P., Srivastava, G., Mehrotra, R. C., *J. Organometal. Chem.* (1974) **65**, 195.
14. Jones, K., Lappert, M. F., *Proc. Chem. Soc.* (1964) 22.
15. Mehrotra, R. C., Bachlas, B. P., *J. Organometal. Chem.* (1970) **22**, 121.
16. Dubukina, O. V., Lutsenko, I. F., *Zh. Obsch. Chim.* (1970) **40**, 2766.
17. Gaur, D. P., Srivastava, G., Mehrotra, R. C., *J. Organometal. Chem.* (1973) **63**, 213.
18. Azerbaev, I. N., Erzhanov, K. B., Polatebekov, N. P., Kochin, D. A., *Izv. Akad. Nauk. Kaj. SSR, Ser. Khim.* (1972) **22**, 50.
19. Voronkov, M. G., Ivanova, N. P., Timilova, L. F., Mirskov, R. G., *Dolk. Vsas, Kouf. Khim. Atsatilana, 4th* (1972) **2**, 188; *Chem. Abs.* (1973) **79**, 78898.
20. George, T. A., Jones, K., Lappert, M. F., *J. Chem. Soc.* (1965) 2157.
21. Mehrotra, R. C., Bachlas, B. P., *J. Organometal. Chem.* (1970) **22**, 129.
22. Tzschach, A., Ponicke, K., *Z. Anorg. Allg. Chem.* (1974) **404**, 121.
23. Wagner, D., Verheyden, J. P. H., Moffat, J. G., *J. Org. Chem.* (1974) **39**, 24.
24. Baur, D. P., Srivastava, G., Mehrotra, R. C., *J. Organometal. Chem.* (1973) **63**, 221.
25. Harrison, P. G., Zuckerman, J. J., *Inorg. Nucl. Chem. Lett.* (1970) **6**, 5.
26. Allan, M., Ianzen, A. F., Willis, C., *Can. J. Chem.* (1968) **46**, 3671.
27. Gaur, D. P., Srivastava, G., Mehrotra, R. C., *J. Organometal. Chem.* (1973) **47**, 95.
28. Harrison, P. G., Zuckerman, J. J., *Inorg. Chem. Acta.* (1970) **4**, 235.
29. Kennedy, J. D., *J. Chem. Soc., Perkin II* (1973) 1785.
30. Lorbert, J., Kula, M. R., *Chem. Ber.* (1964) **97**, 3444.

31. Amberger, E., Kula, M. R., Lorberth, J., *Angew. Chem. Int. Ed.* (1964) 3, 138.
32. Carraher Jr, C. E., Scherubel, G. A., *J. Polym. Sci., Part A1* (1971) 9, 983.
33. Yoshino, N., Kondo, Y., Yoshino, T., *Synt. Inorg. Metal. Org. Chem.* (1973) 3, 397.
34. Mori, F., Sano, K., Matsuda, H., Matsuda, S., *Kogyo. Kagaku. Zasshi.* (1969) 72, 1543.
35. Iwamoto, N., Ninagawa, A., Matsuda, H., Matsuda, S., *Kogyo. Kagaku. Zasshi.* (1971) 74, 1400.
36. Delmond, B., Pommier, J. C., Unpublished data; B. Delmond, Doctoral Thesis, Bordeaux, 1972.
37. Voronkov, M. G., Romadans, J., *Z. Obsch. Khim.* (1969) 39, 2785.
38. Pijselman, J., Pereyre, M., *C. R. Acad. Sci., Ser. C* (1972) 274, 1583.
39. Folli, U., Iarossi, D., Taddei, F., *J. Chem. Soc., Perkin II* (1973) 1284.
40. Domingos, A. M., Sheldrick, G. M., *Acta. Cristallogr., Sect. B* (1974) 30, 559.
41. Marchand, A., Mendelsohn, J., Valade, J., *C. R. Acad. Sci., Ser. C* (1964) 259, 1737.
42. Pommier, J. C., Pereyre, M., Valade, J., *C. R. Acad. Sci., Ser. C* (1965) 260, 6397.
43. Pereyre, M., Bellegarde, B., Valade, J., *C. R. Acad. Sci., Ser. C* (1967) 265, 939.
44. Bellegarde, B., Pereyre, M., Valade, J., *C. R. Acad. Sci., Ser. C* (1967) 264, 340.
45. Davies, A. G., Harrison, P. G., *J. Chem. Soc., Part C* (1971) 1769.
46. Goncharenko, L. V., Tvogorov, R. N., Belavin, I. Yu., Dudukina, O. V., Baukov, Yu. I., Lutsenko, I. F., *Z. Obsch. Khim.* (1973) 43, 1733.
47. Armitage, D. A., *Inorg. Nucl. Chem. Lett.* (1973) 9, 1225.
48. Barrau, J., Massol, M., Stagé, J., *J. Organometal. Chem.* (1974) 71, C 45.
49. Chausov, V. A., Baukov, Yu. I., *Z. Obsch. Khim.* (1975) 45, 1032.
50. Burlachenko, G. S., Baukov, Yu. I., Lutsenko, I. F., *Zh. Obsch. Khim.* (1972) 42, 387.
51. Massol, M., Satge, J., *Synth. Inorg. Metalorg. Chem.* (1973) 3, 1.
52. Chauzov, V. A., Vodolazskaya, V. M., Nitaeva, N. S., Baukov, Yu. I., *Zh. Obsch. Khim.* (1973) 43, 597.
53. Chauzov, V. A., Baukov, Yu I., *Zh. Obsch. Khim.* (1970) 40, 940.
54. Goncharenko, L., Tvogorov, A. N., Belavin, Yu I., Dudukina, O. V., Baukov, Yu I., Lutsenko, I. F., *Zh. Obsch. Khim.* (1973) 43, 1733.
55. Mirskov, R. G., Vlasov, V. M., Torpishchev, Sh. I., *Khim. Atsetilena* (1968) 181.
56. Armitage, D. A., Sinden, A. W., *J. Organometal. Chem.* (1975) 90, 285.
57. Voronkov, M. G., Mirskov, R. G., Ishchenko, O. S., Sitnikova, S. P., *Z. Obsch. Khim.* (1974) 44, 2462.
58. Armitage, D. A., Sinden, A. W., *J. Organometal. Chem.* (1972) 44, C 43.
59. Bellegarde, B., Pereyre, M., Valade, J., *Bull. Soc. Chim. Fr.* (1967) 3082.
60. *Ibid.* (1967) 746.
61. Kurashiki Rayon Co Ltd. French. Patent 1.368.522; *Chem. Abs.* (1965) 62, 2794.
62. Itoi, K., Kumano, S., *J. Chem. Soc. Jap. Ind. Chem. Soc.* (1967) 70, 82.
63. Pijselman, J., Pereyre, M., *J. Organometal. Chem.* (1971) 32, C 72.
64. Pereyre, M., Pijselman, J., *J. Organometal. Chem.* (1970) 25, C 27.
65. Pijselman, J., Pereyre, M., *J. Organometal. Chem.* (1973) 63, 139.
66. Pommier, J. C., Valade, J., *C. R. Acad. Sci., Ser. C* (1965) 260, 4549.
67. Shastokovskii, M. F., Vlasov, V. M., Mirskov, R. G., USSR Patent 173 757, *Chem. Abs.* (1966) 64, 5138.
68. Souvie, J. C., Pommier, J. C., Unpublished data; J. C. Souvie, Doctoral Thesis, Bordeaux, 1972.
69. Valade, J., Pereyre, M., *C. R. Acad. Sci., Ser. C* (1962) 254, 3693.
70. Savel'eva, N. I., Kostyuk, A. S., Baukov, Yu. I., Lutsenko, I. F., *Zh. Obsch. Khim.* (1971) 41, 2339.
71. Dergunov, Yu. I., Kuz'mina, E. A., Vorotyntseva, V. D., Gerega, V. F., Finkel'shtein, A. I., *Zh. Obsch. Khim.* (1972) 42, 372.

72. Paul, Ram Chand, Soni, K. K., Narula, Suraj Prakash, *J. Organometal. Chem.* (1972) **40**, 355.
73. Sheludyakov, V., Khatuntsev, G. D., Mironov, V. F., *Zh. Obsch. Khim.* (1973) **43**, 2697.
74. Pommier, J. C., Delmond, B., Valade, J., *Tet. Letters*, (1967) 5289.
75. Delmond, B., Pommier, J. C., Valade, J., *J. Organometal. Chem.* (1972) **35**, 91.
76. Delmond, B., Pommier, J. C., *Tet. Letters* (1968) 6147.
77. Delmond, B., Pommier, J. C., Valade, J., *J. Organometal. Chem.* (1973) **47**, 337.
78. Delmond, B., Pommier, J. C., Valade, J., *Tetrahedron Lett.* (1969) 2089.
79. Yamaji, Y., Ninomiya, K., Matsuda, H., Matsuda, S., *Kogyo Kagaku Zasshi* (1971) **74**, 1181.
80. Biggs, J., *Tetrahedron Lett.* (1975) 4285.
81. Delmond, B., Pommier, J. C., Valade, J., *J. Organometal. Chem.* (1973) **50**, 121.
82. Delmond, B., Pommier, J. C., Valade, J., *C. R. Acad. Sci., Ser. C* (1972) **275**, 1037.
83. Ratier, M., Delmond B., Pommier, J. C., *Bull. Soc. Chim. Fr.* (1972) 1593.
84. Bats, J. P., Moulines, J., Pommier, J. C., *Tetrahedron Lett.* in press.
85. Odic, Y., Pereyre, M., *J. Organometal. Chem.* (1973) **55**, 273.
86. Pereyre, M., Odic, Y., *Tetrahedron Lett.* (1969) 505.
87. Odic, Y., Pereyre, M., *C. R. Acad. Sci., Ser. C* (1970) **290**, 100.
88. Pommier, J. C., Ratier, M., Chevolleau, D., *J. Organometal. Chem.* (1971) **31**, C 59.
89. Pommier, J. C., Chevolleau, D., *J. Organometal. Chem.* (1974) **74**, 405.
90. Godet, J. Y., Pereyre, M., Pommier, J. C., Chevolleau, D., *J. Organometal. Chem.* (1973) **55**, C 15.
91. Pereyre, M., Godet, J. Y., *Tetrahedron Lett.* (1970) 3653; Godet, J. Y., Pereyre, M., *Bull. Soc. Chim. Fr.*, in press.
92. Godet, J. Y., Pereyre, M., *C. R. Acad. Sci., Ser. C* (1971) **273**, 1183; Godet, J. Y., Pereyre, M., *J. Organometal. Chem.*, (1972) **40**, C 23; Godet, J. Y., Doctoral Thesis, Bordeaux, 1974.
93. Davies, A. G., Muggleton, B., Godet, J. Y., Pereyre, M., Pommier, J. C., *J. Chem. Soc., Perkin II*, in press.
94. David, S., *C. R. Acad. Sci., Ser. C* (1974) **278**, 1051.
95. Saigo, K., Morikawa, A., Mukaiyama, T., *Chem. Lett.* (1975) 145.
96. Jenkins, I. D., Verheyden, J. P. H., Moffat, J. G., *J. Am. Chem. Soc.* (1971) **93**, 4323.
97. Lorberth, J., *J. Organometal. Chem.* (1969) **19**, 435.
98. Mel'nichenko, L. S., Zemlanskii, N. N., Karandi, I. V., Kolosova, N. D., Kocheshkov, K. A., *Dokl. Akad, Nauk. SSSR* (1971) **200**, 346.
99. Wiberg, N., Veith, M., *Chem. Ber.* (1971) **104**, 3176.
100. Cuvigny, T., Normant, H., *C. R. Acad. Sci.* (1969) **269**, 1398.
101. *Ibid.* (1969) **268**, 834.
102. Petterson, D. J., U.S. Patent, 3,794,670, *Chem. Abs.* (1974) **80**, 108 670.
103. Harris, D. H., Lappert, M. F., *J. Chem. Soc., D* (1974) 895.
104. Schaeffer, C. D., Zuckerman, J. J., *J. Am. Chem. Soc.* (1974) **96**, 7160.
105. Cotton, J. D., Cungy, C. S., Harris, D. H., Hudson, A., Lappert, M. F., *J. Chem. Soc., D* (1974) 651.
106. Armitage, D. A., Sinden, A. W., *J. Organometal. Chem.* (1972) **44**, C 43.
107. Manoussakis, G. E., Tossidis, J. A., *Inorg. Nucl. Chem. Lett.* (1969) **5**, 733.
108. Buechel, K. H., Hamburger, B., Klesper, H., Paulus, W., Pauli, O., *British Patent*, 1,319,889, *Chem. Abs.* (1973) **79**, 78965.
109. Buechel, K. H., Hamburger, B., Klesper, H., Paulus, W., Pauli, O., *German Patent*, 2,056,652, *Chem. Abs.* (1972) **77**, 101899.
110. Carraher, C. E., Winter, D. O., *Makromol. Chem.* (1971) **141**, 251.
111. *Ibid.* (1972) **141**, 237.
112. Carraher Jr., C. E., Winter, D. O., *Makromol. Chem.* (1972) **152**, 55.
113. Goetze, H. G., *J. Organometal. Chem.* (1973) **47**, 625.
114. Goetze, H. J., *Angew. Chem. Germ. Ed.* (1974) **86**, 104.

112   ORGANOTIN COMPOUNDS: NEW CHEMISTRY AND APPLICATIONS

115. Cardin, D. J., Lappert, M. F., *J. Chem. Soc.*, D (1967) 1034.
116. Lorberth, J., *J. Organometal. Chem.* (1969) **16**, 235.
117. Roesky, H. W., Vieser, H., *Chem. Ber.* (1974) **107**, 3186.
118. Isida, T., Akiyama, T., Nabika, K., Sisido, K., Kozima, S., *Bull. Chem. Soc. Jap.* (1973) **46**, 2176.
119. Citron, J. D., *J. Organometal. Chem.* (1971) **30**, 21.
120. Sommer, R., Mueller, E., Neumann, W. P., *Justus Liebigs Ann. Chem.* (1968) **718**, 11.
121. Pommier, J. C., Duchene, A., Valade, J., *Bull. Soc. Chim. Fr.* (1968) 4677.
122. George, T. A., Lappert, M. F., *J. Chem. Soc.*, D (1966) 463.
123. Roubineau, A., Pommier, J. C., *C. R. Acad. Sci.* (1973) **277**, 579.
124. Mousseron, M., Julien, J., Jolchine, Y., *Bull. Soc. Chim. Fr.* (1952) 757.
125. Kirrmann, A., Muths, R., Riehl, J. J., *Bull. Soc. Chim. Fr.* (1958) 1469.
126. Kirrmann, A., Duhamel, P., *Bull. Soc. Chim. Fr.* (1960) 3334.
127. Roubineau, A., Pommier, J. C., *C. R. Acad. Sci.* (1975) **281**, 47.
128. Abel, E. W., Crow, J. P., *J. Chem. Soc.*, A, (1968) 1361.
129. Davies, A. G., Kennedy, J. D., *J. Chem. Soc.*, C (1971) 68.
130. Ishii, Y., Itoh, K., *Asaki Garasu Kogyo Gijutsu Shoreikai Kenkyu Hokoku* (1968) **14**, 39.
131. Chandra, G., George, T. A., Lappert, M. F., *J. Chem. Soc.*, C (1969) 2565.
132. Kupchick, E. J., Pisano, M. A., Parikh, D. K., D'ico, M. A., *J. Pharm. Sci.* (1974) **63**, 261.
133. Itoh, K., Katsuura, T., Matsuda, I., Ishii, Y., *J. Organometal. Chem.* (1972) **34**, 63.
134. Itoh, K., Matsuda, I., Ishii, Y., *J. Chem. Soc.*, C (1971) 1870.
135. Harrison, P. G., *J. Chem. Soc.*, Perkin I (1972) 130.
136. Susuki, H., Itoh, K., Matsuda, I., Ishii, Y., *Bull. Soc. Chim. Jap.* (1974) **47**, 3131.
137. Itoh, K., Matsuda, I., Ishii, Y., *Tetrahedron Lett.* (1969) 2675.
138. Dergunov, Yu. I., Gordetsov, A. S., Vostokov S. A., Gerega, V. F., *Z. Obsch. Khim. SSSR* (1974) **44**, 2166.
139. Lappert, M. F., Mc Meekink, J., Palmer, D. E., *J. Chem. Soc., Dalton Trans.* (1973) 151.
140. Itoh, K., Matsuda, I., Kaatsuura, T., Ishii, Y., *J. Organometal. Chem.* (1969) **19**, 347.
141. Jones, K., Lappert, M. F., *J. Organometal. Chem.* (1965) **3**, 295.
142. Pommier, J. C., Roubineau, A., *J. Organometal. Chem.* (1969) **16**, 23.
143. Manoussakis, G. E., Tossidis, J. A., *J. Inorg. Nucl. Chem.* (1972) **34**, 2449.
144. Pommier, J. C., Roubineau, A., *J. Organometal. Chem.* (1969) **17**, 25.
145. *Ibid.* (1973) **50**, 101.
146. Roubineau, A., Pommier, J. C., *C. R. Acad. Sci.* (1974) **279**, 953.
147. Prouvost, B., Pommier, J. C., unpublished data; B. Prouvost, Doctoral Thesis, Bordeaux, 1975.
148. George, T. A., Lappert, M. F., *J. Organometal. Chem.* (1968) **14**, 327.
149. Brocas, J. M., Pommier, J. C., *Int. Symp. Germanium, Tin, and Lead, 1st,* Marseille, 1974; Brocas, J. M., Pommier, J. C., *J. Organometal. Chem.,* in press.
150. Brocas, J. M., De Jeso, B., Pommier, J. C., *J. Organometal. Chem.,* in press.
151. Brocas, J. M., Pommier, J. C., *J. Organometal. Chem.* (1975) **92**, C 7.
152. De Jeso, B., Pommier, J. C., unpublished data.

RECEIVED May 3, 1976.

7

Reactions of Electrophilic Reagents with Tin Compounds Containing Organofunctional Groups

JAMES L. WARDELL

University of Aberdeen, Meston Walk, Old Aberdeen, AB9 2UE, Scotland

This paper is concerned in general with reactions of functionally substituted organotin compounds with electrophiles. Specifically, reactions of sulfur substituted alkyl tin compounds [$R_3Sn(CH_2)_nSR'$ (n = 1-4) and $R_3SnCHXCH_2SR'$ (X = Cl, Br, SCN)] are featured. The types of reactions undergone depend on the situation of the sulfur with respect to the tin; (1) for α-substituted sulfides, R_3SnCH_2SR', the most frequently observed reactions are R–Sn bond cleavage and cleavage of the tin–methylene bond; (2) for β sulfides, R–Sn cleavage and alkene elimination arise, and (3) in more remote situations with respect to the tin, the sulfides give products from both R–Sn and Sn–$(CH_2)_nSR'$ (n ≥ 3) cleavage. The importance of the organic groups, R, and electrophiles, EY, on the course of the particular reaction is discussed.

Tin compounds containing organofunctional groups constitute a small but important proportion of the total number of known organotin species. Their number is steadily increasing as more synthetic routes become established and representative examples of such compounds containing oxygen (e.g., epoxides, ethers, and alcohols), sulfur (e.g., sulfides and sulfones), and nitrogen (e.g., amines and nitriles) functional groups as well as halides have been prepared (1, 2). Prime attention in these studies of functionally substituted compounds has centered on their preparations and physical properties, with less regard paid to reactions and relative reactivities. It is toward the latter aspect of the chemistry of functionally substituted alkyl tin species that this article is concerned. Furthermore, only reactions with electrophiles are considered.

Most reactions involving organotin species are cleavage reactions although modifications of one functional group to another without cleavage of any tin–carbon bond are certainly not rare (1, 2). Reactivity sequences of simple alkyl-

and aryltins toward electrophiles have been established. Aryl–tin bond cleavage is an electrophilic aromatic substitution reaction and so in general the more electron releasing the substituents on the aryl group the easier is the latter's cleavage from tin, i.e., the reactivity correlates with the σ^+ substituent constants. Deviations from a reactivity/σ^+ sequence, however, can arise, e.g. aryl–tin bond cleavage by iodine in carbon tetrachloride solution is an overall third order process with the additional role for the second molecule of iodine modifying the accepted electronic effects of the substituents (3). For cleavage of simple alkyl groups from tin, reactivity sequences obtained have been shown to be variable. One of the more important factors influencing the situation is solvent effects, e.g., iodination in nonpolar solvents results in the cleavage sequence i-Pr > Pr > Et > Me, while the reverse trend holds in polar solvents. This change is a consequence of different mechanisms; in nonpolar solvents, a cyclic four-center transition state is obtained, while in polar solvents, an open transition state occurs (4).

The position of vinyl, allyl, and benzyl groups (all in a sense functionally substituted groups) in electrophilic cleavage sequences has also been established, viz., Ph > PhCH$_2$ > Me; CH$_2$CH=CH$_2$ > Ph; and Ph > CH$_2$=CH > Me [however the relative positions of the vinyl and phenyl groups are apparently reversed in reactions with sulfenyl halides (5)].

Generally for functionally substituted alkyltins, the electronic effects of the substituent, in particular its ability to release or withdraw electrons, and its siting in the organic group relative to the tin atom must be of great importance in determining the reactivity of the organotin bond. In an α position, i.e., R$_3$SnCH$_2$Y, the ability of the group Y to stabilize the carbanion, $^-$CH$_2$Y, or developing carbanion in the transition state, must play the most significant role in determining the cleavage reactivity of the Sn–CH$_2$Y bond. Steric factors could perhaps also be important. In positions other than alpha, i.e., beta and beyond, electronic effects (and steric effects) will have a decreased significance. For β substituents specifically, there is also the possibility of a reaction leading to alkene elimination.

Table I.   Reactions of Cyanoalkyl- and Carboxyalkyltin
Compounds with Halogens[a]

Compound	Reagent	Tin Product, Isolated Yield %
Pr$_3$SnCH$_2$CO$_2$Et	Br$_2$	Pr$_3$SnBr, 90
Bu$_3$SnCH$_2$CH$_2$CO$_2$Me	Br$_2$	Bu$_2$Sn(Br)CH$_2$CH$_2$CO$_2$Me, 90
Bu$_3$SnCH$_2$CN	I$_2$	Bu$_3$SnI
Bu$_3$SnCH$_2$CH$_2$CN	Br$_2$	Bu$_2$Sn(Br)CH$_2$CH$_2$CN, 85
Ph$_3$SnCH$_2$CH$_2$CN	I$_2$	Ph$_2$Sn(I)CH$_2$CH$_2$CN, 62
Ph$_3$SnCH$_2$CH$_2$CH$_2$CN	I$_2$	Ph$_2$Sn(I)CH$_2$CH$_2$CH$_2$CN, 44[b]

[a] Data from Refs. 6, 7, and 8.
[b] Yield quoted is for isolated fluoride, Ph$_2$Sn(F)CH$_2$CH$_2$CH$_2$CN

Reactions of α and β cyanoalkyltin and carboxyalkyltin compounds with halogens, studied by van der Kerk and Noltes (*6, 7, 8*), illustrate the basic situation (Table I); reactivity sequences, $YCH_2 >$ Bu; Ph $> YCH_2CH_2$ (Y = CN, CO_2R) are found. The enhanced reactivity of the $NCCH_2$–Sn bond is a consequence of the (−M) electronic effect of the CN group which can stabilize any developing negative charge on the α-carbon atom. $[\overline{CH}_2{-}C{\equiv}N \leftrightarrow CH_2{=}C{=}N^-]$. The β-CN substituent cannot so stabilize the corresponding carbanion, $\overline{CH}_2CH_2CN$, and so reaction occurs not at the Sn–CH_2CH_2CN bond but at the Bu–Sn bond in the $Bu_3SnCH_2CH_2CN$ reaction. A similar argument holds for the CO_2R substituted compounds.

It should not be inferred from these halogen reactions that consistent cleavage reactivities should be anticipated from electrophilic reactions of organofunctional tin compounds. It has already been pointed out that relative reactivities of simple alkyl– and aryl–tin bonds can vary, and this is clearly true for sulfur-substituted organotin compounds.

α-Substituted Sulfur Organotin Compounds

Reactions of R_3SnCH_2SR' with electrophiles (EY) are shown in Scheme 1. For R_3SnCH_2SR', in which R is an alkyl group such as Me or Bu, primary reaction preferentially occurs at the Sn–CH_2SR bond—e.g., in reactions of $Bu_3SnCH_2SC_6H_4Me$-p with HCl (*9*), of Bu_3SnCH_2SMe with MeI (*10*), and of $Me_3SnCH_2SC_6H_4Me$-p with $HgCl_2$, Br_2, and I_2 (*11*). The electronic effect of sulfur (+M, −I, with the former dominant) results in an overall electron release; however sulfur can stabilize an alpha negative charge—this had been attributed to the use of its empty d orbitals but more recently is ascribed to sulfur's polarizability. Hence cleavage of the Sn–CH_2SR' bond in preference to simple alkyl–tin bonds can be accounted for. The situation of phenyl vs. CH_2SR' is more complicated and is shown in Table II along with some data for the corresponding oxygen and selenium analogues. The data in the table give the percent of the reaction giving Ph–Sn bond cleavage. The remainder, if any, of the total reaction was accounted for and occurred at the Sn–CH_2SR' bond. Primary products detected spectroscopically, however, could not always be isolated (*12*).

Some general findings from the reactions of Ph_3SnCH_2ZR' are: (1) only phenyl–tin bond cleavage occurs in reactions of the oxygen derivative, (2) an approximate balance between cleavage of the Ph–Sn and Sn–CH_2ZR' (Z = S, Se) bonds is obtained in the halogen reactions, (3) sulfenyl halides only react with the substituted alkyl group, and (4) $HgCl_2$ only cleaves the Ph–Sn bonds in all Ph_3SnCH_2ZR' compounds.

The rationale for 1 (above) is that oxygen, unlike sulfur and selenium, is unable to stabilize the appropriate carbanion, $\overline{CH}_2ZR'$, or its precursor in the transition state. All three heteroatoms (O, S, and Se) are electron releasing and, in the absence of any strong polarizability effect (as with oxygen), the operation of this electron releasing effect alone would lead to destabilization of the carb-

Table II. Percent of Reaction of Ph_3SnCH_2ZR' with Electrophiles Leading to Phenyl–Tin Bond Cleavage (Z = O, S, Se)

Reagent/Solvent / Temperature, °C	Ph_3SnCH_2- OC_6H_4Me-p	Ph_3SnCH_2- SC_6H_4Me-p	Ph_3SnCH_2- SeC_6H_4Me-p
Br_2/CCl_4 / 20°	100[a]	10[a]	17[a]
$I_2/CHCl_3$ / 20°	100[a]	64[a]	21[a]
$HgCl_2/EtOH$ / Reflux	90[b]	87[b]	86[b]
$PhSCl/CCl_4$ / 20°	0	0	0

Compound	Reagent/Solvent / 20°C			
	$I_2/CHCl_3$	I_2/CCl_4	$I_2/MeCOMe$	$I_2/MeOH$
$Ph_3SnCH_2SC_6H_4Me$-p	64[a]	22[a]	13[a]	33[a]

$Ph_3SnCH_2SC_6H_4X$-$p/I_2/CCl_4$ / 50°C								
Percent	NH_2	OMe	Me	t-Bu	H	Cl	Br	NO_2
X, %	3	11	22	19	24	20	21	58

[a] Yields calculated from GLC.
[b] Yields based on isolated yields.

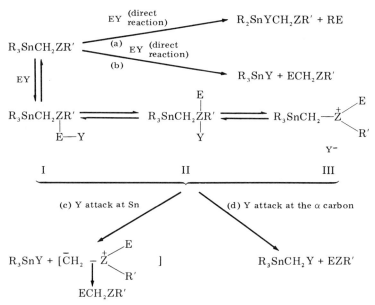

Z = 0, S and Se; EY = electrophilic species; $HgCl_2$, Br_2, I_2; MeI

Scheme 1. Reactions of R_3SnCH_2ZR (Z = O, S, Se) with electrophiles, EY

anion, $\overline{C}H_2ZR'$, and reaction at the phenyl–tin bond would then be anticipated. The balance, observed in the halogen cleavage reactions of Ph_3SnCH_2ZR', is upset in reactions of $Ph_3GeCH_2SC_6H_4Me$-p, $Me_3SnCH_2SC_6H_4Me$-p (only metal–$CH_2SC_6H_4Me$-p cleavage in both cases), and $(p$-$MeC_6H_4)$ $_3SnCH_2SC_6H_4Me$-p (only p-MeC_6H_4–Sn cleavage). Hence a reactivity sequence $R = p$-$MeC_6H_4 >$ Ph $>$ Me in p-$MeC_6H_4SCH_2SnR_3$ is obtained. The variation in products from each of the Ph_3SnCH_2ZR' compounds (Z = S and Se) in these electrophilic reactions clearly means that much more has to be considered than just the charge separations in the ground states of the relevant Ph–Sn and Sn–CH_2ZR' bonds. Solvation effects and electronic demands of the various transition states must also be considered. However, a major impact on the type of reaction results from the coordination of the heteroatom, Z, with the electrophile, EY (Scheme 1). In this scheme, complexes of charge-transfer (I) and oxidative addition (II) types are considered, as is the ionization of the latter to provide III. With each electrophile and solvent, the relative importance of I–III in the reactions will vary. As well as being routes to products proceeding via such complexes, direct routes are also visualized. Both complex types I and II are known. For example, with organic compounds and halogens, crystal structures have been determined for $(CH_2)_4Se{:}I_2$, $(PhCH_2)_2S{:}I_2$; $C_4H_8O_2{:}Br_2$ (all of the molecular complex type), and $[(p$-$MeC_6H_4)_2Se(X)_2$ (X = Cl, Br); $(p$-$ClC_6H_4)_2S(Cl)_2]$ (oxidative addition type) (*13, 14, 15, 16*), while in solution, equilibrium constants have been calculated for iodine charge transfer complexes with sulfides and selenides (*17, 18, 19, 20, 21, 22, 23, 24, 25*). Iodine charge transfer complexes with Ph_3SnCH_2ZR' (Z = S and Se) have λ_{max} in the 350–355 nm range in carbon tetrachloride solution.

The effects of X in $Ph_3SnCH_2SC_6H_4X$-p on the course of the reaction should be somewhat limited as a result of the remoteness of X from the reaction sites. What effects there are would be opposing. An electron-releasing X group would enhance the overall reactivity and render the heteroatom, Z, a more powerful donor species toward electrophiles while at the same time it would decrease the stability of the carbanion, $\overline{C}H_2ZC_6H_4X$-p or its precedent.

The rates of reaction of $Ph_3SnCH_2SC_6H_4X$-p with iodine in carbon tetrachloride solution, in fact, only covered a fairly small overall range and were in the sequence $NO_2 <$ Cl, Br $<$ t-Bu, Me, H $<$ OMe. From Table IIc, no consistent trend can be seen in the amounts of PhI produced, although it appears that more phenyl–tin cleavage does occur with the more electron withdrawing X groups.

Alkyl iodides have been shown not to cause the electrophilic cleavage of aryl– or simple alkyl–tin bonds. However alkyl iodides did react with the α sulfides. Petersen's reaction (*10*) of Bu_3SnCH_2SMe with MeI initially produced the isolable sulfonium salt $\underline{Bu_3SnCH_2}S^+Me_2$, I^- which decomposed spontaneously to Bu_3SnI and the ylide $\overline{C}H_2S^+Me_2$ on attack of I^- at the Sn atom. This gives some basis for step c in Scheme 1. The sulfonium salt $Bu_3SnCH_2S^+Me_2$, $MeSO_4^-$ was considerably more stable, an obvious consequence of the reduced nucleophilicity of $MeSO_4^-$ compared with I^-. In contrast to the reaction of Bu_3SnCH_2SMe with MeI, the products from the $Ph_3SnCH_2SC_6H_4Me$-p/EtI reaction were Ph_3SnCH_2I and $EtSC_6H_4Me$-p, i.e., step (d) in Scheme 1 was followed. In this case, a sulfonium salt could not be isolated. Some precedents for a nucleophilic attack on a carbon alpha to tin are known (26), e.g., Reaction 1.

$$R_3SnCH_2X + I^- \rightarrow R_3SnCH_2I + X^- \qquad (1)$$

$$X = Cl, Br$$

From studies of sulfenyl halides with organotin compounds, Ph–Sn bonds have been shown to be inert to sulfenyl systems (27), and so any reactions of RSCl with Ph_3SnCH_2ZR' (Z = S, Se) would have to occur elsewhere in the molecule, such as with the heteroatom, Z. Interaction of RSCl with the heteroatom, Z, would provide $[Ph_3SnCH_2Z^+(R')SR]$ Cl^-, the subsequent reaction of which can not lead totally to $RSCH_2ZR'$ since the latter compound is stable under the reaction conditions and would not give the complexity of products obtained. It is tempting to involve ylides, $\overline{C}H_2Z^+(R')SR$ (and their subsequent reactions) in the sulfenyl chloride reactions. Some basis for the sulfonium ion intermediate, $Ph_3SnCH_2Z^+(R')SR$, Cl^- comes from reactions of RSX with organic sulfides R^1SR^2 in which the products are R^1SSR and others derived from the carbonium ion $[R^2]^+$ (28, 29, 30, 31). In particular from a kinetic study of the reaction of PhMeCHSMe with PhSCl, the rate limiting step was in fact the formation of the sulfonium salt, $[(PhMeCH)(Me)S^+(SPh)]Cl^-$.

Extensive phenyl–tin cleavage in all $HgCl_2/Ph_3SnCH_2ZR'$ reactions in ethanol solution is difficult to account for. Mercury chloride forms strong unionized complexes with dialkyl sulfide donors. However these are considered to be dissociated extensively in solvents such as ethanol. Weaker complexes with alkyl aryl sulfides would be even more dissociated in solution and so the $HgCl_2$ and $Ph_3SnCH_2SC_6H_4Me$-p can be considered as essentially free rather than complexed in ethanol solution (32, 33, 34). While the amount of reaction occurring via complexes could be small, direct reaction could still, in principle, lead to cleavage of the Sn–CH_2ZR' bond (Z = S, Se) as well as of the phenyl–tin bond, but only the latter occurs.

β Sulfides

The β effect, whereby an alkene is lost from a β-substituted compound, did not occur in reactions of either the β-cyano- or β-carboxyethyltin compounds (6, 7, 8), but it has been demonstrated in a number of other reactions involving electrophilic species (Reactions 2, 3, 4, and 5) including reactions of β sulfides (Reactions 3 and 4). Related reactions (Reactions 6 and 7) may also be considered.

$$Ph_3SnCH_2CH_2OH \xrightarrow{H^+} [Ph_3SnCH_2CH_2\overset{+}{O}\overset{H}{\underset{H}{\diagdown}}] \xrightarrow[-H^+]{(H_2O)} Ph_3SnOH + CH_2{=}CH_2 \quad (2)$$

Ref. 36

$$Ph_3SnCH_2CH_2SR' + EY \rightarrow Ph_3SnY + CH_2{=}CH_2 + R'SE \quad (3)$$
$$R' = p\text{-}MeC_6H_4 \;; EY = Br_2; 2\text{-}NO_2C_6H_4SCl, MeI$$

Ref. 36

$$Ph_3SnCHXCH_2SR' + EY \rightarrow Ph_3SnY + CH_2{=}CHX + R'SE \quad (4)$$
$$R' = 2\text{-}NO_2C_6H_4; X = Cl, Br, SCN; EY = ArSCl$$

Ref. 5

$$Me_3SnCH_2CH_3 + Ph_3C^+BF_4^- \rightarrow Me_3Sn^+BF_4^- + Ph_3CH + CH_2{=}CH_2 \quad (5)$$

Ref. 37

$$R_3SnCH_2CH_2CH{=}CH_2 \xrightarrow{E^+} [R_3SnCH_2CH_2\overset{+}{C}HCH_2E] \xrightarrow{Y^-} R_3SnY +$$

$$\begin{array}{c} CH_2 \\ \diagup \quad \diagdown \\ CH_2{-}CHCH_2E \quad (6) \end{array}$$

$$EY = Br_2, Cl_2, I_2, ArSCl; R = Me, Bu, \text{ but not } R = Ph$$

Ref. 38

$$EY = HOTs; R = Me$$

Ref. 39

$$Bu_3SnCH = CHR \xrightarrow{Pb(OCOCH_3)_4}$$

$$Bu_3SnCHPb(OCOCH_3)_3\overset{+}{C}HR, OAc^- \rightarrow Bu_3SnOCOCH_3 + RCH{=}CHPb(OCOCH_3)_3$$
$$(7)$$

Ref. 40

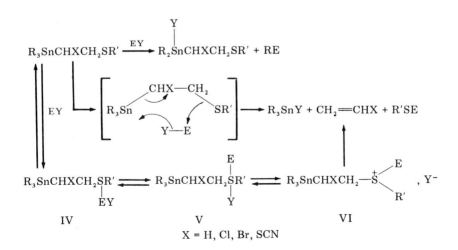

Scheme 2. Reactions of $R_3SnCHXCH_2SR'$ with electrophiles, EY

As well as alkene eliminations, reactions of the two types of β sulfides, $Ph_3SnCH_2CH_2SR'$ and $Ph_3SnCHXCH_2SR'$ (X = Cl, Br), with electrophiles also produced phenyl–tin bond cleavages (Scheme 2). On the other hand no simple cleavage reaction occurred to give YCH_2CH_2SR' or $YCHXCH_2SR'$ products since there is insufficient stabilization for the carbanions $\bar{C}H_2CH_2SR'$ and $\bar{C}HXCH_2SR'$ or their precursors. Comparing the reactions of $Ph_3SnCHXCH_2SR'$ with $Ph_3SnCH_2CH_2SR'$, it was found that alkene elimination occurred less readily with the former. This was considered a consequence of the sulfur atom in $Ph_3SnCHXCH_2SR'$ being less basic or a weaker donor and so could not coordinate as well with the electrophile to give complexes IV and V in Scheme 2. For alkene elimination to occur, a positive character to the β carbon might have to be formed during the reaction and ionization of complex V to give cations VI has been considered in Scheme 2. Whenever the formation of VI is difficult, for example as is likely in $Ph_3SnCH_2CH_2SR'$ reactions with $HgCl_2$ or I_2, ethylene elimination does not result; when it is easier, as in reactions with Br_2, MeI and RSCl, ethylene is evolved in preference to reactions at the phenyl–tin bond. Of course a concerted pathway to alkene formation cannot be totally excluded from the data so far obtained. Of the electrophiles which reacted with $Ph_3SnCHXCH_2SR'$ (proton acids, RSCl, Br_2, I_2, and $HgCl_2$), only RSCl resulted in alkene formation, while MeI proved unreactive.

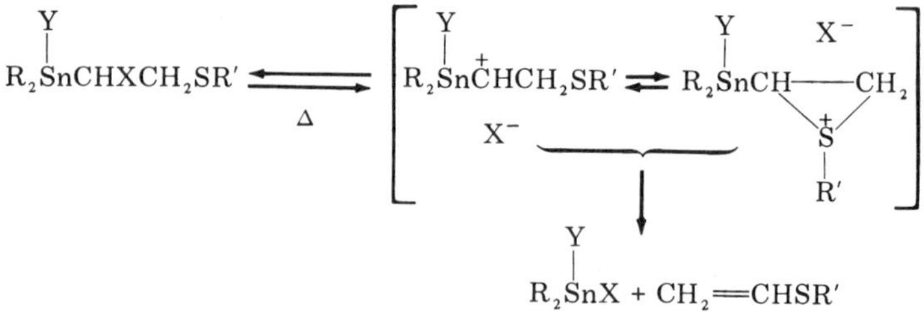

Y = R, Cl, Br, I, SCN

Scheme 3. Thermal decomposition of $R_2YSnCHXCH_2SR'$

$Ph_3SnCHXCH_2SR'$ (X = Cl, Br) and the initial cleavage products, $Ph_2YSnCHXCH_2SR'$, were thermally labile and produced vinyl sulfides. The rates of decomposition of $Ph_2BrSnCHClCH_2SC_6H_3NO_2$-2-X-4 were in the sequence X = Me > H > NO_2 and Scheme 3 is suggested. Other thermally labile β-substituted organotin compounds are known, including $Ph_3SnCH_2CH_2O$-$COCH_3$ (6). In contrast, $Ph_3SnCH_2CH_2SC_6H_4Me$-p was thermally stable.

γ- and δ-Sulfides

With both γ-and δ-sulfides, the position is much clearer. Poller has shown that $Bu_3SnCH_2CH_2CH_2SR$ and gaseous HCl provide $CH_3CH_2CH_2SR$ (R =

XC_6H_4, X = p-Me, H) and Bu_3SnCl as well as Bu_2SnCl_2 (41). On reaction with $Ph_3Sn(CH_2)_nSR$ (n = 3, 4; R = p-MeC_6H_4), Br_2, I_2, $HgCl_2$ and proton acids (EY) provided PhE and $Ph_2YSn(CH_2)_nSR$. The sulfenyl halide, o-$NO_2C_6H_4SCl$, proved inert (42). Methyl iodide and $Ph_3Sn(CH_2)_3SC_6H_4Me$ slowly produced $MeSC_6H_4Me$-p and $Ph_3Sn(CH_2)_3I$. The pathway for this reaction most likely proceeds via a sulfonium salt intermediate. In these compounds, the sulfur is sufficiently remote from the tin, and a situation does not result (i.e., on complex formation) whereby the substituted alkyl–tin bond becomes at least as reactive as the phenyl–tin bond. While complex formation most probably does occur, as seems obvious in the MeI reaction, such complexes do not lead to reaction at the tin–carbon bond.

Acknowledgment

The author thanks R. D. Taylor and D. W. Grant for their help.

Literature Cited

1. Gielen, M., Nasielski, J. in "Organotin Compounds," A. K. Sawyer, Ed., Vol. III, Chap. 9, Dekker, New York, 1971.
2. Luijten, J. G. A., van der Kerk, G. J. M. in "Organometallic Compounds of the Group IV Elements," A. G. MacDiarmid, Ed., Vol. I, Part II, Chap. 4, Arnold, London, 1968.
3. Bott, R. W., Eaborn, C., Waters, A. J., *J. Chem. Soc.* (1963) 681.
4. Gielen, M., Nasielski, J., *Rec. Trav. Chim.* (1963) **82**, 228.
5. Wardell, J. L., *J. Chem. Soc. Dalton* (1975) 1786.
6. van der Kerk, G. J. M., Noltes, J. G., *J. Appl. Chem.* (1959) **9**, 179.
7. van der Kerk, G. J. M., Luijten, J. G. A., *J. Appl. Chem.* (1956) **6**, 93.
8. van der Kerk, G. J. M., Noltes, J. G., Luijten, J. G. A., *J. Appl. Chem.* (1957) **7**, 356.
9. Brasington, R. D., Poller, R. C., *J. Organometal. Chem.* (1972) **40**, 115.
10. Peterson, D. J., *J. Organometal. Chem.* (1971) **26**, 215.
11. Wardell, J. L., unpublished data.
12. Taylor, R. D., Wardell, J. L., *J. Chem. Soc. Dalton* in press.
13. Rømming, C., *Acta. Chem. Scand.* (1960) **14**, 2145.
14. Hassel, O., Hvoslef, J., *Acta. Chem. Scand.* (1954) **8**, 873.
15. McCulloch, J. D., Marsh, R. E., *Acta. Crystallogr.* (1950) **3**, 41.
16. Baenziger, N. C., Buckles, R. E., Maner, R. J., Simpson, T. D., *J. Am. Chem. Soc.* (1969) **91**, 5749.
17. Santini, S., Reichenbach, G., Sorriso, S., Ceccon, A., *J. Chem. Soc. Perkin II* (1974) 1056.
18. Mancini, V., Piovesana, O., Santini, S., *Z. Naturforsch B* (1974) **29**, 815.
19. van der Veen, J., Stevens, W., *Rec. Trav. Chim.* (1963) **82**, 287.
20. McCulloch, J. D., Eckerson, B. A., *J. Am. Chem. Soc.* (1945) **67**, 707.
21. McCulloch, J. D., Barsh, M. L., *J. Am. Chem. Soc.* (1949) **71**, 3029.
22. McCulloch, J. D., Eckerson, B. A., *J. Am. Chem. Soc.* (1951) **73**, 2954.
23. Tideswell, N. W., McCulloch, J. D., *J. Am. Chem. Soc.* (1957) **79**, 1031.
24. McCulloch, J. D., Mulvey, D., *J. Phys. Chem.* (1960) **64**, 264.
25. McCulloch, J. D., Zimmermann, I. C., *J. Phys. Chem.* (1960) **64**, 1084.
26. Bott, R. W., Eaborn, C., Greasley, P. M., *J. Chem. Soc.* (1964) 4804.
27. Wardell, J. L., Ahmed, S., *J. Organometal. Chem.* (1974) **78**, 395.
28. Moor, C. G., Porter, M., *J. Chem. Soc.* (1958) 2890.

29. Oki, M., Kobayashi, K., *Bull. Soc. Chem. Jap.* (1970) **43**, 1223, 1229.
30. Oda, R., Hayashi, Y., *Tetrahedron Lett.* (1967) 3141.
31. Epshtein, G. Yu., Usov, I. A., Ivin, S. Z., *J. Gen. Chem. USSR* (1964) **34**, 2359.
32. Biscarini, P., Fusina, L., Nivellini, G. D., *Inorg. Chem.* (1971) **10**, 2564.
33. Biscarini, P., Nivellini, G. D., *J. Chem. Soc. A* (1969) 2206.
34. Vecera, M., Gasparic, J., Jurecek, M., *Coll. Czech. Chem. Comm.* (1959) **24**, 640.
35. Davis, D. D., Gray, C. E., *J. Org. Chem.* (1970) **35**, 1303.
36. Taylor, R. D., Wardell, J. L., *J. Organometal. Chem.* (1975) **94**, 15.
37. Jerkunica, J. M., Traylor, T. G., *J. Am. Chem. Soc.* (1971) **93**, 6278.
38. Peterson, D. J., Robins, M. D., Hansen, J. R., *J. Organometal. Chem.* (1974) **73**, 237.
39. Pommier, J. C., Kuivila, H. G., *J. Organometal. Chem.* (1974) **74**, 67.
40. Corey, E. J., Wollenberg, R. H., *J. Am. Chem. Soc.* (1974) **96**, 5581.
41. Ayrey, G., Brasington, R. D., Poller, R. C., *J. Organometal. Chem.* (1972) **35**, 105.
42. Wardell, J. L., unpublished data.

RECEIVED May 3, 1976.

Synthesis of Novel Substituted Alkyltin Halides

R. E. HUTTON

Akzo Chemie UK, Ltd., Kirkby Industrial Estate, Liverpool L33 7TH, England

V. OAKES

Akzo Chemie, P.O. Box 247, Stationsstraat 48, Amersfoort, Holland

A new synthetic route to mono- and di-substituted alkyltin compounds is described. The route involves reaction of a carbonyl-activated olefin with a tin(II) halide and a hydrogen halide to produce mono-substituted alkyltin trihalide. Disubstituted alkyltin dihalides are formed by reaction of a carbonyl-activated olefin with metallic tin and hydrogen halide. The reactions are nonhazardous, highly specific, and proceed in high yield at ambient temperatures and atmospheric pressure. Prior formation of intermediate chlorostannanes is believed to be involved. The products have industrial potential as intermediates for PVC stabilizers.

O rganotin compounds have gained considerably in industrial importance in the past decade, and their output is expected to grow at a high rate. In 1965, the world consumption was 5000 tons and was expected to reach 25,000 tons in 1975. A major proportion of this tonnage is used to produce organotin stabilizers for thermal stabilization of PVC.

Four industrial routes are used to produce organotin compounds (Figure 1). The first three routes give high yields, essential for economic reasons because of the expense of tin. However, they suffer the disadvantages of being hazardous, of using up a stoichiometric amount of another metal, and of requiring a further disproportionation stage with $SnCl_4$ to give the major industrial intermediates required, R_2SnCl_2 and $RSnCl_3$.

The fourth route, the direct reaction (*1, 2*) goes part way to solving some of these difficulties, but for alkyl groups higher than methyl it gives a mixture of mono- and dialkyltin halides. The monoalkyltin byproduct is probably formed by reaction of RX on a tin(II) halide which is an intermediate. Also yields de-

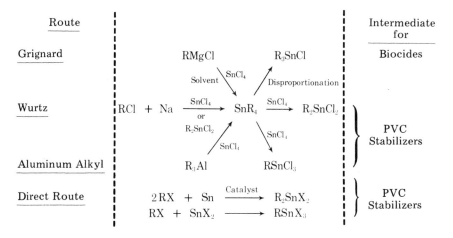

Figure 1.   Methods of organotin production

crease as the alkyl series is ascended, preventing methods to make higher mo-lecular weight products of lower volatility.

Clearly, an incentive exists to find an improved industrial synthesis for producing tin–carbon bonds.   Of the laboratory routes available, the addition of tin–hydrogen bonds across a carbon–carbon double bond seemed to have the most potential.   Excellent work on adding stable organotin hydrides to double bonds had been carried out at the T.N.O. Institute at Utrecht, starting in 1956 (3) and culminating in a detailed mechanistic study (4).

$$R_3SnH + R'CH{=}CH_2 \longrightarrow R_3SnCH_2CH_2R'$$

Since this route involved preparation of the stable organotin hydrides by reduction of organotin halides with lithium aluminium hydride, it would not be economically viable for industrial use.   The possibility of using inorganic tin hydrides was, therefore, considered.

Discussion

Inorganic hydrides are difficult to prepare.   The chlorostannanes are par-ticularly difficult since they are very thermally unstable.   Stannane, SnH_4, was prepared as early as 1919 by Paneth and Furth, (5) in very low yield by electro-lytic reduction of tin(II) sulfate in sulfuric acid.   A typical modern preparation involves low temperature reduction of $SnCl_4$ by lithium aluminum hydride in ether (6).   The chemistry of stannane has not received much attention owing to its instability, and most work has been confined to its synthesis, isolation, and decomposition (7, 8).   In 1969, Reifenberg and Considine studied the reaction with olefins to give tetraalkyltin compounds (9).   It is interesting to note in view of the work reported in this paper that Reifenberg and Considine found that acrylate esters were unreactive.

The reaction of HCl with stannane at $-70°C$ gives rise to chlorostannane, H_3SnCl (*10*). A mechanism was suggested for its decomposition on warming to room temperature.

$$SnH_4 + HCl \longrightarrow H_2 + H_3$$

$$2H_3SnCl \longrightarrow 2HCl + 2SnH_2 \longrightarrow Sn + SnH_4$$

$$SnH_4 + 2HCl \longrightarrow 2H_2 + H_2SnCl_2 \longrightarrow SnCl_2 + H_2$$

Dichlorostannane is suggested as an intermediate in this decomposition, and this seems to be the only literature reference to this species.

Some controversy has existed about the formation of trichlorostannane which is believed to exist as the dietherate from reaction of tin(II) chloride and excess HCl in diethyl ether (*11*). Trihalostannane etherates have also been prepared by reaction of tin(II) halides with lithium aluminum hydride in diethylether (*12*). Trichlorostannane is reported to decompose above $30°C$ with the reformation of tin(II) chloride and HCl.

We decided to investigate the reactions of tin(II) chloride and anhydrous HCl in ether as a potential source of trichlorostannane and to attempt hydrostannation reactions with various olefins. Surprisingly, almost quantitative yields of organotin trichloride could be obtained with activated olefins such as acrylate esters under mild reaction conditions.

$$SnCl_2 + HCl \xrightarrow[\substack{20\,°C}]{\substack{diethyl \\ ether}} [HSnCl_3] \xrightarrow{\substack{O \\ \| \\ ROC-CH=CH_2}} \overset{O}{\underset{\|}{ROC}}-CH_2CH_2SnCl_3 \quad (1)$$

The trichlorostannane intermediate may be preformed with the olefin added subsequently, or tin(II) chloride may be suspended in the solvent with activated monomer and hydrogen chloride bubbled into the suspension. The upper temperature limit is governed by the reflux temperature of the acrylate ester and by polymerization starting to occur. The scope of the reaction with respect to the olefinic reactant is shown in Table I.

Monomers such as styrene which add HCl or polymerize cannot be used. It can be seen from Table I that it is necessary to have a carbonyl group adjacent to the double bond for reaction to occur. Notable exceptions to this rule are maleic acid esters and isophorone. The significance of this will be discussed later. Hydrobromic and hydroiodic acids will undergo the reaction also with the corresponding tin(II) halide.

Any solvent which does not interact with HCl may be used. With certain monomers, for example butyl acrylate, the monomer acts as its own solvent and complexing agent and no other solvent is necessary. Even aqueous hydrochloric

Table I. Scope of the Reaction

$$SnCl_2 + HCl + \rangle C{=}C\langle \rightarrow \rangle CH{-}\overset{|}{\underset{|}{C}}{-}SnCl_3$$

Reactive Monomers		*Unreactive Monomers*	
Acrylates	$CH_2{=}CH\ COOR$		
Methacrylates	$CH_2{=}CMe\ COOR$	Acrolein	$CH_2{=}CH\ CHO$
Crotonates	$CH_3CH{=}CH\ COOR$	Alpha Olefins	$CH_2{=}CH{-}R$
Acrylic Acid	$CH_2{=}CH\ COOH$	Allyl Chloride	$CH_2{=}CH\ CH_2Cl$
Methacrylic Acid	$CH_2{=}CMe\ COOH$	Allyl Alcohol	$CH_2{=}CH\ CH_2OH$
Acryloyl Chloride	$CH_2{=}CH\ COCl$	Styrene	$Ph{-}CH{=}CH_2$
Vinyl Ketones	$CH_2{=}CH\ COR$	Isophorone	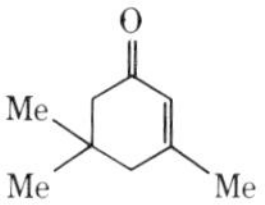
Phorone	$Me_2C{=}CH\ COCH{=}CMe_2$		
Propiolic Acid	$CH{\equiv}C\ COOH$	Vinyl Acetate	$CH_3COOCH{=}CH_2$
Acrylamide	$CH_2{=}CH\ CONH_2$	Acetylene	$CH{\equiv}CH$
Diethyl Ethylidene Malonate	$MeCH{=}C(COOEt)_2$	Furan	
		Maleic Acid Esters	$ROOC\ CH{=}CH\ COOR$
		Phenyl Acetyene	$Ph{-}C{\equiv}CH$

acid has been used to prepare β-carbomethoxyethyltin trichloride from methyl acrylate. The yield of organotin was lower however, because of hydrolysis and hydrochlorination of the monomer. If the trichlorostannane is preformed, however, an oxygenated solvent must be used.

In organotin stabilizers for PVC, the major intermediate is the dialkyltin dichloride, R_2SnCl_2. We considered, therefore, how this reaction could be extended to produce such compounds. The reaction between metallic tin and hydrogen halides produces tin(II) halide and hydrogen. Does this reaction, however, involve the transient formation of dichlorostannane?

Table II. Reactions of Powdered

Methyl Acrylate, g	*Solvent, ml*	*Reaction Time, hr*	*Hydrogen Halide, g*
87.4	Et_2O, 140	3	HCl 87
95.7	Et_2O, 110	14	HCl 42
37.1	Hexane, 140	12.5	HCl 46
95.7	Et_2O, 140	10.5	HBr 110
174.2	No solvent	15	HCl 40

$$Sn + 2HX \longrightarrow [H_2SnX_2] \longrightarrow SnX_2 + H_2$$

If such an intermediate is formed could it be trapped with an activated olefin before decomposition?

$$Sn + 2HX + 2ROCOCH{=}CH_2 \longrightarrow (ROCOCH_2CH_2)_2SnX \qquad (2)$$

We therefore investigated a reaction in which hydrogen chloride was passed into a suspension of tin in methyl acrylate at room temperature and atmospheric pressure with diethyl ether as solvent. The product after working up was shown to be predominantly bis(β-carbomethoxyethyl)tin dichloride, $Cl_2Sn(CH_2CH_2COOMe)_2$, with small amounts of β-carbomethoxyethyltin trichloride. The yield was 98% based on tin consumed. Table II shows a series of reactions in which powdered tin (60 g) was allowed to react with methyl acrylate at 20°C.

Depending on the conditions, minor amounts of the β-carbomethoxyethyltin trichloride are formed. This is minimized by slow passage of hydrogen chloride which prevents formation of tin(II) chloride, followed by Reaction 1. A variety of olefins were examined and the scope of the reaction is shown in Table III.

A list of monomers similar to that in the first reaction is shown to be reactive. The more sterically hindered monomers tend to have a reduced rate of reaction but still give high yields. Again, other hydrogen halides undergo the reaction.

Many of the monomers shown as unreactive in Table III are destroyed by other reactions such as preferential addition of hydrogen chloride across the double bond or polymerization. In every case, only the β adduct was isolated as shown by NMR spectroscopy. The α adduct

$$\left[X{-}\overset{\overset{\displaystyle O}{\|}}{C}{-}\underset{\underset{\displaystyle SnCl_2}{|}}{CH}{-}CH_3 \right]_2$$

was not detected.

Tin with Methyl Acrylate at 20°C

Weight of Unreacted Tin, g	$Cl_2Sn(CH_2CH_2-COOMe)_2$, g	$Cl_3SnCH_2CH_2-COOMe$, g	Yield Based on Tin Consumed
0.5	125.6	46.4	98%
3.7	161.35	5.85	98%
1.5	146	27	99%
9.5	157.4	38.6	100%
5.0	158.5	—	95%

Table III. Scope of the Reaction

$$Sn + 2HCl + 2 \ce{>C=C<} \rightarrow \left(\ce{>CH-C|-} \right)_2 SnCl_2$$

Reactive Monomers		Unreactive Monomers	
Acrylates	$CH_2{=}CH \cdot COOR$		
Methacrylates	$CH_2{=}CH \cdot MeCOOR$	Alpha Olefins	$CH_2{=}CH{-}R$
Crotonates	$CH_3CH{=}CH \cdot COOR$		
Acrylic Acid	$CH_2{=}CH \cdot COOH$	Styrene	$PhCH{=}CH_2$
Acryloyl Chloride	$CH_2{=}CH \cdot COCl$		
Methyl Vinyl Ketone	$CH_2{=}CH \cdot COCH_3$	Isophorone	
Phorone	$Me_2C{=}CH \cdot COCH{=}CMe_2$		
Propiolic Acid	$CH{\equiv}C \cdot COOH$		
Acrylamide	$CH_2{=}CH \cdot CONH_2$	Maleic Acid Esters	$ROOC\ CH{=}CH\ COOH$
Diethyl Ethylidene Malonate	$MeCH{=}C(COOEt)_2$		$ROOC\ CH{=}CH\ COOR$

Disproportionation Reactions

The chemical properties of functionally substituted alkyltin halides show some interesting differences from conventional alkyltins. For example, in disproportionation reactions, the functionally substituted alkyltin halides are much more labile. In Table IV, the first reaction to give an organotin trichloride from an organotin dichloride shows high reactivity in the case of the bis(β-carbobutoxyethyl)tin dichloride compared with dibutyltin dichloride. Similarly, higher reactivity is shown in the second reaction with (β-carbobutoxyethyl)tin trichloride compared with butyltin trichloride. The increased reactivity of the functionally substituted alkyltins in these reactions may be caused by coordination of the carbonyl group to tin.

Table IV. Disproportionation Reaction Yields

	Yield	
	$R = Alkyl$	$R = CH_2CH_2COOBu$
$R_2SnCl_2 + SnCl_4 \xrightarrow{150°C} 2RSnCl_3$	2% After 10 hr	100% After 1 hr
$2RSnCl_3 + Bu_4Sn \xrightarrow[4\ hr]{170°C} R_2SnCl_2$	65%	95%

Mechanism of Reaction

The two new synthetic routes to functionally substituted organotin halides may proceed by an ionic mechanism. Both reactions proceed in a highly polar medium, and the addition of free radical inhibitors such as hydroquinone has no effect on the rate of reaction.

A concerted mechanism as shown in Figure 2 is favored as a tentative suggestion. This mechanism is favored against a simple four-center mechanism since isophorone, with a trans configuration, is unreactive.

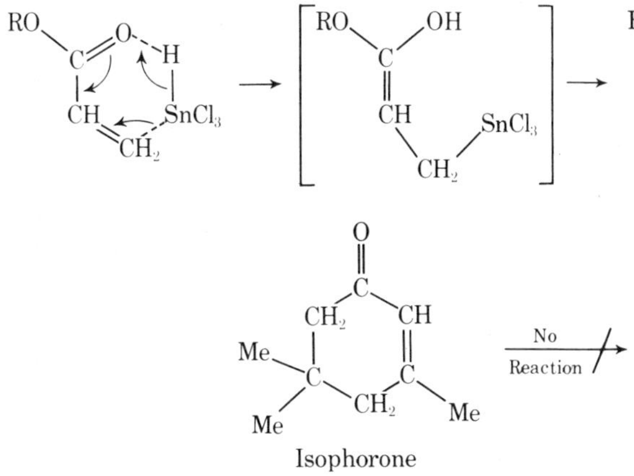

Isophorone

Figure 2. Tentative mechanism for functionally substituted organotin halides.

The polarity of the tin–hydrogen bond may be reversed in the case of chlorostannanes compared with stable organotin hydrides where hydride ion transfer is favored in those cases where an ionic mechanism operates (4).

Clearly, more work is required to confirm the existance of the chlorostannane intermediates, particularly dichlorostannane Cl_2SnH_2 (or ClSnH?). However, this leads to the intriguing thought that these chlorostannane intermediates may be involved in the long list of reduction reactions of either metallic tin and HCl or tin(II) chloride and HCl.

Experimental

A typical procedure is presented in each case, followed by a tabular summary of results.

Reaction of Tin(II) Chloride and Anhydrous Hydrogen Chloride with Activated Olefins. *β-Carbomethoxyethyltin Trichloride.* Anhydrous tin(II) choride (80 g), dimethoxy ethane (150 ml), and methyl acrylate (36.3 g) were placed in a 500-ml, three-necked flask. The flask was fitted with a stirrer thermometer, reflux condenser, and gas inlet tube. To the stirred suspension, gaseous

Table V.   Monomers in Diethyl Ether

Monomers	Temp, °C	HCl × Theory
Methyl Acrylate	25	1.3
bEthyl Acrylate	15	2.2
Methyl Acrylate	25	5.0
Methyl Crotonate	20–25	4.1
Mesityl Oxide	25	1.5
bAcrylic Acid	15	2.0
Acryloyl Chloride	15	1.9
Crotonic Acid	15	2.4
bMethyl Vinyl Ketone	15	2.4
Phorone	20–25	2.2
bDiethyl Ethylidene Malonate	20–25	2.2
bPropiolic Acid	20–25	2.4
Acrylamide	20	1.7

a Within the limits of experimental determination, $SnCl_2$ is quantitatively converted to organotin product.
b Monomer added to $HSnCl_3$.
c Product obtained by crystallization from the reaction medium: the actual conversion is probably higher.
d Accurate estimation of the conversion is hampered by the elimination of HCl from the product $(Cl_3SnCH_2CH_2CO_2H \xrightarrow{-HCl} -Cl_2SnCH_2CH_2CO_2-)$. This reaction also accounts for the anomalous elemental analyses.

anhydrous hydrogen chloride (36 g) was added over 2 hr.   The temperature was maintained at 20°C by an ice/salt bath.

The solvent was removed on a film evaporator, and the residue was extracted with 100 ml toluene.   Volatile matter was distilled up to 100°C at 4 mm Hg pressure to leave a residue (117 g) which crystallized on cooling.   IR, NMR, and elemental analysis were consistent with the composition: $CH_3O-COCH_2CH_2SnCl_3$, mp, 70°C; bp, 174°C/4 mm.   Tabular details of other monomers (1 mole) in diethyl ether solvent reacting with tin(II) chloride (1 mole) are given in Table V.

Reaction of Metallic Tin and Anhydrous Hydrogen Chloride with Activated Olefins.   *Bis(β-carbomethoxyethyl)tin Dichloride.*   Into a 500-ml, three-necked flask equipped with cooling bath, stirrer, condenser, thermometer, and gas inlet tube was placed powdered tin (60 g), methyl acrylate (87.4 g), and diethyl ether (140 ml).

Solvent Reacting with Tin(II) Chloride

Reaction Time, hr	$SnCl_2$[a] Conversion, %	Analysis of Product		
		Cl,[g] %	Sn,[g] %	Mp, °C
3.5	98	34.6 (34.1)	38.1 (38.3)	70
1	79	32.2 (32.6)	37.5 (36.3)	68
3.5	62	32.4 (32.6)	36.3 (36.3)	86
4	30	31.7 (32.6)	36.6 (36.3)	81
2	80[c]	32.9 (32.8)	38.1 (36.6)	123
1	~75[d]	31.2 (35.7)	43.5 (39.8)	—
3	90	34.7 (33.6)	38.0 (37.4)	42
5	~40	e		—
2.5	70	e		70
4	<20	28.7 (29.0)	32.7 (32.6)	77
2.5	<20	e		—
2	~60[f]	29.4 (40.1)	44.7 (40.1)	—
10	<20	e	—	

[e] Purification of the organotin compound was difficult—characterization was by NMR spectroscopy.
[f] Accurate estimation of conversion is hampered by HCl addition to the organotin compound ($Cl_3SnCH=CHCO_2H \xrightarrow{HCl} Cl_3SnCH—CH_2CO_2H$) and HCl elimination as in [d].
|
Cl

[g] Values in parenthesis indicate theoretical values.

Dry HCl (87 g) was bubbled into the stirred suspension at 20°C over 3 hr. The ether was evaporated off, and the residue was extracted with hot chloroform (300 ml). Unreacted tin (0.5 g) and traces of tin(II) chloride were removed. The chloroform was evaporated to yield 177.2 g of a white solid. Nuclear spin resonance spectroscopy showed the product was $(MeOCOCH_2CH_2)_2SnCl_2$ with 27 wt% $MeOCOCH_2CH_2SnCl_3$. The yield was 98% based on tin converted.

Pure $(MeOCOCH_2CH_2)_2SnCl_2$ was obtained by washing the mixture with diethyl ether in which the trichloride is soluble. The white crystalline substance which remained was shown by elemental analysis, IR, and NMR spectroscopy to be pure $(MeOCOCH_2CH_2)_2SnCl_2$, mp 132°C.

Tabular details of reactions of other monomers (2 moles) with metallic tin (1 mole) and dry hydrogen chloride in diethyl ether solvent are given in Table VI.

Table VI.   Reactions

Monomer	$HCl \times$ Theory	Time, hr	Yield, %
Methyl Acrylate	1.4	13	99 $(90)^b$
Ethyl Acrylate	1.5	25	94
n-Butyl Acrylate	1.4	20	99
Methyl Methacrylate	1.7	11	85
Methyl Crotonate	1.1	17	55^c
Mesityl Oxide	1.9	10	80 $(46)^b$
Methyl Vinyl Ketone	1.4	14	~80
Acryloyl Chloride	1.6	20	Low
Acrylic Acid	1.6	15	50^f
Propiolic Acid	1.9	12	80^g
Dimethyl Glutaconate	1.8	13	62^c
Dimethyl Ethylidene Malonate	2.0	9	66
Acrylamide	1.6	23	79

[a] Values in parenthesis indicate theoretical values for the dialkyltin dichloride products. Values quoted are for mono/di mixture.
[b] Yields of di material isolated by crystallization from the reaction medium. Other values are yields of mono/di mixture based on tin charged.
[c] Sn was largely converted to $SnCl_2$ in these reactions.
[d] The product was contaminated with resinous material.
[e] Purification of the organotin compound was difficult and characterization was by NMR of the product mixture.

Literature Cited

1. Oakes, V., Hutton, R. E., *J. Organometal. Chem.*, (1965) **3**, 472–477.
2. Ibid. (1966) **6**, 133–140.
3. van der Kerk, G. J. M., Luijten, J. G. A., Noltes, J. G., *Chem. Ind.* (1956) 352.
4. Leusink, A. J., "Hydrostannation, A Mechanistic Study," Organisch Chemisch Instituut, TNO, Utrecht, Holland, 1966.
5. Paneth, F., Furth, K., *Ber.* (1919) **52**, 2020.
6. Bontschev, Z. et al., *Z. Anorg. Algem. Chem.* (1966) **347** (3–4), 199.

of Other Monomers

Analysis of Product

Cl,[a] %	Sn,[a] %	% Mono[h]	mP[i], °C
—	—	9	
19.5	32.1		132
(19.5)	(32.7)		
20.2	31.31	~15	
(18.04)	(30.15)		
17.34	26.38	Not determined	Liquid
(15.86)	(26.54)		
24.31	32.53	57.5	111
(18.20)	(30.44)		
17.95	30.17	None detected	
(18.20)	(30.44)		
—	—	22	
18.81	30.94		158
(18.32)	(30.60)		
24.79[d]	31.2[d]	47	—
(21.4)	(35.8)		
[a]—	—	Mainly mono	—
21.13	33.37	Not determined	
(21.25)	(35.58)		
19.77	31.40	Not determined	
(21.40)	(35.78)		
15.50	22.53	None detected	
(14.00)	(23.40)		
14.46		~40	—
(12.50)			
20.96	35.21	None detected	240–250
(21.3)	(35.60)		

[f] Product isolated from reaction residue by $CHCl_3$ extraction.
[g] Elemental analyses and yield affected by secondary processes involving HCl.
[h] Determined by NMR.
[i] Melting point of purified di compound.

7. Emeléus, H. J., Kettle, S. F. A., *J. Chem. Soc.* (1958) 2444.
8. Tamaru, K., *J. Phys. Chem.*, (1956) **60**, 612.
9. Reifenberg, G. H., Considine, W. J., *J. Am. Chem. Soc.* (1969) **92**, 2401.
10. Amberger, E., *Angew. Chem.*, (1960) **72**, 78.
11. Nefedov, O. M., Kolesnikov, O. P., *Izv. Akad. Nauk, SSSR Ser. Khim.*, (1966) **2**, 201.
12. Nametkin, N. S., Kurjmin, O. V., Korolev, V. K., Chernysheva, T. I., *Izv. Akad. Nauk, SSSR Ser Khim.* (1972) **4**, 985.

RECEIVED May 12, 1976.

9

The Use of Estertin Stabilizers in PVC

D. LANIGAN
Akzo Chemie, Liverpool, England

E. L. WEINBERG
Interstab Chemicals Inc., New Brunswick, N.J.

This paper describes the evaluation of first-generation estertin stabilizers in the major applications areas of extrusion, calendering, injection molding, blow molding, plastisol coating, and rotational casting. Comparisons were made with octyl, butyl, and methyltin stabilizers as well as barium–cadmium–zinc and calcium–zinc stabilizer systems. Estertins are a new class of organotin stabilizers for use in PVC and are based on the discovery of a novel route to produce intermediates. The unique structure of estertins, compared with alkyltins, gives low volatility and extractability while retaining all the virtues of conventional alkyltins (e.g., excellent heat stability and clarity). The results to date show that the performance of estertins is equivalent to or better than available, commercial alkyltin stabilizers in all applications tested.

Since the commercial acceptance of organotin compounds as stabilizers for PVC in 1939, their use has increased considerably, and today they command a major share of the world stabilizer market. Initially, dibutyltin dilaurate and dibutyltin maleate established themselves, and these were followed by other dibutyltin compounds and by dioctyltin and dimethyltin derivatives. All these compounds belong to the alkyltin class and have gained rapid acceptance in the market because of their inherent technical advantages over other products despite their relatively high cost.

Although considerable improvements have been made in the effectiveness of alkyltins, there has been no major breakthrough in tin chemistry for PVC stabilizers until 1974 when an entirely new process for organotin intermediates was discovered (*see* Chapter 8). Moreover the process is simple and does not involve the complicated or hazardous routes used for making alkyltin intermediates.

This has led to the development of a completely new class of organotin stabilizers known collectively as estertins. These new stabilizers have the general structure:

$$
\begin{array}{c}
\text{R}_1-\text{O}-\overset{\overset{\text{O}}{\|}}{\text{C}}-\text{CH}_2-\text{CH}_2 \\
\text{R}_1-\text{O}-\underset{\underset{\text{O}}{\|}}{\text{C}}-\text{CH}_2-\text{CH}_2 \\
\text{R}_1\text{OC}-\text{CH}_2-\text{CH}_2-\text{Sn} \\
\underset{\text{O}}{\|}
\end{array}
$$

Initial technical evaluation of these stabilizers showed that they were similar in performance to the existing alkyltins. Furthermore, primary tests to assess the health and safety aspects of the estertins showed them to be only mild skin irritants and to cause no eye irritation. Extraction tests in different media showed estertins to be less extractable than an octyltin. In addition, acute oral toxicity values (LD_{50}) in rats showed estertin compounds to give higher values than butyltin and methyltin compounds but generally lower values than octyltin compounds (*see* Table I). As a result of these very promising data, the first generation of estertin stabilizers was developed. These consist of four liquid thiotin stabilizers coded Stanclere T208, T209, T215, and T217.

Naturally with this range of products there is a possibility that more than one of the estertin stabilizers can be used for applications other than for those they

Table I. Acute Oral Toxicity of Chloride and Isooctyl
Thioglycollate (IOTG) Derivatives

		LD_{50} *Values, mg/kg*	
R		*X = Cl*	*X = IOTG*
RSnX$_3$	methyl	1370	920
	butyl	2300	1063
	octyl	3800	3400
	ester	5500	1230
R$_2$SnX$_2$	methyl	75	1210
	butyl	126	510
	octyl	7000	1975
	ester	2350	1430
R$_3$SnX	methyl	9	20
	butyl	349	1350
	octyl	29200	26550
	ester	a	a

[a] Note that no detectable amount of tri-compounds could be found in the compounds prepared by estertin process.

were originally developed. In general, however, the following general classification can be made. Stanclere T208 is a liquid thiotin stabilizer with low odor and is recommended for use in bottles and plastisols. However, it has shown promise in the area of calendering. Stanclere T209 is a more powerful stabilizer than T208. It is also a liquid thiotin compound and was specifically developed for use in demanding applications such as rigid calendering and injection molding. Stanclere T215 is a liquid thiotin stabilizer giving a low odor and is specifically designed for plastisols. Stanclere T217 is a liquid thiotin stabilizer specifically developed for use at low levels in rigid pipe extrusion. This paper describes the work carried out to evaluate these products in the many different applications in which tin stabilizers are used.

Multiple Screw Extrusion of Water Pipe

The use of organotin stabilizers in this application is almost exclusive to the American pipe market and is in contrast to the other major technological areas

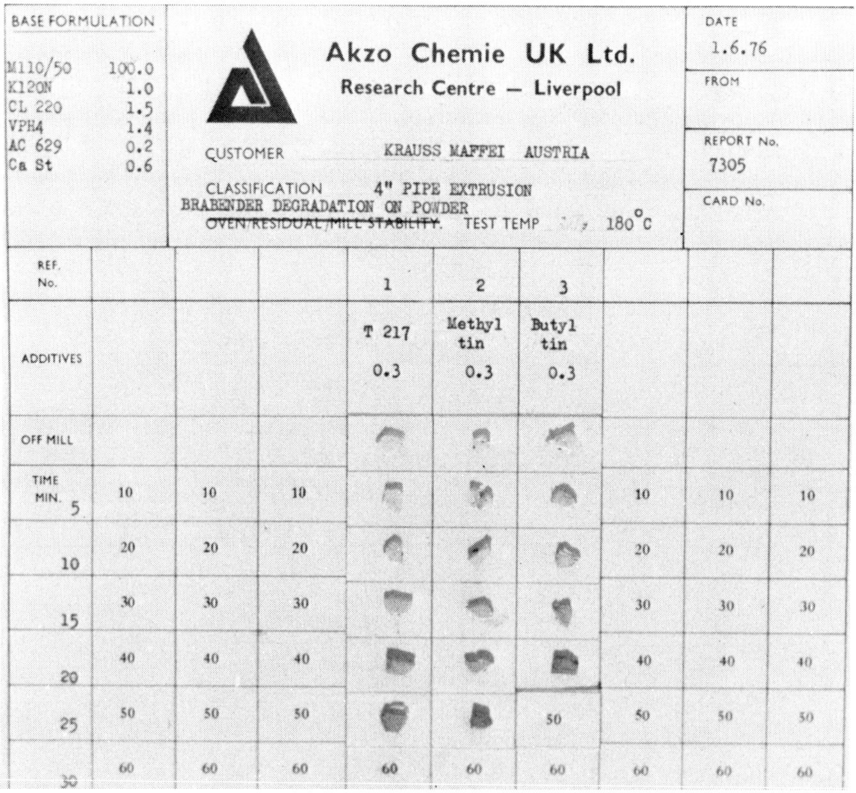

BASE FORMULATION		Akzo Chemie UK Ltd.		DATE 1.6.76
M110/50	100.0	Research Centre — Liverpool		FROM
K120N	1.0			
CL 220	1.5			
VPH4	1.4			
AC 629	0.2	CUSTOMER KRAUSS MAFFEI AUSTRIA		REPORT No. 7305
Ca St	0.6			
		CLASSIFICATION 4" PIPE EXTRUSION BRABENDER DEGRADATION ON POWDER OVEN RESIDUAL MILL STABILITY. TEST TEMP , 180°C		CARD No.

REF. No.				1	2	3			
ADDITIVES				T 217 0.3	Methyl tin 0.3	Butyl tin 0.3			
OFF MILL									
TIME MIN 5	10	10	10				10	10	10
10	20	20	20				20	20	20
15	30	30	30				30	30	30
20	40	40	40				40	40	40
25	50	50	50			50	50	50	50
30	60	60	60	60	60	60	60	60	60

Figure 1. Torque viscometer degradation of powder for a 4-in. pipe extrusion

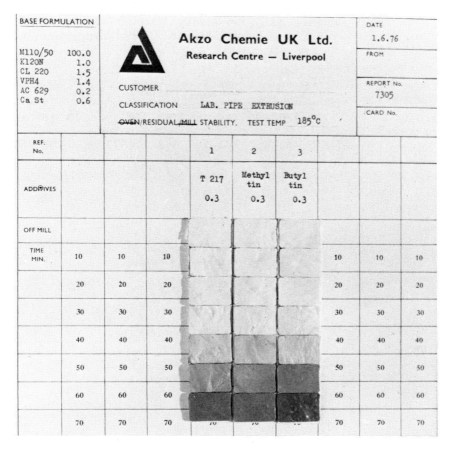

Figure 2. Residual stability laboratory extrusion

in Europe and Japan where mainly lead, and to a lesser extent calcium–zinc, stabilizers are used. The type of stabilizers currently used are butylthiotin, methylthiotin, and mixed metal compounds. In recent years there have been significant reductions in the price of tin stabilizers coupled with major improvements in heat stability performance, and today it is normal for levels of ca. 0.3 phr of stabilizer to be used in formulations for manufacturing water pipe using twin or multiple screw extruders. The use of very low levels of stabilizer differs significantly from the levels used in other applications such as bottles, film, and sheet and, therefore, it has been necessary to develop a specific estertin stabilizer for this application. This is coded Stanclere T217. Important technical properties required from the stabilizer system are: (a) the ability to produce pipe with a good base initial color and color hold, and (b) the ability to enable regrind to be processed satisfactorily. These two properties are best assessed on the large-scale plant, but as costs prohibit this during the development of a stabilizer,

Table II. Extrusion Results Obtained with Formulation
Krauss Maffei KMD 90 Extruder Set up to

Stabilizer	Extrusion Temperatures, °C										
Estertin	170	180	180	165	155	162	170	170	170	200	190
Stanclere T217 Alkyltin	170	180	180	165	160	162	170	170	170	200	188

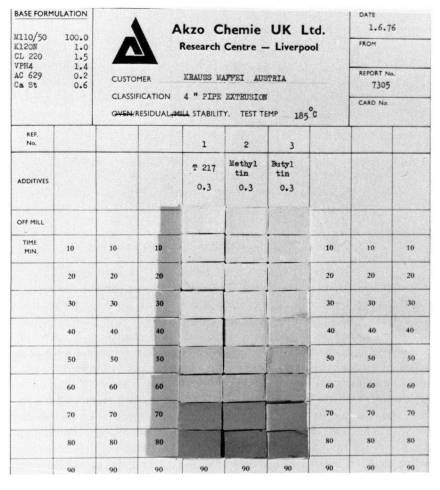

Figure 3. Residual stability production extrusion

Containing Estertin T217 and That on a
Manufacture 111 × 5.3 mm Pipe

Motor Speed, rpm	Motor Load, %	Back Pressure, tons	Feeder Screw Speed, rpm	Output, kg/hr
2200	62	18	51	310
2200	62	18	51	314

small scale tests have been devised to simulate the extrusion process, the most commonly used being the Brabender plastograph as well as small scale laboratory extruders.

The Brabender plastograph can be used to assess the degree of lubrication imparted to the formulation as well as dynamic heat stability. Comparison of the performance of Stanclere T217 with major tin stabilizers used by the American pipe industry is shown in Figure 1.

Figure 1 shows that Stanclere.T217 is essentially non-lubricating and very similar to the market leaders. Comparison of the dynamic heat stability results obtained on the Brabender shows that T217 over a range of shear rates gives excellent white coloration and slightly longer time to decomposition than the standard formulation. Figure 1 confirms that the initial color and color hold of the formulation based on T217 is equivalent to both alkyltins. Confirmation of these results has been obtained by producing pipe on an AGM CT 40 conical twin screw laboratory extruder and a production size machine. Comparison of the oven heat stability of the pipe made from Stanclere T217 showed it was equivalent to that given by the control (*see* Figures 2 and 3).

Confirmation of the Brabender results was also obtained during large-scale trials on a Krauss Maffei KMD 90 extruder. The comparative results are given in Table II. Very similar output, back pressure, and motor load were recorded for both formulations, and the quality of pipe produced (whiteness, outside and inside surface quality) was equivalent. Light stability test carried out on pressure pipe formulations in the Xenotest apparatus has shown little difference between estertins, butylthiotins, and methylthiotins in this application after 3000 hr exposure.

Clear Extruded Sheet

This product is mainly used outdoors and, therefore, demands the use of organotin carboxylates rather than organothiotins because of the superior light-stabilizing power given to the sheet by the former compounds. As the first generation of estertins are all sulfur containing, it is thought that they may not impart good light stability. Nevertheless, efforts are currently underway to assess

Table III. Light Stability Results of Estertins and Organotin
Carboxylate in Listed Base Formulation[a]

		Time to Initial Discoloration, hr		
Method	Type of Exposure	T208	T209	Organotin Carboxylate
Xenotest	artificial	>3500	>3500	>3500

[a] Base Formulation: PVC (K55 suspension) 100, stabilizer 2.5, Interstab G8215 1.5, processing aid 1.5, UV absorber 0.2, blue pigment 0.001.

Table IV. Light Stability Results of Estertin Thio vs. Alkyltin Thio
and Carboxylate Stabilizers

		Time to Initial Discoloration, hr		
Method	Type of Exposure	Estertin T209	Alkylthiotin	Organotin Carboxylate
Xenotest	artificial	600	400	1000

the light stability of estertins in this application. So far, Xenotest results have shown that there is little difference between estertins T208 and T209 and organotin carboxylates (*see* Table III).

With formulations containing no UV absorbers, the results (*see* Table IV) show that estertin stabilizers give a better result than alkylthiotins but worse than organotin carboxylates. These results are, however, academic since it is not recommended that any clear product is used outdoors without a UV absorber.

Rigid Foamed Profile

Production-scale trials carried out to compare the performance of estertins and alkylthiotins in this application have shown that a product of superior appearance can be obtained using estertins. The machine chosen for the evaluation was an EC-200A-6-15 Aragon vertical, single-screw extruder. The extrusion

Table V. Results of Trial Comparing Stanclere T208 and an
Alkylthiotin on an Aragon Extruder Making Rigid Foamed Profile

Stabilizer	Extrusion Temperatures, °C					Main Drive Speed, rpm	Motor Load, amps	Out-put, kg/hr	Foam Density, gl.
Estertin Stanclere T208	160	170	178	195	180	1050	20	29	0.75
Alkyl-thiotin	160	170	178	195	180	1050	20	30	0.68

conditions used were optimized for the control formulation, stabilized with alkylthiotin, and the performance of the estertin formulation was compared under the same conditions (*see* Table V). The result obtained was a profile of superior surface gloss and slightly higher density at the same output.

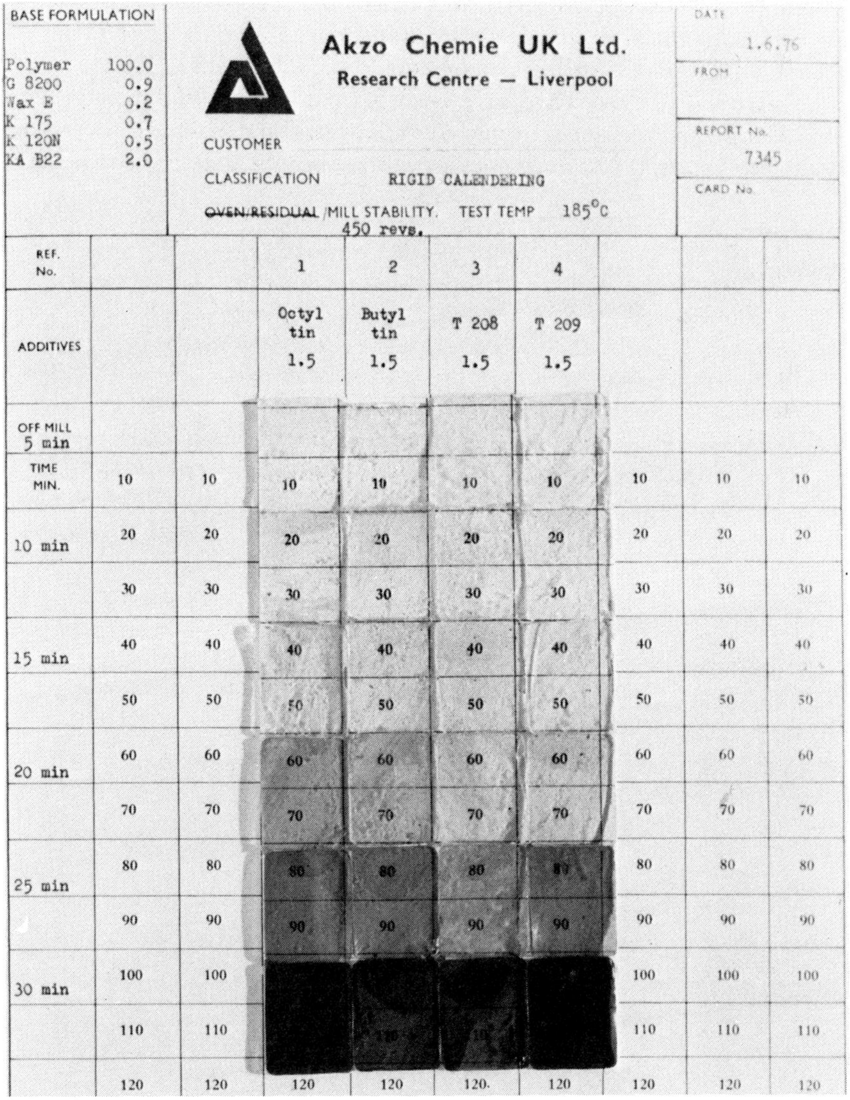

BASE FORMULATION				
Polymer	100.0			
G 8200	0.9			
Wax E	0.2			
K 175	0.7			
K 120N	0.5			
KA B22	2.0			

Akzo Chemie UK Ltd.
Research Centre — Liverpool

DATE 1.6.76
REPORT No. 7345

CUSTOMER

CLASSIFICATION   RIGID CALENDERING

~~OVEN/RESIDUAL~~/MILL STABILITY.   TEST TEMP   185°C
450 revs.

REF. No.			1	2	3	4			
ADDITIVES			Octyl tin 1.5	Butyl tin 1.5	T 208 1.5	T 209 1.5			
OFF MILL 5 min									
TIME MIN.	10	10	10	10	10	10	10	10	10
10 min	20	20	20	20	20	20	20	20	20
	30	30	30	30	30	30	30	30	30
15 min	40	40	40	40	40	40	40	40	40
	50	50	50	50	50	50	50	50	50
20 min	60	60	60	60	60	60	60	60	60
	70	70	70	70	70	70	70	70	70
25 min	80	80	80	80	80	80	80	80	80
	90	90	90	90	90	90	90	90	90
30 min	100	100					100	100	100
	110	110					110	110	110
	120	120	120	120	120	120	120	120	120

Figure 4. Heat stability—lab calendar

Rigid Calendered Film

In 1974 the use of tin stabilizer in the rigid calendering market was estimated as 1300 M ton/year in Europe. Octylthiotins are used mainly for non-toxic applications, and butylthiotins are used for general purpose film. The application is demanding and relies very much on obtaining the correct lubrication system as well as having a good heat stabilizer. For this reason, high-speed mill tests are considered the most meaningful small-scale method of comparing different systems before any large scale trials are carried out. It is also true, as with most PVC processes, that the comparative results can vary according to conditions of test and other variables such as polymer grade and lubricant system.

Estertins T208 and T209 have been compared with leading octyl and butylthiotin stabilizers and shown to give equivalent, and in some cases better, results (see Figure 4). As a consequence of these results, large-scale trials were

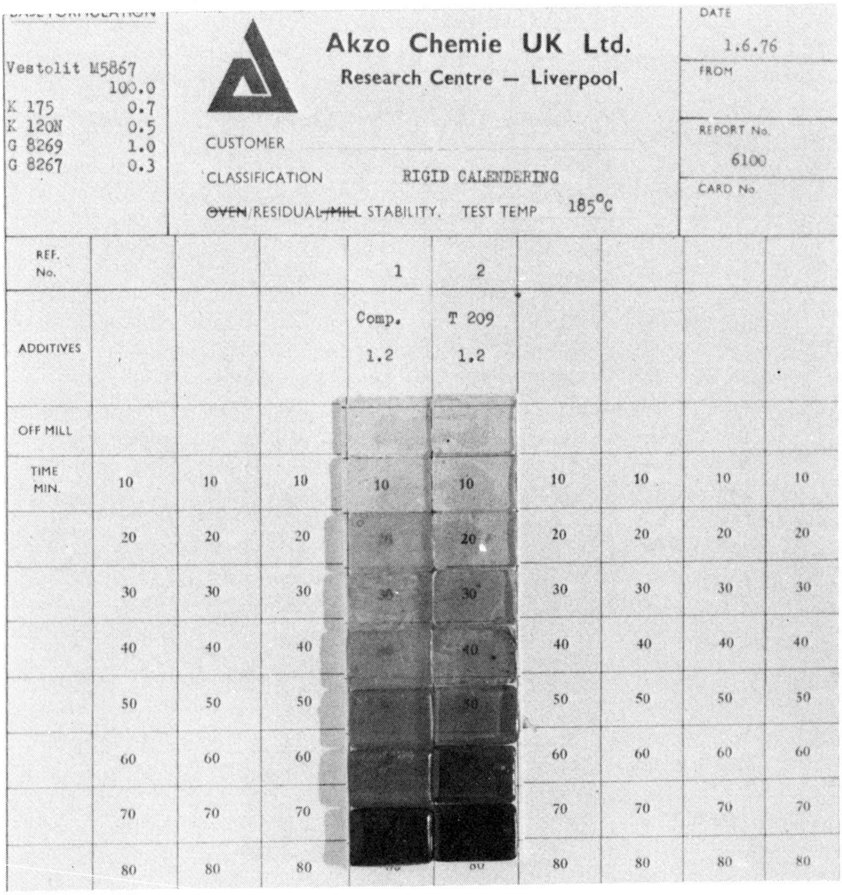

Vestolit M5867	100.0		Akzo Chemie UK Ltd.	DATE 1.6.76
K 175	0.7		Research Centre — Liverpool	FROM
K 120N	0.5			
G 8269	1.0		CUSTOMER	REPORT No. 6100
G 8267	0.3		CLASSIFICATION RIGID CALENDERING	
			OVEN/RESIDUAL/MILL STABILITY. TEST TEMP 185°C	CARD No.

REF. No.			1	2				
ADDITIVES			Comp. 1.2	T 209 1.2				
OFF MILL								
TIME MIN.	10	10	10	10	10	10	10	10	10
	20	20	20		20	20	20	20	20
	30	30	30		30	30	30	30	30
	40	40	40		40	40	40	40	40
	50	50	50		50	50	50	50	50
	60	60	60		60	60	60	60	60
	70	70	70		70	70	70	70	70
	80	80	80		80	80	80	80	80

Figure 5. Residual heat stability—commercial

Table VI.   Processing Conditions for Rigid Calendered
Film Commercial

Extruder Compounder Conditions, °C	Mill Temperatures, °C	Calender Temperatures, °C	Take-Off Roll Temperatures, °C
125 120 165	155/160	180/185/190/185	116/95/85

performed on a Berstoff calender with Stanclere T209.   The formulation used
is confidential, but the trial compared a leading octyltin stabilizer with Stanclere
T209 with the same lubrication system at a level of 1.2 phr: the conditions used
are given in Table VI.   The film produced was of excellent quality and at least
equivalent to that given by the formulation based on the octyltin stabilizer.
Comparative oven heat stability tests carried out on the film are given in Figure
5 and show a greater reserve of heat stability and surface gloss from the estertin
formulation.

Extruded Rigid Foil

A large quantity of the foil used by vacuum-forming compounds for sale
to the package industry is now made by an extrusion/calendering process.   This
is technically an interesting application since it involves a closed extrusion operation during which dryblend is converted to a melt followed by a small open
calendering process.   Laboratory trials showed that both Stanclere T208 and T209
compared favorably with an octylthiotin stabilizer and were suitable for further
study on the larger scale (*see* Figure 6).

The trials carried out on a Kleinewefer Kalendrette showed that optimum
results were obtained with a mixture of T208 and T209 in the ratio of 1.0:0.5 phr
against an octylthiotin at 1.5 phr.   Three very important observations were made
during the production operation.   First, the foil produced had excellent color,
clarity and surface gloss; secondly, the machine was much easier to clean with
the formulation based on the estertin, and finally the odor of the foil after a short
storage period was less than standard production.

Vacuum forming trials carried out on the foil confirmed its excellent quality
both in a deep draw application and in special tests to highlight the quality of
the surface of the film.   Production trials showed that the film vacuum formed
easily and quickly giving moldings with a greater brilliance and sparkle than given
by the foil stabilized with octylthiotin.   The odor produced during foaming was
also low and completely acceptable.

Bottles

Estertins in this application must have restricted use since they are not yet
approved for use in products in contact with food.   Nevertheless two estertins
T208 and T209 have been compared with both octylthiotin and methylthiotin

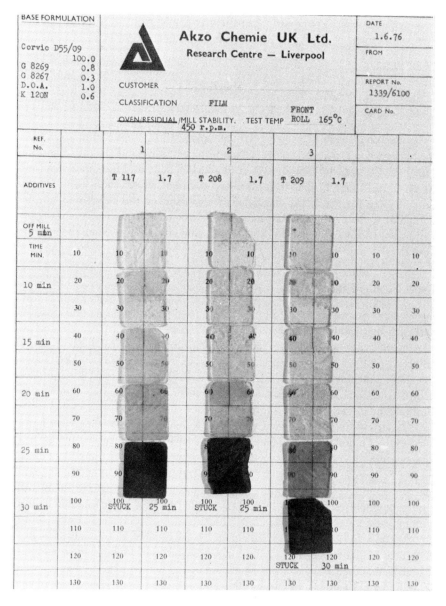

Figure 6. Dynamic mill stability test results

stabilizers for this application. Initial laboratory results indicated that both gave good heat stability performance against the market leaders (*see* Figure 7) and that bottles produced from T208 were of very low odor.

To obtain confirmation of these results, large scale trials were performed on the following blow molding machines: (a) Sidel DSL 3M rotary blow molder,

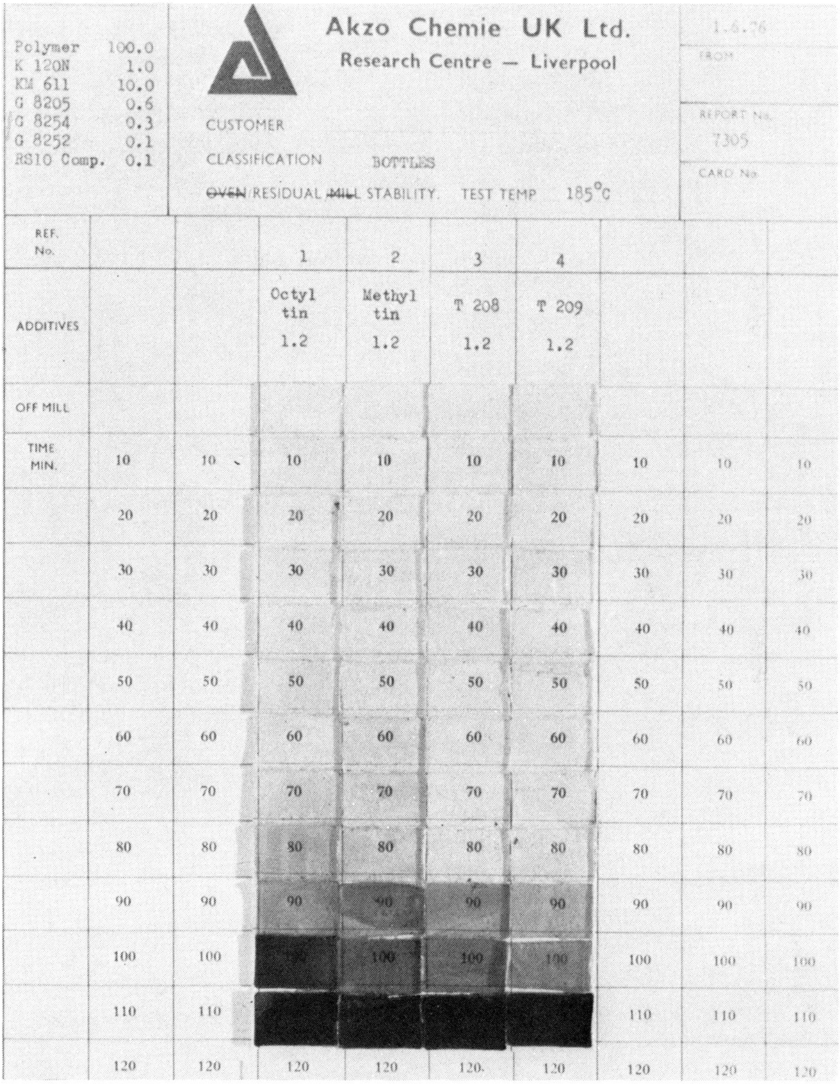

Polymer	100.0
K 120N	1.0
KM 611	10.0
G 8205	0.6
G 8254	0.3
G 8252	0.1
RS10 Comp.	0.1

Akzo Chemie UK Ltd.
Research Centre — Liverpool

CUSTOMER

CLASSIFICATION   BOTTLES

OVEN RESIDUAL MILL STABILITY.   TEST TEMP $185^{\circ}C$

REF. No.			1	2	3	4			
ADDITIVES			Octyl tin 1.2	Methyl tin 1.2	T 208 1.2	T 209 1.2			
OFF MILL									
TIME MIN.	10	10	10	10	10	10	10	10	10
	20	20	20	20	20	20	20	20	20
	30	30	30	30	30	30	30	30	30
	40	40	40	40	40	40	40	40	40
	50	50	50	50	50	50	50	50	50
	60	60	60	60	60	60	60	60	60
	70	70	70	70	70	70	70	70	70
	80	80	80	80	80	80	80	80	80
	90	90	90	90	90	90	90	90	90
	100	100		100	100	100	100	100	100
	110	110					110	110	110
	120	120	120	120	120.	120	120	120	120

Figure 7. Residual heat stability of blown bottles (lab)

(b) Fischer 90/20D blow molder fitted with twin parison head, and (c) Bekum blow molders. All the trials confirmed that estertin T208 could be substituted in a typical octylthiotin formulation without making significant adjustments to lubrication or machine conditions. The most demanding trial carried out was on the Sidel DSL 3M machine producing 3000 1-l. bottles per hr. Extrusion conditions for this trial are given in Table VII.

Table VII.   Sidel Bottle

Stabilizer	Machine Temperatures, °C						
Octylthiotin	170	184	166	166	154	180	203
Estertin T208	169	180	165	160	154	177	196

[a] Base formulation: Lacqvyl S071S 100, Kane Ace B28 10, Paraloid

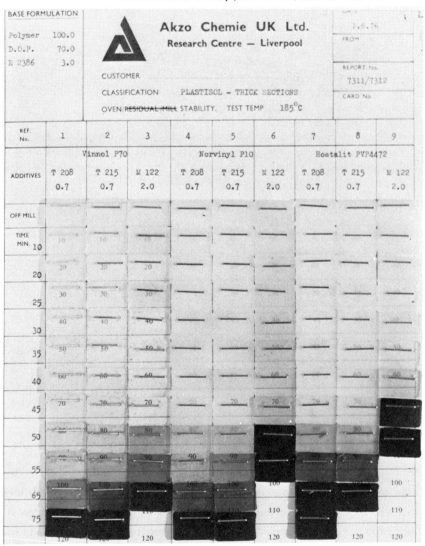

Figure 8.   Heat stability—plastisols comparing estertins vs. metal soap stabilizers in various resins

Trial Conditions[a]

Speed, rpm	Amperes	Output, bottle/hour	% Regrind
32	59	3000 (139 kg/hr)	20
32	60	3000 (139 kg/hr)	20

K120N 1.0, stabilizer 1.5, Interstab EC55 1.0, Reckitts RS10 0.01.

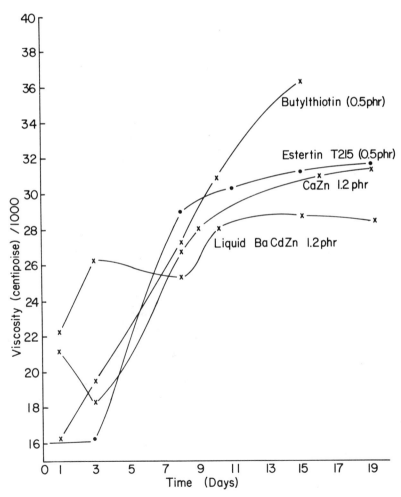

Figure 9. Viscosity build up of a plastisol containing different stabilizers against time

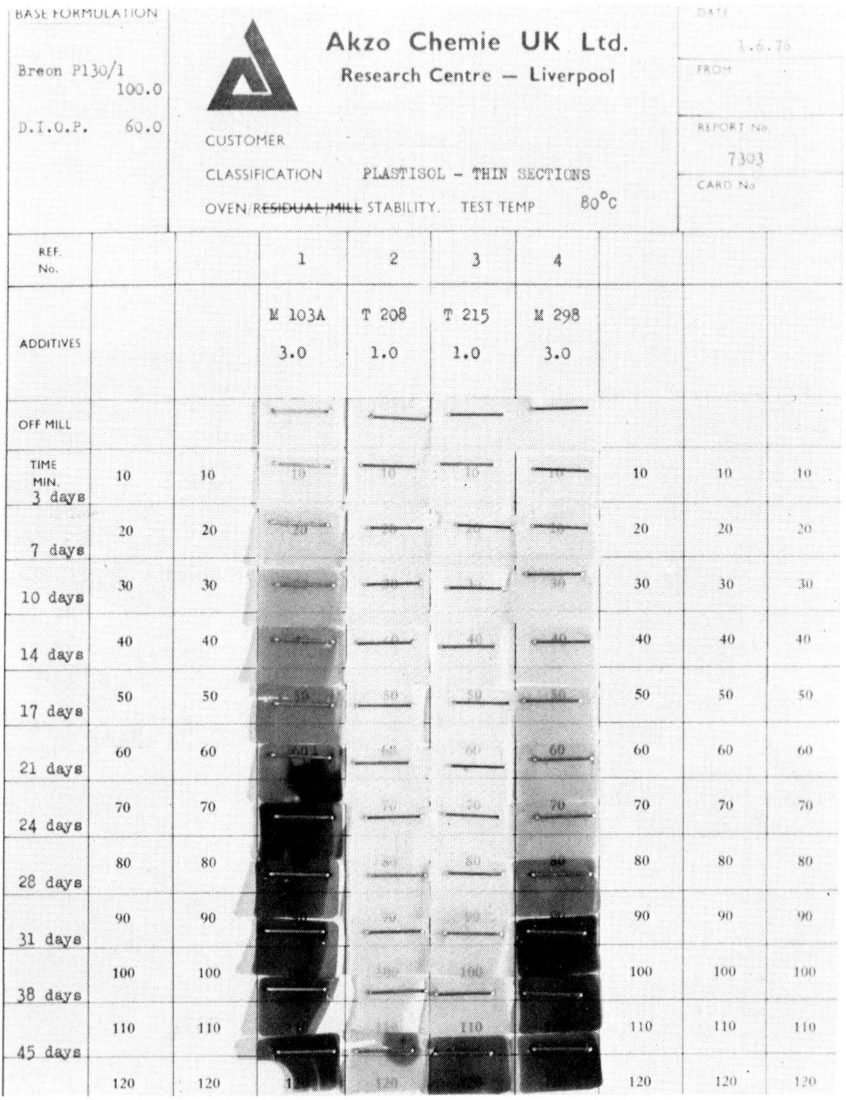

BASE FORMULATION

Akzo Chemie UK Ltd.
Research Centre — Liverpool

Breon P130/1
            100.0

D.I.O.P.   60.0

CUSTOMER

CLASSIFICATION   PLASTISOL - THIN SECTIONS

OVEN/RESIDUAL/MILL STABILITY.   TEST TEMP $80^{\circ}C$

DATE
FROM
REPORT No.   7303
CARD No.

REF. No.			1	2	3	4			
ADDITIVES			M 103A	T 208	T 215	M 298			
			3.0	1.0	1.0	3.0			
OFF MILL									
TIME MIN. 3 days	10	10	10	10	10	10	10	10	10
7 days	20	20	20	20	20	20	20	20	20
10 days	30	30	30	30	30	30	30	30	30
14 days	40	40	40	40	40	40	40	40	40
17 days	50	50	50	50	50	50	50	50	50
21 days	60	60	60	60	60	60	60	60	60
24 days	70	70	70	70	70	70	70	70	70
28 days	80	80	80	80	80	80	80	80	80
31 days	90	90	90	90	90	90	90	90	90
38 days	100	100	100	100	100	100	100	100
45 days	110	110	110	110	110	110	110	110
	120	120	120	120	120	120	120	120

*Figure 10. Low temperature heat stability
plastisol*

Reversed Roll Coated Wall Covering

The stabilization of plastisols is mainly achieved with metal soap stabilizers with only a few specialized applications using tin stabilizers. In general, stabilization of different emulsion resins with metal soap stabilizers requires extensive knowledge of the particular resin as most resins have a different sensitivity to zinc.

As a result many metal soap stabilizers exist because they were developed for specific polymer systems.

Following laboratory tests, Stanclere T215 and T208 were developed for use in plastisols, and their properties are best described by reference to Figures 9–12.

They show that T208 and T215 in this application have the following advantages: (a) they show less sensitivity than metal soaps to polymer change (Figure 8), (b) they are efficient at low levels of use (Figure 8), (c) the viscosity build-up characteristics and air release properties of the plastisol are satisfactory (Figure 9), (d) they have a low odor, (e) they show excellent long term, low temperature (80°C) heat stability compared with that given by liquid Ba–Cd–Zn and Ca–Zn stabilizers (Figure 10), and (f) they show good light stability on exposure in the Xenotest apparatus (Figure 11).

Large scale trials have been carried out on a modern Bone Craven reverse roll coating equipment to confirm laboratory results. Conditions used are given in Table VIII. Particular attention was paid to the odor produced during processing and in the final product, and they were both found to be completely satisfactory.

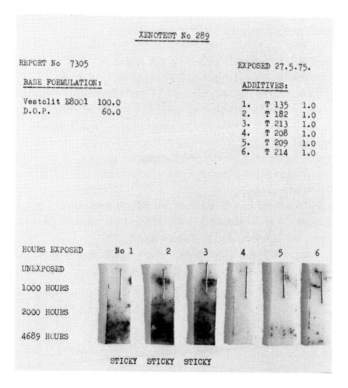

Figure 11. Light stability (Xenotest) plastisol. Removed from Xenotest 9-3-76 after 4689 hr exposure

Table VIII.   Processing Conditions Used for Manufacture of Vinyl
Wallpaper on Bone Craven Reverse Roll Coating Equipment

Processing Conditions

Coating weight of vinyl covering (g/m^2) dry weight	90–110
Line speed, m/min	120
Oven temperatures, °C	170 190 200 205

Rigid Injection-Molded Fittings

This demanding application requires an efficient heat stabilizer as well as a well-balanced lubrication system.   Trials carried out on Stork Reed 500-ton injection molding machine with granulate stabilized with Stanclere T209 gave 6-in. diameter couplers of equivalent appearance to material stabilized with a liquid butylthiotin stabilizer.   Such initial results are considered excellent since no change in lubrication was required and, therefore, it will most likely be possible to develop improved systems with T209.

Health and Safety Aspects

Increasing interest and pressure are being shown in this subject from the following points of view: (a) potential risks to industrial plant personnel manufacturing basic raw materials such as PVC stabilizers, (b) potential risks to industrial plant personnel converting basic raw materials into products for the market, and (c) potential risks to the users of the final products.   As already mentioned, the route to produce estertin intermediates does not involve the complicated or hazardous routes used for making alkyltin intermediates.   Work has also been carried out to assess the risks relative to alkyltin stabilizers of estertins, and the results are given below.

Primary Skin and Eye Irritation.   Results of tests on rabbits, following the U.S. Code of Federal Regulation, Title 16, section 1500.41 (skin) and section 1500.42 (eye), are summarized in Table IX.

Migration into Food Simulants.   The migration of estertins is compared with an octyltin, commonly used in bottle formulations.   All stabilizers were used

Table IX.   Primary Skin and Eye Irritation Results

	Skin Irritation	Eye Irritation
RSnCl$_3$[a]	mild	mild
R$_2$SnCl$_2$	mild	severe
RSn(IOTG)$_3$	mild	non
R$_2$Sn(IOTG)$_2$	mild	non

[a] R = R'OCOCH$_2$CH$_2$–.

at an addition level of 1.2 phr, and the bottles were sealed immediately after blowing. The stabilizer contents of the simulants were measured at 40°C after an incubation time of 10 days. The figures mentioned in Table X are the highest

Table X. Extraction Data from Rigid Bottles

Simulant	Octyltin	Stanclere ET208	Stanclere ET209
Distilled water	0.19	0.15	0.07
Acetic acid (3% aq.)	0.39	0.23	0.13
Ethanol (10% aq.)	0.26	0.19	0.15
Cooking oil	0.20	0.17	0.11

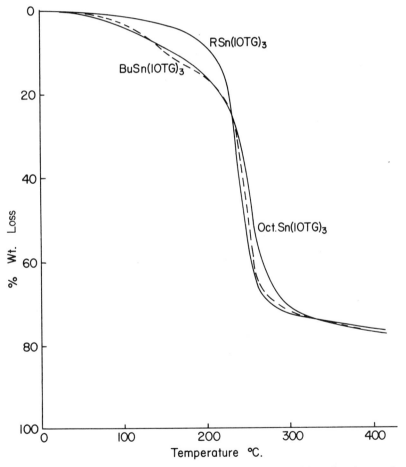

Figure 12. T.G.A. curves for monalkyltin trisisooctylthioglycolate salts (heating rate 10°/min in air)

of three separate determinations, and they give the amount of stabilizer in the extract in ppm.

Acute Oral Toxicity.   A comparison of the acute oral toxicity values (LD_{50}) in rats between conventional alkyltin compounds and estertin compounds is given in Table I.

Volatility.   Although the TGA curves for alkyltin and estertin stabilizers are in the same order (see Figures 12 and 13), the chloride compounds are shown to be less volatile (Figures 14 and 15).   The lower volatility of the estertin chlorides could be an important safety factor for plant personnel operating open conversion processes such as calendering.

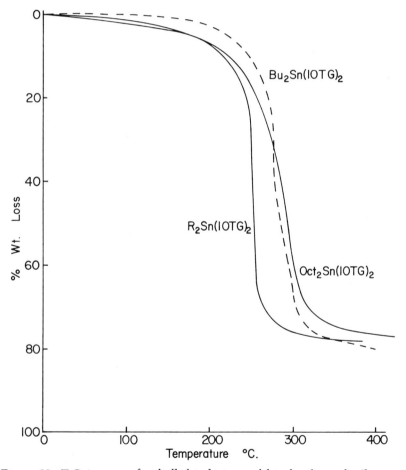

Figure 13.   T.G.A. curves for dialkyltin bisisooctylthioglycolate salts (heating rate 10°/min in air)

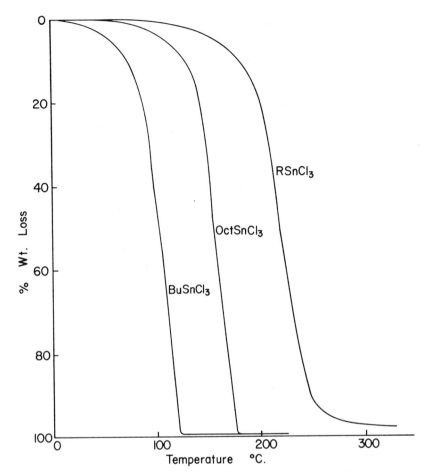

Figure 14. T.G.A. curves for monoalkyltin trischlorides (heating rate 10°/min in air)

Mutagenicity. Monoestertin trisisooctylthioglycolate and diestertin bisisooctylthioglycolate compounds have given negative results when examined by the direct plating mutagenicity test using *Salmonella typhimurium* at five dose levels per compound.

Chronic Toxicity. On the basis of acute oral toxicity, four-week studies were undertaken on the estertin stabilizers to determine a suitable dietary level for a 13-week study. A similar study was carried out on the corresponding chlorides to provide additional product safety data. At the satisfactory conclusion of these studies, a 13-week study on estertin stabilizers was commissioned. All animals have now been sacrificed, and petitions for approval as food packing additives will be submitted when the final reports are available.

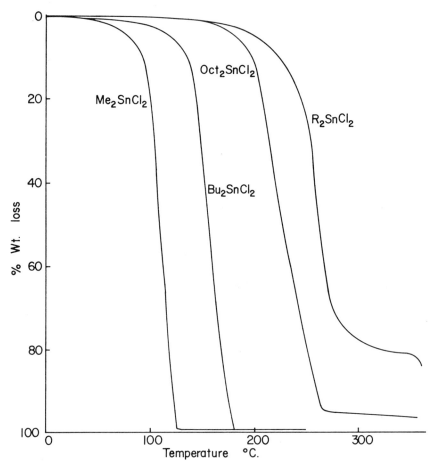

Figure 15. T.G.A. curves for dialkyltin dichlorides (heating rate 10°/min in air)

Conclusions

The first generation of estertin stabilizers has been shown to give at least equivalent and in some cases improved performance vs. that of alkylthiotin stabilizers in the major applicational areas.   Against liquid Ba–Cd–Zn and Ca–Zn stabilizer, estertins have definite advantages in plastisols.   Added to their excellent heat stabilizing performance is the benefit that the intermediate chlorides are of low volatility and toxicity.   These data point to estertins being major stabilizers for PVC in the future.

RECEIVED May 12, 1976.

Structure and Bonding in Organotins by Gamma Resonance Spectroscopy

ROLFE H. HERBER and MICHAEL F. LEAHY

Rutgers University, New Brunswick, N. J. 08903

The recently developed effective vibrating mass (EVM) model, which provides a semi-theoretical framework within which the molecular weight of covalent organometallic compounds can be determined from Raman spectroscopic and temperature-dependent Mössbauer effect experiments, has been extended to two organotin thiol compounds. In $[(CH_3)_2SnS]_3$, although the Mössbauer active atom does not occupy the center of mass of the molecule, the agreement between the trimeric formula weight (542) and the calculated effective vibrating mass (552) demonstrates further the validity of the basic assumptions of the EVM model. In the distorted, nearly tetrahedral molecule $Sn(SCH_2CH_2S)_2$, the EVM model calculations suggest an appreciable interaction between proximal molecules in the solid through S-atom bridging between adjacent tin atoms. This gives rise to quasi-linear Sn-S-Sn- chains in consonance with the chemical and physical properties of this compound, as well as the results of a recent crystal structure determination.

Shortly after the first results were published concerning the chemical applications of gamma resonance spectroscopy (Mössbauer Effect) (*1, 2*) making use of the 23.8 keV excitation in [119]Sn, it became clear that this technique would be valuable in elucidating the structure and bonding in organotin compounds, and a very large literature has grown up in this field (*3, 4*). The earliest data to be extracted from such spectra focussed attention principally on the isomer shift and quadrupole splitting parameters, since it became evident that these hyperfine interactions could elucidate the formal oxidation state of the metal atom [e.g., Sn(II) or Sn(IV)] and its coordination number more readily than most other available techniques. In addition, it was observed (*5, 6*) that there appeared to be a qualitative relationship between the temperature dependence of the [119]Sn recoil-free fraction, $f(T)$, and the coordination number of the metal atom and/or intermolecular bonding. The observation of a resonance effect at room temperature has frequently been interpreted in terms of an extended network-associated or polymeric structure for the organotin compound in question. In fact, the temperature dependence of the recoil-free fraction observed in ^{119}Sn Mössbauer spectroscopy of organotin compounds can be used to provide quantitative information concerning the structure and bonding in such compounds. A model has been developed (7, 8, 9) which permits the exploitation of the data obtained by $f(T)$ and vibrational spectroscopic investigations to obtain a clearer understanding of the association of organotin compounds in the solid state. This model has been called the effective vibrating mass (EVM) model, and the fundamental assumptions of this approach are reviewed briefly.

A Brief Review of the EVM Model

The principal parameter of interest in the context of this discussion is the recoil free fraction, f, observed in a Mössbauer experiment. This parameter—which can have values ranging from 0 to 1—is the probability of emitting (f_s) or absorbing (f_a) a gamma ray without recoil. Although in principle it is possible to evaluate this parameter quantitatively under a given set of experimental conditions, it is more convenient to examine the temperature dependence of the recoil free fraction $f(T)$ by extracting the temperature dependence of the area under the resonance curve $A(T)$ in such spectra. The definition of f and A are given by

$$f = \exp(-\langle x^2 \rangle \lambda^{-2})$$ (1)

and

$$A = \frac{\pi}{2} f_s \Gamma L(t)$$ (2)

in which $\langle x^2 \rangle$ is the expectation value of the mean square displacement of the Mössbauer atom, λ is the wavelength (divided by 2π) of the Mössbauer transition, Γ is the natural line width of the gamma transition (i.e., $\Gamma = h/2\pi\tau$, where τ is the mean lifetime of the excited state) and $L(T)$ is a saturation function (10) of the absorber thickness, t_a. The latter parameter is defined by the relationship

$$t_a = \left(\frac{\pi}{2}\right) \sigma_0 \Gamma n f_a$$ (3)

in which n is the number of Mössbauer-active nuclei per unit area of the absorber.

The total experimental line width, Γ_e, in turn is given by

$$\Gamma_e = \Gamma_a + \Gamma_s + 0.27\Gamma\sigma_0 f_a \approx \Gamma (2 + 0.27\sigma_0 f_a)$$ (4)

where the a and s subscripts refer to absorber and source respectively and σ_0 is the cross-section for recoilless ((resonant) scattering and has the value 1.403 ×

10^{-18} cm^2 for the 23.8 keV transition of [119]Sn. The temperature dependence of the recoil-free fraction of the absorber f_a is given from the Debye model of solids by an expression of the form

$$f_a(T) = \exp - \frac{3}{2}\frac{E_R}{k\theta}\left[1 + 4\left(\frac{T}{\theta}\right)^2 \int_0^{\theta/T} \frac{x}{e^x - 1}dx\right] \qquad (5)$$

In the high temperature limit, the integral in Equation (5) goes to θ/T so that the slope of the $f(T)$ vs. temperature curve is given by

$$\frac{d\ln f_a}{dT} = -\frac{6E_R}{k\theta_M^2} \qquad (6)$$

where θ_M is a characteristic temperature which is (for an ideal, monatomic isotropic cubic solid) equivalent to the Debye temperature, θ_D, calculated from low temperature specific heat data. The recoil energy, E_R, can be evaluated from momentum conservation considerations

$$E_R = \frac{E_\gamma^2}{2Mc^2} \qquad (7)$$

so that

$$\frac{d\ln f}{dT} = \frac{-3E_\gamma^2}{M_{eff}c^2k\theta_M^2} \qquad (8)$$

where M_{eff} is the effective recoiling mass which contains the Mössbauer active atom.

In general, except for certain solids such as intercalation compounds and ionic lattices, the effective recoiling mass is not known, so Equation 8 contains two unknown experimental parameters, M_{eff} and θ_M. In molecular solids it may be assumed that because of both energetic and symmetry considerations the intra- and intermolecular motions do not couple, and hence that as a first approximation the organotin compound condensed phase can be thought of as an array of hard sphere particles which interact within the unit cell by van der Waals intermolecular forces. The motion corresponding to the intra-unit cell vibrations of these molecules against each other can be probed by Raman spectroscopy, and the relevant frequencies are normally observed in the lattice mode region of the spectrum lying generally below ~200 cm^{-1}. If one of the lattice region vibrational modes corresponds to the unique intermolecular intraunit cell vibration of two molecules against each other, then the frequency of this mode, ω_L, can be used to calculate an effective Debye temperature by the relationship

$$\theta = \frac{\hbar\omega_L}{k} \qquad (9)$$

Inserting this value of θ into Equation 8 leads to the relationship (7):

$$\frac{-d \ln A}{dT} = \frac{3E_\gamma^2 k}{M_{eff} c^2 h^2 \omega_L^2} \tag{10}$$

in which the only unknown quantity is the effective vibrating mass M_{eff}. It is thus possible to use the temperature dependence of the [119]Sn recoil-free fraction to elucidate the possible association of organotin molecules in the solid state, since it should clearly be feasible to distinguish between monomers, dimers, and other low molecular weight polymeric units by means of the above formalism. A number of such studies (13, 14) have previously been published for a variety of molecular geometries and coordination numbers of tin, and the general validity of this approach appears to be reasonably well established. However, because of the predominance of sp^3, sp^3d, and sp^3d^2 hybridization of the tin atom in which there are, respectively, four, five, and six identical or similar ligands bonded to the metal, most molecules which have been subjected to temperature dependent Mössbauer effect studies are those in which the tin atom occupies the center of mass of the molecule. Clearly, if the assumptions of the EVM model are correct, this fact should be unimportant in the applicability of this model to molecular systems, and thus a wider test would be to make certain that the details of the intramolecular geometry do not influence the interpretation of the $f(T)$ data.

In the present study we have extended the EVM model to the investigation of two organotin compounds in which the metal atom does not occupy the center of mass of the molecule as a whole and in which molecular association by bridging ligands has either been shown by x-ray diffraction data or is presumed to exist on the basis of other spectroscopic and chemical information.

Experimental

The experimental details of the Mössbauer spectroscopic methodology used in these studies has been described (7, 8, 9). All isomer shifts cited in the present work are with respect to a room temperature (295 °K) BaSnO$_3$ absorber spectrum. Velocity calibration of the spectrometer was effected by measuring the Fe(0) magnetic hyperfine spectrum as reported earlier (16).

The organotin compounds were prepared by literature methods (17, 18, 19), and their purity was ascertained by standard analytical techniques including IR, mass spectrometric, and Mössbauer methods as appropriate.

Discussion of Results

[(CH$_3$)$_2$SnS]$_3$. The Mössbauer parameters for this compound are summarized in Table I, and the low frequency Raman spectrum is shown in Figure 1. In the latter, it will be noted that there are six distinct bands in the region below ~100 cm^{-1}, and these are assumed to arise from either discrete lattice modes of the covalent solid or low energy librational and/or torsional modes of the indi-

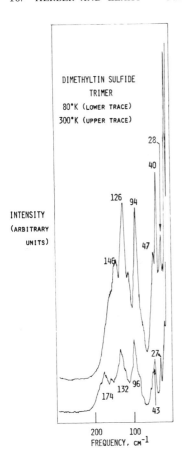

DIMETHYLTIN SULFIDE
TRIMER
80°K (LOWER TRACE)
300°K (UPPER TRACE)

INTENSITY
(ARBITRARY
UNITS)

200   100
FREQUENCY, CM^{-1}

Figure 1. Low frequency (lattice mode region) Raman spectrum of [(CH$_3$)$_2$SnS]$_3$ at 300° and 80°K. The mode at ~27.5 cm^{-1} has been identified as the intermolecular, intra-unit cell vibration of two trimeric molecules against each other on the basis of the present work.

vidual molecules. Replacing θ in Equation 8 by the value calculated using the 27.5 cm^{-1} band, and using the temperaturre dependence of the ^{119}Sn recoil-free fraction extracted from the Mössbauer experiments, that is, ln $A = -0.02464T - 35.84T^{-1}$ (multiple regression coefficient 0.999), leads to a calculated value of M_{eff} of 552. This is in good agreement with the trimeric mass of 542 and in consonance with the x-ray diffraction evidence (20) for the six-membered ring structure of this compound in the solid state. The temperature dependence of the area under the resonance curve is summarized graphically in Figure 2.

It is thus clear that the EVM model assumption that molecules in covalent (organometallic) solids could be considered as hard sphere entities in the context of understanding the temperature dependence of the recoil-free fraction extracted from Mössbauer data is valid even when the center of mass of the molecule is not coincident in space with the lattice position of the Mössbauer-active metal atom. The EVM model thus provides an important and useful approach to the determination of the extent of association of molecular monomers into low weight polymeric arrays in covalent solids, and the basic assumption that the local mi-

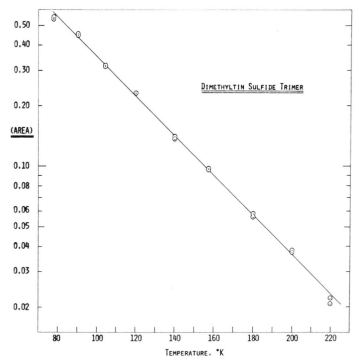

Figure 2.   Temperature dependence of the area under the ^{119}Sn
resonance curve for $[(CH_3)_2SnS]_3$ *in the temperature range 78*
$\leq T \leq 218°K$. *The temperature dependence is given by* $\ln A$
$= -0.02464T - 35.847/T$

crostructure of the molecular units is unimportant in this model is well borne out. The application of this model to molecules which cannot, for one reason or another, be studied by x-ray diffraction techniques (e.g., they are liquids at room temperature, they do not readily form single crystals, they decompose in the x-ray beam, etc.) is thus limited only by the provisions that an otherwise suitable Mössbauer-active atom can be built somewhere into the molecular structure and that the lattice-mode region of the vibrational spectrum is accessible by appropriate (Raman or far-IR) spectroscopic techniques.

Table I. ^{119}Sn

Compound	I.S. (78°K) mm/sec	Q.S. (78°K) mm/sec
$[Sn(SCH_2CH_2S)_2]_n$	1.401 ± 0.010	1.026 ± 0.008
$[(CH_3)_2SnS]_3$	1.341 ± 0.008	1.824 ± 0.010

$[Sn(SCM_2CH_2S)_2]_n$. The Mössbauer parameters for this compound are summarized in Table I and a typical Mössbauer spectrum is shown in Figure 3. The low frequency region of the Raman spectrum of this compound is shown in Figure 4. Among the physicochemical properties of this compound which must be considered in an interpretation of these data are the relatively high melting point *(21)* (182°–183°C), low solubility in nonpolar solvents, appreciable

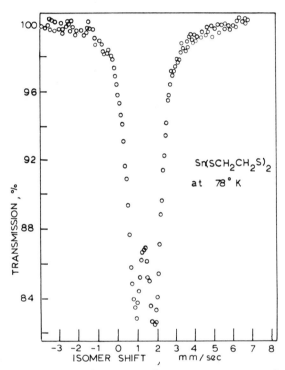

Figure 3. ^{119}Sn Mössbauer spectrum of $Sn(SCH_2CH_2S)_2$ at 78°K. The isomer shift standard is the center of a room temperature $BaSnO_3$ absorber spectrum run using the same $Ba^{119m}SnO_3$ source used in the sample spectra. A geometrical base line correlation has been applied to all data.

Mössbauer and Raman Data

$\ln f\ (T)$	Raman Modes cm^{-1}	(<100 cm^{-1})				
$-0.01180T - 16.63T^{-1}$	~80°K:31.5	41.5		75.5		
	~300°K:28					
$-0.02464T - 35.84T^{-1}$	~80°K:16	20.5	27.5	43	49	96
	~300°K:15.5	22	28	40	47	94

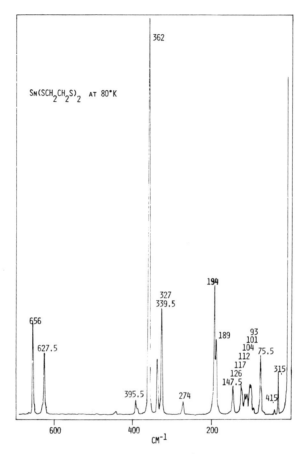

Figure 4.   Low frequency (lattice mode region)
Raman spectrum of $[Sn(SCH_2CH_2S)_2]_n$

solubility in polar (coordinating) solvents, monomeric molecular weight in benzene, absence of mass spectral fragments heavier than P^+, and the nondegeneracy of the IR and Raman-active bands observed in the Sn–S regions of the vibrational spectra.

The temperature dependence of the recoil-free fraction over the temperature range $78 < T < 240°K$ is given by $\ln A(T) = -0.01180T - 16.63T^{-1}$ with a multiple regression correlation coefficient of 0.997.   This result, together with the proper ω_L value extracted from the low temperature Raman spectrum mode at 41.5 cm^{-1}, leads to a calculated molecular weight of 506 ± 32 which is in reasonably good agreement with the assumption that this molecule is a dimer in the solid state (twice the formula weight is 606).   However, this interpretation is not completely consistent with the mass spectral data which show no fragment heavier than P^+ when the inlet temperature is held to ~10°–20° below the melting point of the solid (just adequate to provide a sufficient vapor pressure to permit the recording of a meaningful mass fragmentation spectrum) or with the monomeric weight observed in benzene.

These data can, however, be understood in terms of a structure in the solid in which two of the four sulfur atoms are coordinated, respectively, to tin atoms in the two adjacent molecular layers giving the metal atoms effectively a coordination number of 6 (22). Under this hypothesis, the tin-orbital hybridization becomes d^2sp^3 and allows the four sulfur atoms to occupy the equatorial positions in a somewhat distorted octahedral structure. The quadrupole hyperfine interaction observed in the ^{119}Sn Mössbauer spectrum arises from the fact that the two axial sulfur atoms are inequivalent to the four equatorial sulfur atoms (which concomitantly belong to two different types), thus giving rise to a nonvanishing electric field gradient tensor at the metal atom lattice position.

The low solubility of the one-dimensional polymeric solid in nonpolar solvents (such as benzene and n-butyl benzene) is consistent with the integrity of the polymeric structure under these conditions. Treatment with donor (polar) solvents, on the other hand, results in the depolymerization of the solid state structure, leading to a solvated monomeric species in solution in which the two axial positions are occupied by two donor solvent molecules, as shown in III. Moreover the mass spectral results suggest that under thermal activation at ~165 °C, the solid ⇌ vapor equilibrium also results in the disruption of the one-dimensional polymeric structure, this time giving rise to a (presumably quasitetrahedral) monomeric structure in which the tin atom is bonded to only the four sulfur atoms of the di-thiol ligands.

The Raman data show both the symmetric and asymmetric Sn–S modes, but it is not possible to interpret these data further with respect to the presumed anisobidentate (23) nature of the ligands. Clearly the presumed structure in the solid represented by II should give rise to two distinct equatorial plane Sn–S interactions and hence to two distinct Sn–S bond distances in the SnS$_4$ portion of the molecule. The vibrational spectroscopic data are not sufficient to permit an unambiguous assignment of a cis or trans configuration to this structure, but the latter would seem to be preferable on purely symmetry arguments. Such structural details will have to await an x-ray diffraction elucidation of the architecture of this molecule in the solid state.

Finally, it is appropriate to return to the interpretation of the effective vibrating mass calculated from the EVM model applied to the ^{119}Sn Mössbauer data. It had been observed earlier (24) in the case of $(CH_3)_3SnF$, which has been shown both by IR and x-ray diffraction data (25, 26) to consist of one-dimensional polymeric chains made up of not quite planar $(CH_3)_3Sn$ fragments bonded by nonlinear fluorine bridges, that an effective vibrating mass, corresponding to $[(CH_3)_3SnF]_2$ portions of the chain giving rise to a Raman active vibration at 41 cm^{-1}, could account for the temperature dependent ^{119}Sn Mössbauer recoil-free fraction data. In this case, the temperature dependence of the recoil-free fraction is given by the expression

$$\ln f(T) = -1.75 \times 10^{-2}T - 59.5\ T^{-1}$$

in the range $72 < T < 296\ °K$. The value of M_{eff} calculated from these data and $\nu = 41\ cm^{-1}$, substituted into Equation 10 yield an effective vibrating mass of 349 which is reasonably good agreement with 2MW of 366. Thus, in the case of this one-dimensional polymeric chain structure, agreement between the calculated effective vibrating mass and the assumption that two adjacent dimeric fragments of the chain account for the intra-unit cell motion of the metal atom leads to an internally self-consistent view of the experimental data.

The present data on the Sn(VI)bis(1,2-ethane dithiol) species suggest that a quite analogous situation obtains in this case. The structure indicated schematically above (II) can be viewed as that of a quasi-linear tin–sulfur array in which the cyclic moieties of the ligands interact only weakly by normal covalent solid van der Waals forces with proximal chains lying on either side. The solubility, mass spectral, and molecular weight data, are consistent with this interpretation of the results of the ^{119}Sn temperature-dependent, recoil-free fraction Mössbauer resonance data.

[Note added in proof: After this paper had been presented at the New York ACS Symposium, it was pointed out to us by A. G. Davies that the crystal structure of $Sn(SCH_2CH_2S)_2$ had been determined by C. A. Mackay who then generously made available to us a copy of his unpublished data relevant to this compound. Mackay's study shows that each tin atom is surrounded by four sulfur atoms arranged in a distorted tetrahedron with Sn–S distances of 2.405(1)A and 2.388(1)A, showing the anisobidentate nature of the 1,2 ethane dithiol ligand as predicted. Moreover, the intermolecular Sn–S distances are 3.764A and 3.811A, considerably smaller than the 4.05A deduced for the sum of the Sn and S van der Waals radii, indicative of the intermolecular bonding which had been postulated from the temperature-dependent Mössbauer data. Finally, it should be noted that the crystallographic data show that there are two tin–sulfur distances of ∼3.76A and two of 3.81A between adjacent monomer units, forming a chain of linked molecules with closest interaction between the sulfur atom of one molecule and one of the carbon atoms of the next unit.]

Summary and Conclusion

The present study has dealt with the extension of the EVM model to low molecular weight, polymeric, solid-state structures in which the Mössbauer-active atom does not occupy the center of mass of the molecule. Good agreement has been obtained between the calculated effective vibrating mass and the formula molecular weight for $[(CH_3)_2SnS]_3$ making the assumption that the 27.5 cm^{-1} band observed in the Raman spectrum of this compound can be assigned to an intra-unit cell, intermolecular vibration of two trimeric units against each other.

Application of the EVM model to the compound $[Sn(SCH_2CH_2S)_2]_n$ has led to a value of $n = 2$, but this result, in conjunction with cryoscopic molecular weight data, mass spectral results, and the interpretation of low temperature Raman data, must be understood in terms of the presence of an extended linear, one-dimensional, polymeric structure in the solid. In the proposed structure, the two five-membered 1,2-ethane dithiol ligands, which are anisobidentate (23) in the solid, asssme an almost equatorial position around the metal atom with two of the four sulfur atoms interacting with metal atoms in adjacent molecular chains to give rise to a six-coordinate metal atom in a distorted octahedral array.

Acknowledgments

The authors are indebted to H. A. Schugar for the sample of $[Sn(SCH_2CH_2S)_2]_n$ and to C. Weston for his careful mass spectral examination of this compound. We are very grateful to C. A. Mackay for making a copy of his thesis available to us prior to publication.

Finally, the authors are especially indebted to J. J. Zuckerman, for providing the opportunity to present the results of the present study and to discuss outstanding problems with many of the active workers in this field.

Literature Cited

1. Greenwood, N. N., Gibb, T. C., "Mössbauer Spectroscopy," Chapman and Hall, London, 1971.
2. Gol'danskii, V. I., Herber, R. H., "Chemical Applications of Mössbauer Spectroscopy," Academic, New York, 1968.
3. Zuckerman, J. J., *Adv. Organomet. Chem.* (1970) **9**, 21.
4. Bancroft, G. M., Platt, R. H., "Advances in Inorganic Chemistry and Radiochemistry," H. J. Emeleus, A. G. Sharpe, Eds., Academic, New York, 1972.
5. Gol'danskii, V. I., Makarov, E. F., Stukan, R. A., Sumarokova, T. N., Trukhanov, V. A., Khrapov, V. V., *Dokl. Akad. Nauk SSSR* (1964) **156**, 400.
6. Herber, R. H., Stöckler, H. A., "Application of the Mössbauer Effect in Chemistry and Solid State Physics," *Int. Atomic Energy Agency, Tech. Repts.* Ser. No. **50**, Vienna, 1966.
7. Herber, R. H., Leahy, M. F., Hazony, Y., *J. Chem. Phys.* (1974) **60**, 5070.

8. Hazony, Y., Herber, R. H., "Mössbauer Effect Methodology," Vol. 8, I. Gruverman, Ed., Plenum, New York, 1973.
9. Herber, R. H., Hazony, Y., *J. de Physique* (1974) **C6**, 131.
10. Hafemeister, D. W., Shera, E. B., *Nucl. Instr. Methods* (1966) **41**, 133.
11. Muir, Jr., A. H., *Atomics International Report* **AI-6699**, Canoga Park, Calif.
12. Wertheim, G. K., "Mössbauer Effect, Principles and Applications," Academic, New York, 1962.
13. Herber, R. H., Fischer, J., Hazony, Y., *J. Chem. Phys.* (1973) **58**, 5185.
14. Herber, R. H., *J. Chem. Phys.* (1970) **52**, 6045.
15. Rein, A. J., Herber, R. H., *J. Chem. Phys.* (1975) **63**, 1021.
16. Spijkerman, J. J., deVoe, J. R., Travis, J. C., *Nat. Bur. Stds. Spec. Pub.* **260-20**, 1970.
17. Reichle, W. T., *J. Org. Chem.* (1961) **26**, 4634.
18. Reichle, W. T., *Inorg. Chem.* (1962) **1**, 650.
19. Poller, R. C., Spillman, J. A., *J. Chem. Soc. A* **958** (1966).
20. Menzebach, B., Bleckmann, P., *J. Organomet. Chem.* (1975) **91**, 291.
21. Finch, A., Poller, R. C., Steele, D., *Trans. Farad. Soc.* (1965) **61**, 2628.
22. Epstein, L. M., Straub, D. K., *Inorg. Chem.*, (1965) **4**, 1551.
23. DeVries, J. L. K. F., Herber, R. H., *Inorg. Chem.* (1972) **11**, 2458.
24. Herber, R. H., Chandra, S., *J. Chem. Phys.* (1971) **54**, 1847.
25. Clark, H. C., O'Brien, R. J., Trotter, J., *Proc. Chem. Soc.* (1964) **85**.
26. Clark, H. C., O'Brien, R. J., Trotter, J., *J. Chem. Soc.* (1964) 2332.

RECEIVED May 7, 1976. Work supported by National Science Foundation grant DMR 76-00139 and the Research Council of Rutgers University.

11

Organotins in Agriculture

MELVIN H. GITLITZ

M&T Chemicals Inc., P.O. Box 1104, Rahway, N.J. 07065

*Although organotin compounds were suggested as having bio-
logical activity as early as 1929, it is only relatively recently
that they have found a place in agriculture. The evolution of
organotin compounds as commercially important agricultural
fungicides and insecticides both here and abroad is discussed.
Our limited knowledge of the relationship of structure to biological activity is considered and recent attempts to fill this
void are described. Steric effects seem to be important. In
vitro tests of fungicidal activity have limited value to the organotin pesticide chemist since they disregard phytotoxicity
which is a problem with many organotins. General toxicological properties are described.*

Between one-half and one-third of the world's agricultural food
crops are lost annually to pests (1). We continually battle against the
competition from insects, fungi, bacteria, and weeds for our food. The most
effective weapons that we have been able to devise thus far to assist us in this
struggle are the chemical pesticides. Organotins are a relatively new class of
compounds receiving increasing interest as agricultural pesticides.

The use of organometallic compounds as pesticides and disease-control agents
is not a recent phenomenon. Organoarsenic compounds, for example, have had
a long and colorful history from Ehrlich's Salvarsan [4,4'-diarsenobis(2-amino-
phenol) dihydrochloride trimer] to the methylarsonic acid salts still used today
as crabgrass herbicides and to control other grassy weeds.

Another group of biologically active organometallics are the organomercurials which until recently were used widely as seed treatments to control
seed-borne fungus diseases in many areas of the world and are still important as
medicinals. There is at least one distinctive difference between tin and arsenic
or mercury. Unlike these latter two elements, tin's inorganic compounds are
orders of magnitude less toxic than its most toxic organometallic compounds.

167

This property gives tin a unique place among the heavy metals and is an important factor in its agricultural utility. Although there were reports of the biological activity of organotin compounds as early as 1929 (2), it was not until the early 1950's that van der Kerk and Luijten at TNO in Utrect systematically explored the in vitro fungicidal and antibacterial properties of organotin compounds (3, 4). Their research laid the foundation for further studies which ultimately led to the commercial application of organotins as bio-control agents. Table I shows the diversity of biocidal applications for which organotins are being used or for which they have been suggested.

Table I. Biocidal Applications of Organotin Compounds (5)

Fungicidal
Control of fungi on potatoes and sugar beets[a]
Control of scab on pecans and peanuts[a]
Control of rice blast and pine needle blight
Preservation of wood (from fungi and insects)[a]
Paint additive to prevent mold growth

Bacteriostatic
Control of slime in paper and wood pulp production
Fabric disinfectant[a]
Antimicrobial activity in synthetic fibers

Insecticidal
Antifeedant against insect larvae
Chemosterilant (preventing reproduction)
Arachnidicide against phytophagous mites[a]

Other
Tapeworm and helminthes eradication in poultry[a]
Protection of surfaces (ships, piers, etc.) from attack by marine organisms[a]
Plankton control in reservoirs
Molluscicide for bilharzia control

[a] Commercially significant usage.

Journal of Organometallic Chemistry

In the early 1960's the first organotin agricultural fungicide, triphenyltin acetate, was introduced in Europe commercially by Farbwerke Hoechst A.G. as Brestan. Brestan was recommended for the control of Phytophthora (late blight) on potatoes and Cercospora on sugar beets at rates of a few ounces per acre (6). Shortly thereafter, triphenyltin hydroxide was introduced as Duter by Philips-Duphar N.V. with about the same spectrum of disease control as Brestan. Duter is presently registered in the United States as a fungicide for potatoes, sugar beets, pecans, and peanuts. Both materials are extremely effective protectant fungicides. Incidentally, both triphenyltin hydroxide and acetate exhibit the unusual and interesting property of deterring insects from feeding

(7). This antifeedant effect is also shown by other triphenyltin compounds (8). Several triphenyltin compounds including the hydroxide have also been shown to be effective fly sterilants at sub-lethal concentrations (9).

Although research on the biological activity and toxicology of organotins progressed at a rapid pace in the interim, it was not until 1968 that the next major agricultural application of organotins was realized. In that year, tricyclohexyltin hydroxide, the result of a joint research effort by M&T Chemicals and Dow Chemical Co., was introduced under the name Plictran (10). It was recommended for the control of phytophagous (plant-feeding) mites on apples and pears. It is now also registered for use on citrus, stone fruits, and hops. Plictran is very selective and gives good control of harmful arachnids with little toxicity towards other insects such as honeybees. The compound also shows antifeedant activity against some insect larvae (11).

Tricyclohexyltin hydroxide can be prepared by the two routes shown in Figure 1. The first process, termed the butyl transfer route, involves two steps to prepare tricyclohexyltin chloride (12). In the first step, butyltin trichloride is allowed to react with 3 moles of cyclohexylmagnesium chloride to form butyltricyclohexyltin which is then allowed to react with tin tetrachloride in an inert solvent to give tricyclohexyltin chloride and butyltin trichloride. The butyltin trichloride is extracted with water, distilled, and recycled into the process.

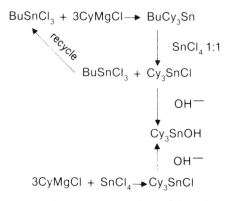

Figure 1. Routes to tricyclohexyltin hydroxide. Data from Refs. 12 and 13.

An alternate route to tricyclohexyltin chloride involves the reaction of 3 moles of cyclohexyl Grignard and 1 mole of tin tetrachloride to form the organotin directly (13). Although this route seems economically more attractive, it requires very close control of reaction conditions to minimize byproducts. The commercial success of Plictran miticide stimulated research efforts on organotins in other companies. There are now three other organotin miticides which have been disclosed in the patent literature. The structures are shown in Figure 2.

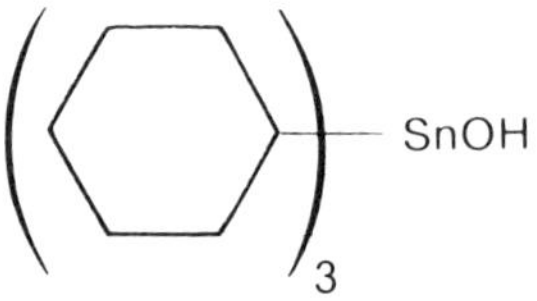

Plictran (Dow)

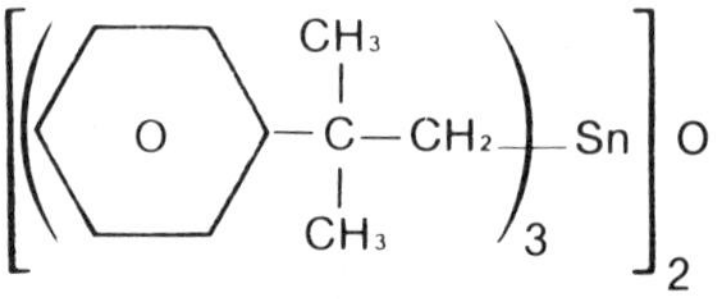

Vendex (Shell)

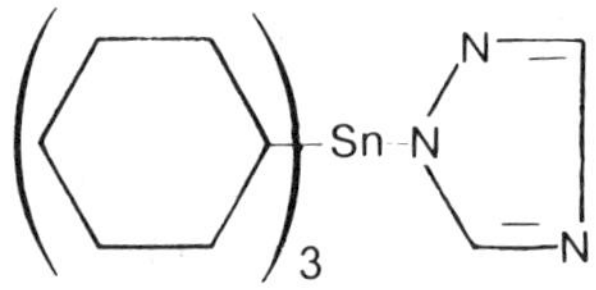

Tricyclazol (Bayer, Chemagro)

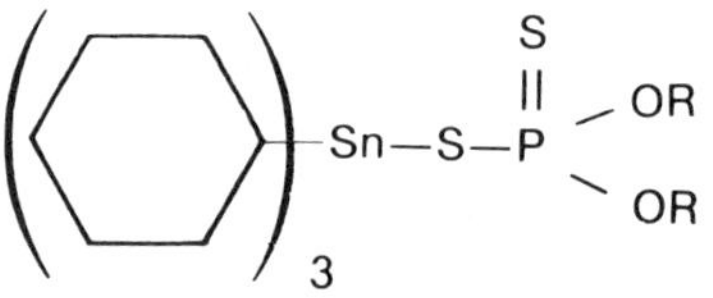

R-28627 (Stauffer)

Figure 2. Organotin acaricides. Plictran, Ref. 21; *Vendex, Ref.* 26; *Tricyclazol, Ref.* 48; *R-28627, Ref.* 49.

The lower tri-*n*-alkyltins from trimethyl to tri-*n*-pentyl show high biological activity. The trimethyltins are highly insecticidal and the tripropyls, tributyls, and tripentyls have a high degree of fungicide and bactericide activity. The dialkyltin compounds are generally less active than the trialkyltin compounds. The typical relationship between chain length of di- and trisubstituted organotin

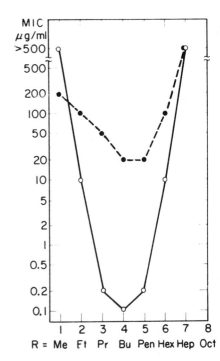

Figure 3. Influence of chain length of di- and trisubstituted tin compounds on minimum concentration inhibitory to Mycobacterium phlei. *Legend:* O——O, *trialkyltin compounds;* ●---●, *dialkyltin compounds* (14)

compounds and their minimum inhibitory concentration for a particular bacteria, *Mycobacterium phlei* is shown in Figure 3.   Note that diorganotins are less active than the analogous triorganotins.

The response of fungi to triorganotins is generally similar to that exhibited by bacteria in the foregoing example and is typically illustrated in Figure 4 below for the fungus *Aspergillus niger*.   Although the lower trialkyltins show high fungicide activity, they are unlikely candidates for agricultural fungicide use because of their high phytotoxicity.   Over the years, various attempts have been made to moderate the phytotoxicity of the lower alkyltins by changes in the anion

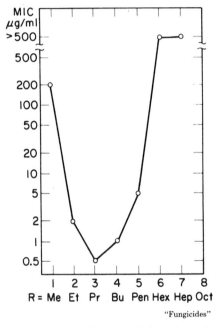

"Fungicides"

Figure 4.   Influence of chain length of trialkyl-substituted tin acetates on minimum concentration inhibitory to Aspergillus niger (15)

group.   These have not been very successful since the nature of the anion group has little influence on the spectrum of biological activity provided that the anion is not biologically active in its own right and that it confers a sufficient minimal solubility on the compound.

While the higher (C_8 and above) trialkyltins are non-phytotoxic, they are neither fungicidal nor insecticidal.   Tricyclohexyl and tricycloheptyltins are also essentially non-phytotoxic.   Attempts to moderate the phytotoxicity of the trialkyltins by synthesizing asymmetric organotins of the form R_2CySnX or RCy_2SnX where R = lower alkyl group, Cy = cyclohexyl, and X = chloride,

1. $R_2Cy_2Sn + SnCl_4$ $\xrightarrow{\text{Hydrocarbon Solvent}}$ $RSnCl_3 + RCy_2SnCl$

2. $R_3CySn + SnCl_4$ $\xrightarrow{\text{Hydrocarbon Solvent}}$ $RSnCl_3 + R_2CySnCl$
    R = Me, Et, n-Pr, n-Bu, n-Pent

Yields: 89-97%
Purity: >95% as isolated (GLC)
Reaction Time: 10-20 min.
$RSnCl_3$ removed by extraction with aqueous HCl

Figure 5. Process for asymmetric cyclohexyltin chlorides. Data from Ref. 16.

Table II. Typical Asymmetric Cyclohexyltin Chlorides[f]

Compound	Tetraorganotin Reagent	Melting Point, °C	Yield,[a] %	Purity,[b] %
Me₂CySnCl[c]	Me₃CySn	33–35	89	99
MeCy₂SnCl	Me₂Cy₂Sn	52–54	97	98
Pr₂CySnCl	Pr₃CySn	d	97	99
PrCy₂SnCl	Pr₂Cy₂Sn	33–35	95	—
Bu₂CySnCl	Bu₃CySn	e	92	97

[a] Based on tetraorganotin.
[b] GLC area %.
[c] Cy = cyclohexyl. Sn and Cl analyses satisfactory and agree with GLC results.
[d] Liquid $n_D^{22} = 1.5142$.
[e] Liquid, $n_D^{24} = 1.5103$.
[f] Adapted from Ref. 16.

hydroxide, or acetate have met with only limited success. Although these compounds are less phytotoxic than the trialkyltins, this is attained at the expense of fungicidal activity. The asymmetric dicyclohexylalkyl and cyclohexyldialkyltin compounds were prepared by the convenient high-yield synthesis shown in Figure 5 (16). This is an extrapolation of the butyl transfer process shown in Figure 1. Typical asymmetric cyclohexyltins prepared in this manner are shown in Table II.

Although the original work on the alkyltins was published more than 20 years ago, little has appeared in the chemical or biological literature since then on the relationship of structure to biological activity in organotin compounds. That which has been reported in this area is usually confined to in vitro fungicidal activity which is of limited value to the agricultural pesticide chemist because it circumvents the important consideration of phytotoxicity. These published reports generally confirm the initial observation of the relative unimportance of the nature of the anionic portion of the triorganotin molecule to the overall biological activity (17, 18, 19), although there are occasional unsubstantiated reports of dependence (20).

It is the nature of the organic groups bonded to tin through carbon and their number which govern the biological activity of organotins. Table III shows some examples of the specificity of the structure of organotins to their acaricidal and fungicidal activity. It can readily be seen that both the number and arrangement of carbon atoms around tin have a profound effect on the acaricidal and fungicidal activity. The activity of the molecule is very sensitive to even the slightest change. Tricyclohexyltin hydroxide is an extremely good acaricide. The insertion of a methylene group between the cyclohexyl ring and tin causes a marked diminution in miticidal activity (Table III, example 3). The insertion of a propylene group renders the compound inactive (example 4).

Attempts to combine the good acaricidal properties of the tricyclohexyltins with the high fungicidal activity of the triphenyltins in one molecule have been successful to a degree. Dicyclohexylphenyltin hydroxide is almost as good an acaricide as the tricyclohexyltins and slightly poorer than the phenyltins as a fungicide (example 5).

Although interposing a methylene group between the cyclohexyl ring and tin led to a decrease in acaricidal and fungicidal activity (example 3), the internally methylene-bridged compound, tris(2-norbornyl)tin hydroxide, is comparable in activity with the tricyclohexyltin compound. Other examples of the specificity of structure to biological activity are shown in Table IV.

Table III.   Specificity of Related Triorganotin Structures

Example	Organotin [a]	Acaridical Activity[a]	Fungicidal Activity[a]	Reference
1	Cy_3SnOH	+ + + +	+ +	21
2	Φ_3SnOH	+ +	+ + + +	22
3	$(CyCH_2)_3SnBr$	+ +	+	23
4	$[(Cy(CH_2)_3]_3SnBr$	−	−	23
5	$Cy_2(\Phi)SnOH$	+ + +	+ + +	24
6	−SnOH	+ + + +	+ +	25

[a] + + + + = highly active; − = inactive; Φ = phenyl.

Table IV.   Relative Acaricidal Activity of Related Triorganotins

Example	Organotin	Acaricidal Activity[a]	Reference
1	$(\Phi CH_2)_3SnCl$	−	26
2	$[(\Phi CH_2CH_2)_3Sn]_2O$	+	26
3	$[(\Phi CH(CH_3)CH_2)_3Sn]_2O$	+ +	26
4	$[(\Phi C(CH_3)_2CH_2)_3Sn]_2O$	+ + + +	26
5	$[(2\text{-Thienyl-}C(CH_3)_2CH_2)_3Sn]_2O$	+ + +	27
6	$[(\Phi CH_2CH_2CH_2)_3Sn]_2O$	−	26
7	$[(4\text{-Me}\Phi C(CH_3)_2CH_2)_2Sn]_2O$	+ +	26

[a] + + + +, highly active; −, inactive.

From the striking increase in activity with the increase in the degree of branching between the phenyl ring and tin, it is obvious that steric effects about the tin atom are of major importance in determining acaricidal activity, but this is not the only factor. To understand why this is so, one has only to examine the results for the methyl-substituted neophyltin in example 7. Perhaps mode-of-action studies such as those with the cyclohexyltins and phenyltins (28, 29) will help us understand what is happening here and point the directions for future syntheses, but at the present almost any conclusion from these results is conjecture. They are presented here to show how difficult it is to predict the activity of new compounds.

The mechanism of action or metabolism of organotin compounds is not discussed here since this topic has been covered in the recent literature (28, 29, 30, 31, 32) and is discussed in other chapters of this volume. However, no discussion of agricultural pesticides is complete without some general toxicological considerations. Table V shows the acute oral toxicities of some representative triorganotins. As a general rule, the triorganotins are more toxic than their diorganotin analogs.

As evidenced by these data, the toxicity is strongly dependent on the nature of the organic groups bonded to tin. The toxicity varies from the highly poisonous trimethyl and triethyltins (which are not used in any commercial applications) to the essentially non-toxic trioctyltins which also have no commercial utility. The triphenyltins and tricyclohexyltins are intermediate in mammalian oral toxicity and the former show evidence of species dependence and perhaps even

Table V.   Acute Oral Toxicities of Triorganotin Compounds

Compound	LD_{50},[a] mg/kg	Test Animal	Reference
Me_3SnOAc	9	rat	33
Et_3SnOAc	4	rat	33
Bu_3SnOAc	133	rat	34
Bu_3SnOAc	380	rat	33
Ph_3SnOAc	125–150	rat	35
Ph_3SnOAc	429–491	rat	36
Ph_3SnOAc	23–41	guinea pig	36
Ph_3SnOH	108	rat	37
Ph_3SnOH	108	mouse	37
Ph_3SnOH	27	guinea pig	37
Cy_3SnOH	540	rat	38
Cy_3SnOH	780	guinea pig	38
$[(Neoph)_3Sn]_2O$[b]	2630	rat	39
$[(Neoph)_3Sn]_2O$	1450	mouse	39
$(C_6H_{13})_3SnOAc$	1000	rat	33
$(C_8H_{17})_3SnOAc$	>1000	rat	33

[a] Single dose (in mg of chemical/kg body weight) causing death in 50% of the test animals.
[b] Neoph = neophyl = β, β-dimethylphenethyl.

strain dependence.. Bis(trineophyltin)oxide appears to be the least toxic triorganotin compound yet considered for agricultural pesticide use.

To put the entire subject of acute oral toxicity in its proper perspective it is interesting to note that allyl alcohol has an LD_{50} for rabbits of 52 mg/kg; hydroquinone, 70 mg/kg (cats); methyl iodide, 150–220 mg/kg (rats); triethylamine 460 mg/kg (rats); and phenol, 530 mg/kg (rats). Thus organotins other than the trimethyls, triethyls, and possibly the tripropyls appear to be no more toxic than many common laboratory and commercial chemicals. Those interested in pursuing the matter of toxicity further are referred to the review by Klimmer (*40*) and other recent references (*41, 42, 43*).

Organotins rest on a sound environmental footing because of the relatively weak nature of the tin–carbon bond. The degradation of triorganotin compounds to non-toxic inorganic tin under environmental stresses such as UV irradiation has been documented (*44, 45, 46, 47*). Persistence should not be a problem for organotin pesticides. In fact, organotin compounds too frequently do not persist long enough in the laboratory for the synthesis chemist to isolate and to characterize them. This is a real problem in the synthesis of new structures.

The future for organotin pesticides is bright. To feed the world's population of the year 2000, it has been estimated that we will have to double our current food output if present rates of population growth continue (*1*). To achieve this goal, we are going to need all the help we can get.

Literature Cited

1. Miller, Paul R., *Agr. Chem.* (1970) Dec. 16.
2. Hartmann, E., Hardtmann, M., Kümmel, P., U.S. Patent **1,744,633**, (Jan. 21, 1930).
3. van der Kerk, G. J. M., Luijten, J. G. A., *J. Appl. Chem.* (1954) **4**, 314.
4. van der Kerk, G. J. M., Luijten, J. G. A., *J. Appl. Chem.* (1956) **6**, 56.
5. Thayer, J. S., *J. Organometal. Chem.* (1974) **76**, 265.
6. Hartel, K., *Agr. Vet. Chem.* (1962) **3**, 19.
7. Ascher, K. R. S., Nissim, S., *Int. Pest Control* (1965) **7**, 21.
8. Byrdy, S., *Rocz. Nauk. Roln.*, (1968) **A 93**, 789. *Chem. Abs.* (1968) **69**, 51184.
9. Kenaga, E. E., *J. Econ. Entomol.* (1965) **58**, 4.
10. Allison, W. E., Doty, A. E., Hardy, J. L., Kenaga, E. E., Whitney, W. K. *J. Econ. Entomol.* (1968) **61**, 1254.
11. Ascher, K. R. S., Avdat, J., Kamhi, J., *Int. Pest Control* (1970) **33**, 11.
12. Kushlefsky, B. G., Reifenberg, G. H., Considine, J., Hirshman, J. L., U.S. Patent **3,607,891** (Sept. 21, 1971).
13. Hirshman, J. L., Natoli, J. G., U.S. Patent **3,355,468** (Nov. 28, 1967).
14. Sijpesteijn, A. K., Luijten, J. G. A., van der Kerk, G. J. M., "Fungicides," D. C. Torgeson, Ed. Vol. 2, p. 352, Academic, New York, 1969.
15. *Ibid*, p. 340.
16. Reifenberg, G. H., Gitlitz, M. H., U.S. Patent **3,789,057** (Jan. 29, 1974).
17. Ison, R. R. Newbold, G. T., Saggers, D. T., *Pesticide Sci.* (1971) **2**, 152.
18. Srivastava, T. N., Rupainwar, R., *Ind. J. Chem.* (1972) **9**, 1411.
19. McIntosh, A. H., *Ann. Appl. Biol.* (1971) **69**, 43.
20. Czerwinska, E., Eckstein, Z., Ejmocki, Z., Kowalik, R. *Bull. Acad. Polon. Sci.* (1967) **15**, 335.

21. Kenaga, E. E., U.S. Patent **3,264,177**, (Aug. 2, 1966).
22. Taylor, J. L., U.S. Patent **3,268,395** (Aug. 23, 1966).
23. Gitlitz, M. H., Kushlefsky, B. G., U.S. Patent **3,790,611** (Feb. 5, 1974).
24. Gitlitz, M. H., U.S. Patent **3,923,998** (Dec. 2, 1975).
25. Gitlitz, M. H., U.S. Patent **3,781,316** (Dec. 25, 1973).
26. Horne, C. A., U.S. Patent **3,657,451** (April 18, 1972).
27. Foster, J. P., Soloway, S. B., U.S. Patent **3,736,333** (May 29, 1973).
28. Ahmad, S., Knowles, C. O., *Comp. Gen. Pharmacol.* (1972) **3**, 125.
29. Wulf, R. G., Ph.D. Thesis University of Missouri, Columbia, 1973.
30. Wulf, R. G., Byington, K. H., *Arc. Biochem Biophys.* (1975) **167**, 176.
31. Desaiah D., Cutkomp L. K., Koch R. B., *Life Sci.* (1973) **13**, 1693.
32. Fish, R. H., Kimmel, E. C., Casida, J., *J. Organometal. Chem.* (1975) **93**, C1.
33. Barnes, J. M., Stoner, H. B., *Brit. J. Indust. Med.* (1958) **15**, 5.
34. Klimmer, O. *Tin and its Uses* (Tin Research Inst.) (1964) **61**, 6.
35. Ascher, K. R. S., Nissim, S. *World Rev. Pest Control* (1964) **3**, 188.
36. Stoner, H. B., *Brit. J. Ind. Med.*, (1966) **23**, 222.
37. "Pesticide Manual," H. Martin, C. R. Worthing, Eds. p. 274, *Brit. Crop Prot. Council 4th Ed., 1974.*
38. *Ibid.*, p. 144.
39. *Ibid.*, p. 228.
40. Klimmer, O. Arzneimittel-Forsch. (1969) **19**, 934.
41. Neumann, W. P., "The Organic Chemistry of Tin," p. 230 Wiley-Interscience, London, 1970.
42. Byington K., Yeh, R. Y., Forte L. R. *Toxicol. Appl. Pharmacol.* (1974) **27**, 230.
43. Luijten J. G. A., Klimmer O., "Toxicological Evaluation of Organotin Compounds," *18th German Tin Meetg. Nov. 15, 1973*, Tin Research Institute.
44. Freitag, K-D., Bock, R., *Pesticide Sci.* (1974) **5**, 731.
45. Barnes, R. D., Bull, A. T., Poller, R. C., *Pesticide Sci.* (1973) **4**, 305.
46. Getzendaner, M. F., Corbin, H. B., *J. Agr. Food Chem.* (1972) **20**, 881.
47. Chapman, A. J., Price, J. W., *Int. Pest Control* (1972) **1**, 11.
48. Buchel, K. H., Hammann, I., U.S. Patent **3,907,818** (1975).
49. Baker, D. R., U.S. Patent **3,919,418** (1975).

RECEIVED May 3, 1976.

Some Recent Chemistry Related to Applications of Organotin Compounds

R. C. POLLER

Queen Elizabeth College, Campden Hill Rd., London W8 7AH, England

Studies of the interactions between PVC and $Bu_2Sn(^{35}SBu)_2$ indicate that, although there is a small amount of exchange between ^{35}SBu groups and reactive chlorine atoms in the polymer, the principal reaction is $Bu_2Sn(^{35}SBu)_2 + 2HCl \rightarrow Bu_2SnCl_2 + 2Bu^{35}SH$. The thiol then adds to the polyene systems in the degrading polymer. The effect of changes in the organic groups bound to tin on the stabilizing efficiency is discussed as is the catalytic effect on dehydrochlorination of the various tin products. The nature of organotin biocides R_3SnX is considered and the preparation of compounds containing sucrose residues in the X group is described. These products have enhanced biocidal activity compared with compounds such as tributyltin oxide, even though they have only a fraction of the tin content.

The principal applications of organotin compounds are as thermal stabilizers for PVC and as biocidal agents. The technology was developed before there was any substantial understanding of the chemical and biological processes involved. Much has been learned of the chemistry of PVC stabilization since this topic was reviewed in 1974 (1), and most of this paper is devoted to this topic. The structure of biocidal agents will be more briefly referred to at the end of this report.

PVC Stabilizers

Although other species are used, a typical organotin stabilizer is a dialkyltin compound. It has the general formula: R_2SnX_2, $R = Me$, Bu, C_8H_{17}, $X = SR'$, $OCOR''$. A typical compound is $Bu_2Sn(SCH_2COOC_8H_{17}\text{-}i)_2$. This structure was evolved empirically but we are now able to describe the function of these compounds in chemical terms.

The X Groups. Frye and his co-workers (2) studied reactions between PVC and organotin stabilizers carrying radioactive labels at various molecular positions. They concluded that exchange reactions occurred between chlorine atoms in the polymer and tin-bound X groups:

$$2 \overset{|}{\underset{\underset{Cl}{|}}{C}} + R_2SnX_2 \longrightarrow 2 \overset{|}{\underset{\underset{X}{|}}{C}} + R_2SnCl_2$$

Subsequent work showed (3, 4) that secondary chlorine atoms are inert to this reaction though allylic chlorines undergo exchange. When we examined uptake of radioactive sulfur by PVC treated with $Bu_2Sn(^{35}SBu)_2$ in vacuo, we obtained the results (5) shown in Figure 1. [These are quite different from the results obtained by Frye (2) who worked with thin films of polymer exposed to air.] There appears to be a slow uptake of radioactivity during an induction period.

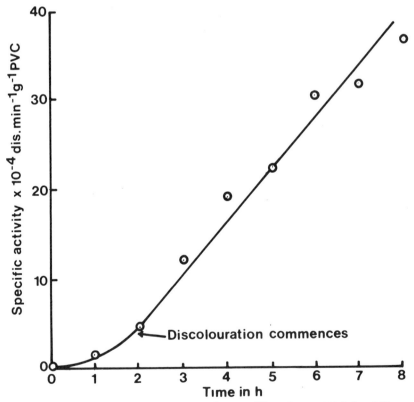

Figure 1.   Interaction between PVC and $Bu_2Sn(^{35}SBu)_2$ at 180°C for different periods of time

With the onset of discoloration the rate of absorption of radioactivity increases. During the induction period the stabilizer has a preventive role, and it is only at this stage that exchange reactions, probably accompanied by rearrangement (*4*) with the small number of allylic chlorine atoms present, are important. If this interpretation is accepted, then the radioactivity at the end of the induction period gives an approximate measure of the number of allylic chlorines present, and this is 3 per 100 polymer molecules. Subsequent experiments (*6*) which exploited differences in reactivity of secondary, tertiary, and allylic chlorine atoms toward labelled ethoxide ions confirmed this figure.

The second and more important reaction of the X groups is with hydrogen chloride:

$$Bu_2SnX_2 \ + \ 2HCl \ \longrightarrow \ Bu_2SnCl_2 \ + \ 2HX$$

At least as important as hydrogen chloride absorption is the generation of HX, usually a thiol which has a healing function by adding to the polyene systems and thereby reducing the unsaturation and the color. This is the reason for the more rapid uptake of radioactivity which occurs after the induction period in Figure 1. We obtained clear evidence for this (*5*) by treating degraded PVC with $Bu^{35}SH$; not only was the color discharged but the uptake of radioactivity was similar to that obtained by interaction between PVC and the labelled stabilizer.

If this interpretation is correct, we would expect to see some correlation between the efficiency of a stabilizer $R_2Sn(SR')_2$ and the ease of addition of the thiol R'SH to alkenes. It is reasonable to assume that the latter process will be a radical one, and reactivity toward alkene addition is then given by the chain transfer constants (*7, 8*). The values given in Table I indicate that, at least toward styrene addition, there is a correlation, and this may account for the fact that the organotin thioglycolates are the most effective stabilizers.

In further studies (*9*) of the interactions between PVC samples and labelled stabilizers, we observed an inverse relationship between the amount of radioactivity taken up and the molecular weight of the polymer (Tables II, III). Although we have not specifically investigated the mechanism of the degradation process, these results, and those reported by other workers (*10, 11, 12, 13, 14, 15, 16*) are consistent with a predominantly radical process initiated at chain ends:

$$Cl\cdot \ + \ \sim\sim CH_2CHClCH_2CHCl\sim\sim \ \longrightarrow$$
$$HCl \ + \ \sim\sim \dot{C}HCHClCH_2CHCl\sim\sim \ \longrightarrow$$
$$Cl\cdot \ + \ \sim\sim CH{=}CHCH_2CHCl\sim\sim$$

A high degree of chain transfer is postulated together with long kinetic chains leading to the observed, relatively short polyene sequences (*17*).

Table I.   Chain Transfer Constants for Thiols with Styrene at 60° (7, 8)

Thiol	Constant
t-BuSH	3
$CH_3(CH_2)_{11}SH$	19
$EtOCH_2CH_2CH_2SH$	21
BuSH	22
$EtOCOCH_2SH$	58

Table II.   Molecular Weights of PVC Samples

Sample No.	$\overline{M}_n{}^a$	$\overline{M}_n{}^b$
1	28000	27500
2	38000	31400
3	45500	39800
4	55000	45000
5	64000	56000
6	82000	79800

a Determined in the ICI (Plastics Division) labs GPC.
b Determined in this laboratory by osmometry.

Table III.   Retention of Radioactivity by PVC Samples after Interaction with $Bu_2Sn(^{35}SBu)_2$

Sample No.	Retained Radioactivity × 10^{-3},a (disintegration min^{-1} g^{-1} PVC) After 6 cycles	After 8 cycles
1	114	109
2	97.8	96.5
3	69.7 (73.3)	66.7
4	76.9 (78.7)	71.5
5	54.8	57.4
6	26.2	35.0

a Figures in parenthesis are repeat experiments.

Table IV.   PVC Containing 2% of Stabilizer $R_2Sn(SCH_2COOC_8H_{17}$-$i)$ Heated at 195°

R	Approx. Time for Complete Discoloration, min.
CH_3	90
CH_3CH_2	90
$CH_3CH_2CH_2CH_2$	70
$CH_3(CH_2)_7$	70
$C_6H_5CH_2$	40
$BrCH_2$	0
C_6H_5	60
p-$CH_3OC_6H_4$	10

The R Groups. There remains the possibility that organotin stabilizers may react directly with chain-carrying chlorine atoms. Experiments in progress (*18*) in which organotin stabilizers are exposed to chlorine atoms derived from the photolysis of carbon tetrachloride or from N-chlorosuccinimide indicate that such reactions are not important.

This leaves the reaction between the stabilizer and hydrogen chloride as the dominant process. In an attempt to shed some light on this reaction, we examined (*18*) the effect on stabilizing efficiency of systematically varying the organic groups in organotin bis(isooctyl thioglycolates) (cf. Ref. *19*). Some results are shown in Table IV. The stabilizers were milled into the PVC at 190°C and the resulting film heated at 195°C; samples were withdrawn at intervals and examined for color. Complete discoloration was taken as the point when the sample became uniformly black or very dark brown. Although the results must be regarded as approximate they show that the methyl- and ethyltin compounds are the most effective. Lengthening the alkyl chain causes a small reduction in stabilizing efficiency suggesting that electron release decelerates the stabilization reactions. It is unlikely that these differences are simply caused by lowering of the molar concentrations since the largest increase in molecular weight is between the dibutyl and dioctyl compounds which have equal stabilizing ability. Introduction of the electron withdrawing groups phenyl or bromine onto the tin-bound methyl groups causes a much bigger reduction in stabilization. Diphenyltin bis(isooctylthioglycolate) is distinctly less efficient than the simple dialkyltin compounds but better than the dibenzyl analogue. Introducing the electron-donating *p*-methoxyl substituent onto the phenyl group reduces stabilization. The contradictions implied by these results can be resolved by recalling that two distinct reactions between stabilizer and hydrogen chloride may occur (*19*). These are tin–sulfur bond cleavage which is a pro-stabilization reaction and tin–carbon cleavage which results in reduced stabilization since the final product will be stannic chloride, a powerful catalyst for dehydrochlorination (Figure 2). Simple dialkyltin stabilizers become converted in the polymer to dialkyltin chlorides but benzyl, bromomethyl, and phenyl groups are more readily

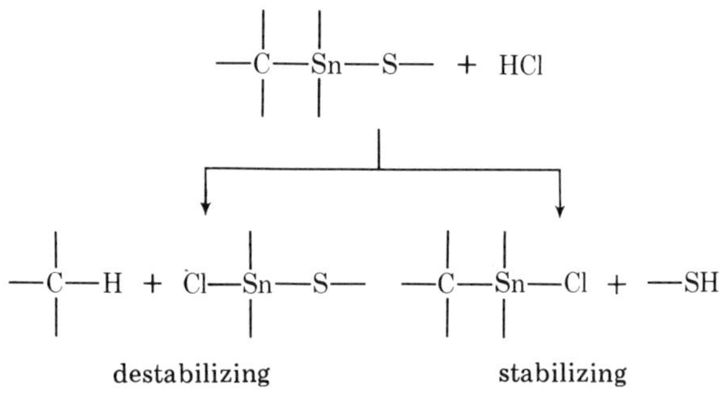

$$
\begin{array}{c}
-\overset{|}{\underset{|}{C}}-\overset{|}{\underset{|}{Sn}}-S- \ + \ HCl \\
\end{array}
$$

$$-\overset{|}{\underset{|}{C}}-H \ + \ Cl-\overset{|}{\underset{|}{Sn}}-S- \qquad -\overset{|}{\underset{|}{C}}-\overset{|}{\underset{|}{Sn}}-Cl \ + \ -SH$$

destabilizing   stabilizing

Figure 2.   Reaction of organotin dithioglycollates with hydrogen chloride

cleaved from tin than alkyl, and the p-methoxyl substituent would be expected (20) to increase the rate of cleavage.

Lewis Acidity Considerations.   The dialkyltin dichloride formed from the stabilizer (either by absorption of hydrogen chloride or by exchange) is itself a Lewis acid catalyst for further dehydrochlorination, though much less active than stannic chloride.

We set out to design a stabilizer in which the Lewis acidity in the ensuing organotin dichloride was suppressed.   We prepared (21, 22) di(4-ketopentyl)tin di(isooctylthioglycolate) in which it was expected that the keto groups would be free in the stabilizer but coordinated to tin in the corresponding dichloride.

$$(CH_3COCH_2CH_2CH_2)_2Sn(SCH_2COOC_8H_{17}\text{-}i)_2$$

The IR spectra (Table V) showed that the C=O band in the dichloride at 1680 cm^{-1} (i.e., a position characteristic of coordinated carbonyl) was displaced to the "free" position of 1705 cm^{-1} when the more powerful donor, 2,2'-bipyridyl was present.   This band is masked in the isooctylthioglycolate, but the NMR spectra (22) (Table VI) demonstrate clearly that coordination of the keto groups

Table V.   IR Spectra of 4-Ketopentyl Compounds[a]

	$\nu(C{=}O)$ (cm^{-1})	Phase
RH	1708	liq. film
R$_2$SnCl$_2$	1680	mull
R$_2$SnCl$_2$·2,2'-bipyridyl	1705	mull

[a] R = CH$_3$COCH$_2$CH$_2$CH$_2$.

Table VI.   NMR Spectra of 4-Ketopentyl Compounds
$$R = CH_3\overset{a}{C}OCH_2CH_2\overset{b}{C}H_2 \text{ in CDCl}_3 \text{ or CCl}_4$$

	τ	
	a	b
RH	7.97	7.68
R$_2$SnCl$_2$	7.74	7.23
R$_2$SnCl$_2$·2,2'-bipyridyl	7.89	7.42
R$_2$Sn(SCH$_2$COOC$_8$H$_{17}$-i)$_2$	7.93	7.52

is absent in the stabilizer but present in the dichloride.   Subsequent tests (22) showed that the ketopentyltin compounds are more than twice as effective in stabilization than the corresponding butyltin derivatives.

Conclusions Concerning the Structure of Stabilizers

Although in the early stages of stabilization, exchange with the small number of allylic chlorine atoms present may be important, the dominant reaction is absorption of hydrogen chloride.   The latter occurs by tin–sulfur bond cleavage, and structural changes which favor this but which do not promote tin–carbon bond cleavage will improve the stabilizer.   The HX molecules liberated must be effective in radical additions to alkenes, i.e., the chain transfer constant should be high.   Finally, reducing the Lewis acidity of the resulting organotin chloride will improve the efficiency of the stabilizer.

Organotin Biocides

The compounds of highest activity are those with three Sn–C bonds; those in commercial use follow the formula: R_3SnX (R= Bu, Ph, C_6H_{11}).   Emphasis is usually placed on the nature of the R groups with the X groups considered to be of little importance.   Once the R_3Sn group gets to the site of biochemical reaction i.e., at cell mitochondria, it may not matter what the X group is but the latter can be significantly involved in transporting the biocide to the reactive site.

Although these compounds appear to pose little threat to the environment (23, 24, 25), their use in agriculture is limited by lack of specificity, and the host suffers as well as the predator.

We considered that if an X group was designed so that the solubility of the resulting R_3SnX compound was substantially increased this might have a considerable effect on biological activity and, hopefully, specificity.   Since nature frequently uses glycoside formation to solubilize compounds and hence promote transport in living tissue, it was decided to incorporate the disaccharide sucrose into the X group.

The most successful method used (26) was to treat the product obtained from the reaction between a cyclic anhydride and sucrose with an organotin hydroxide.

(similarly for succinic and maleic anhydrides)

No attempt was made to purify the mixture of products obtained from the reaction between the anhydride and sucrose before the introduction of the organotin residue.   We do not yet have information about the specificity of these products

Table VII.   Compounds Tested against *Enteromorpha* in Sea Water
Modified with Algal Nutrients

	Concentration[a]	
	1 ppm	0.1 ppm
Bu$_3$SnOCOC$_6$H$_4$COO-sucrose	+	+
Ph$_3$SnOCOC$_6$H$_4$COO-sucrose	+	+
(C$_6$H$_{11}$)$_3$SnOCOC$_6$H$_4$COO-sucrose	—	—
Bu$_3$SnOCOCH$_2$CH$_2$COO-sucrose	+	+
Ph$_3$SnOCOCH$_2$CH$_2$COO-sucrose	+	+

[a] + = effective; — = not effective.

Table VIII.   Action of Organotin Compounds Against Paint-
Destroying Fungi

	No. of Fungi Showing Inhibition of Spore Germination (max 16)		
	100 ppm	10 ppm	1 ppm
Bu$_3$SnOCOC$_6$H$_4$COO-sucrose	16	16	6
Ph$_3$SnOCOC$_6$H$_4$COO-sucrose	16	14	4
(C$_6$H$_{11}$)$_3$SnOCOC$_6$H$_4$COO-sucrose	3	0	0
Bu$_3$SnOCOCH$_2$CH$_2$COO-sucrose	16	16	10
Ph$_3$SnOCOCH$_2$CH$_2$COO-sucrose	16	15	9
(Bu$_3$Sn)$_2$O	13	13	12

Table IX.   Tin Content of Biocidal Organotin Compounds

Compound	% Sn
Tributyltin oxide	39.8
Tributyltin fluoride	38.4
Bu$_3$SnOCOC$_6$H$_4$COO-sucrose	15.2
Ph$_3$SnOCOC$_6$H$_4$COO-sucrose	14.1

but their biological activity is much higher than that of compounds in commercial use.

As potential antifoulants, the compounds are at least three times more effective against the alga *Enteromorpha* than tributyltin oxide where a minimum concentration of 0.3 ppm for a complete kill is necessary under these conditions (Table VII).   The results of screening against the paint-destroying fungi are shown in Table VIII.   Although the tricyclohexyl derivative shows low activity, the other organotin–sucrose compounds inhibit spore germination in all 16 fungi at 100 ppm, and the two tributyltin compounds maintain this at 10 ppm whereas tributyltin oxide is effective against only 13 of the fungi at these concentrations.

The high activity of these compounds is remarkable considering their very low tin content compared with commercial pesticides (Table IX).   These results show that the nature of the X group is important, and work is continuing on other organotin derivatives of carbohydrates.

Acknowledgments

I thank my co-workers, particularly S. Z. Abbas, F. Alavi-Moghadam, G. Ayrey, F. P. Man, and Ann Parkin. I am grateful to the Paint Research Association (UK) for carrying out the biological tests.

Literature Cited

1. Ayrey, G., Head, B. C., Poller, R. C., *Macromol. Rev.* (1974) **8**, 1.
2. Frye, A. H., Horst, R. W., Paliobagis, M. A., *J. Polym. Sci.* (1964) **A2**, 1765, 1785, 1801.
3. Klemchuk, P. P., *Prog. Chem.* (1968) **85**, 1.
4. Ayrey, G., Poller, R. C., Siddiqui, I., *J. Polym. Sci. A1* (1972) **10**, 725.
5. Alavi-Moghadam, F., Ayrey, G., and Poller, R. C., *Eur. Polym. J.* (1975) **11**, 649.
6. Alavi-Moghadam, F., Ayrey, G., Poller, R. C., unpublished data.
7. Gregg, R. A., Alderman, D. M., Mayo, F. R., *J. Am. Chem. Soc.* (1948) **70**, 3740.
8. Walling, C., *J. Am. Chem. Soc.* (1948) **70**, 2561.
9. Alavi-Moghadam, F., Ayrey, G., Poller, R. C., *Polymer* (1975) **16**, 833.
10. Bamford, C. H., Fenton, D. F., *Polymer* (1969) **10**, 63.
11. Yousufzai, A. H. K., Zafar, M. M., Shabik-ul-Hasan, *Eur, Polym. J.* (1972) **8**, 1231.
12. Guyot, A., Bert, M., Michel, A., McNeill, I. C., *Eur. Polym. J.* (1971) **7**, 471.
13. Guyot, A., Bert, M., Michel, A., Spitz, R., *J. Polym. Sci., A1* (1970) **8**, 1596.
14. Michel, A., Bert, M., Guyot, A., *J. Appl. Polym. Sci.* (1969) **13**, 929, 945.
15. McNeill, I. C., Neil, D., Guyot, A., Bert, M., Michel, A., *Eur. Polym. J.* (1971) **7**, 453.
16. McNeill, I. C., Neil, D., *Eur. Polym. J.* (1970) **6**, 143, 569.
17. Braun, D., Thallmaier, M., *Makromol. Chem.* (1966) **99**, 59.
18. Ayrey, G., Man, F. P., Poller, R. C., unpublished data.
19. Rockett, B. W., Hadlington, M., Poyner, W. R., *J. Appl. Polym. Sci.* (1974) **18**, 745.
20. Eaborn, C., *J. Organometal. Chem.* (1975) **100**, 43.
21. Abbas, S. Z., Poller, R. C., *J. Chem. Dalton* . (1974) 1769.
22. Abbas, S. Z., Poller, R. C., *Polymer* (1974) **15**, 543.
23. Barnes, R. D., Bull, A. T., Poller, R. C., *Pest. Sci.* (1973) **4**, 305.
24. Bock, R., Freitag, K. D., *Naturwissenschaften* (1972) **59**, 165.
25. Getzendaner, M. E., Corbin, H. B., *J. Agr. Food Chem.* (1972) **20**, 881.
26. Parkin, A., Poller, R. C., British Provisional Patent **12168/76**.

RECEIVED May 3, 1976. Work is supported in part by the International Sugar Research Foundation.

13

The Influence of Organotin Compounds on Mitochondrial Functions

W. N. ALDRIDGE

Medical Research Council Toxicology Unit, MRC Laboratories, Woodmansterne Rd., Carshalton, Surrey, England

The biological activity of tetraorganotins and diorganotins is considered briefly, and triorganotins are considered in more detail. Tetraorganotin compounds have no direct selective action on mitochondria. In vivo they are rapidly converted to triorganotins and their biological effects are caused by these products. Diorganotins have a property which triorganotins do not and they inhibit the oxidation of α-ketoacids via an affinity for dithiols. Triorganotins derange mitochondrial function in three different ways (a) by secondary responses caused by discharge of a hydroxyl–chloride gradient across mitochondrial membranes, (b) by interaction with the basic energy conservation system involved in the synthesis of ATP, and (c) by an interaction with mitochondrial membranes to cause swelling and disruption. Experiments are described which illustrate the processes which lead to the inhibition of ATP synthesis.

Mono-, di-, tri-, and tetraalkyltins are chemically stable substances. Tin and its inorganic salts have little or no toxicity (*1*). Monoethyltin has a very low toxicity (*2*) whereas the higher alkylated derivatives are toxic (*1*). Both dialkyltin and trialkyltin compounds are inhibitors of oxygen uptake in tissues or mitochondria whereas the tetraalkyltins are inactive (Table I). A rise in the ratio [lactate]:[pyruvate] has been shown to be a sensitive index of effects of triethyltin and is brought about by a rise in lactate associated with a fall in pyruvate (*5*).

In contrast to the inactivity of the tetraethyltin in vitro on intermediary metabolism of brain cortex tissue, it is toxic in vivo (*4, 6*) as are many other tetraalkyltins (*6, 7, 8*). Tetraethyltin is rapidly converted in vivo to triethyltin, and this conversion occurs particularly in the liver (*4*). The toxicity of tetraethyltin appears to be caused entirely by the triethyltin produced from it.

Table I.   Effect of Tetra-, Tri-, and Diethyltin Compounds on the Oxidative Metabolism of Preparations from Rat Brain[a]

Compound	Oxygen uptake	[Pyruvate]	Ref.
	% of Control		
Rat Brain Brei			
Phenylarsenious acid ($30\mu M$)	16	490	3
Diethyltin (810 μM)	34	443	3
Triethyltin ($44\mu M$)	46	48	3
	Oxygen Uptake	$\dfrac{[Lactate]}{[Pyruvate]}$	
Rat Brain Slices			
Triethyltin ($15.8\mu M$)	72	655	4
Tetraethyltin ($150\mu M$)	104	123	4

[a] For brain brei the substrate was lactate and for slices it was glucose.

The results in Table I show quite clearly that the action of diethyltin and triethyltin on rat brain cortex slices is different; both result in an inhibition of oxygen uptake, but diethyltin leads to an increase in the concentration of pyruvate whereas triethyltin produces a decrease (3).   Diethyltin behaves in a similar way to phenylarsenious acid, and this would agree with the observations that both have a high affinity for dithiols (3).   The pyruvic dehydrogenase complex requires lipoic acid as coenzyme which contains vicinal dithiol groups (9).   Further experiments have shown that dialkyltins other than diethyltin also inhibit α-keto acid oxidation.   The results shown in Table II illustrate this point.   All the dialkyltins studied cause inhibition of respiration and also a block in the oxidation of the α-oxoglutarate formed from 1-glutamate.   This increase in α-oxoglutarate

Table II.   Inhibition of α-Oxoglutarate Oxidation of Rat Liver Mitochondria by Dialkyltins and Phenylarsenious Acid[a]

Compound	pI_{50}	R
Phenylarsenious acid	5.8[b]	38
Dimethyltin	3.5	9.1
Diethyltin	4.4	4.5
Di-n-propyltin	4.8	3.4
Diisopropyltin	4.8	9.5
Di-n-butyltin	4.8	5.5
Di-n-pentyltin	3.8	12
Di-n-hexyltin	3.8	8.3

[a] 1-Glutamate was the substrate in a medium as described previously (10). pI_{50} is $-\log$ molar concentration for 50% inhibition of oxygen uptake and R is the ratio of the concentrations of α-oxoglutarate in the medium ± compound.
[b] This concentration caused 94% inhibition.

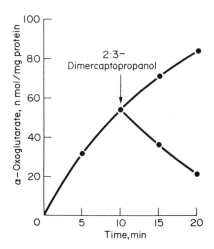

Figure 1.   Accumulation of α-oxoglutarate in the presence of di-n-butyltin
and its prevention by 2,3-dimercap-
topropanol. The substrate was 10 mM
glutamate, and the concentrations of
di-n-butyltin were 10μM and 2,3-dim-
ercaptopropanol 11.2μM.

Table III.   Macromolecules Which Do Not Bind Triethyltin[a]

Phosvitin   Hemoglobins:
Clupeine   Guinea pig
Salmine   Horse
Cytochrome c   Rabbit
Bovine plasma albumin   Human
DNA   Horse myoglobin
RNA   Chymotrypsin
Glycogen   Bovine pancreatic ribonuclease
Dextran
Dextran sulfate

[a] Data from Refs. 13 and 14.

Table IV.   Inhibition of ATP Synthesis of Rat Liver Mitochondria by Trialkyltins (14)

	Concentrations Causing 50% Inhibition, μM
Trimethyltin	4.0
Triethyltin	0.1
Tri-n-propyltin	0.32
Triisopropyltin	0.64
Tri-n-butyltin	0.64
Tri-n-hexyltin	2.5

is progressive and may be reversed by adding 2,3-dimercaptopropanol at equi-molar concentration (Figure 1). It is therefore clear that dialkyltins affect mitochondrial functions by inhibiting, at low concentration, the oxidation of α-keto acids, and that the mechanism of this inhibition is almost certainly caused by its combination with coenzymes or enzymes possessing two thiol groups in the correct conformation. This could be lipoic acid as stated above, but perhaps it is more likely that the lipoyl dehydrogenase, an enzyme possessing two thiol groups close together (11), is the site of inhibition as has been suggested for arsenite (12).

Trialkyltins are stable substances which have little or no affinity for mono- or dithiols (3), an unusual situation for substances containing heavy metals. They do not combine with many macromolecules (Table III) and therefore appear to be unreactive. This lack of reactivity with large molecules which offer an almost unlimited surface for interaction is in sharp contrast to the low concentration of trialkyltins which inhibit mitochondrial functions (Table IV). Oxygen uptake stimulated by ADP (state 3) or by 2,4-dinitrophenol (14), oxidative phosphorylation and ATP hydrolysis stimulated by 2,4-dinitrophenol (14), or other uncouplers (15) and the $^{32}P_i$-ATP exchange reaction (16) are all inhibited by the concentrations listed in Table IV.

Binding to Mitochondria

The opportunity of measuring the binding of triethyltin and trimethyltin to mitochondria was presented because of the availability of highly labeled compounds containing ^{113}Sn, a γ-emitting isotope. By conventional binding experiments and by plotting the results using the method of Scatchard, two classes of sites were found; one of low concentration but of high affinity, and the other of a large number of sites with much lower affinity (17). Only the high affinity site (hereafter called site 1) was of interest with respect to inhibition of mitochondrial functions. For both trimethyltin and triethyltin the same concentration of 0.8 nmole/mg mitochondrial protein was obtained with affinity constants of 1.2×10^4 and $4.7 \times 10^5 M^{-1}$, respectively. An insoluble fraction from the mitochondria containing approximately 15% of the mitochondrial protein contained this class of binding site. Correlation of inhibition of oxidative phosphorylation of these two tin compounds with the binding to this site (calculated from the experimentally determined constants) showed that different results were obtained depending on which substrate was being oxidized (Figure 2). When the substrate was pyruvate, percent inhibition was up to 10 times more than the corresponding percent of site 1 complexed, whereas when phosphorylation was linked to the oxidation of reduced cytochrome c, a one-to-one relationship was obtained (15). At the time, these findings were brought together into one hypothesis for the mechanism of oxidative phosphorylation (18). This has now been shown to be incorrect since the initial assumption that trialkyltins are specific reagents exerting

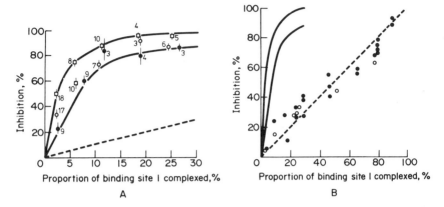

Figure 2.   Relationship between inhibition of mitochondrial functions in relation to binding of triethyltin to a site in rat liver mitochondria.   ATP synthesis was linked to oxidation of A, pyruvate (O, ●), succinate (□), and B, reduced cytochrome c (● at 25°, O at 37°).   The medium contained 100mM chloride (15).   Dashed line indicates a 1:1 relationship, and the solid lines in B are those in A.

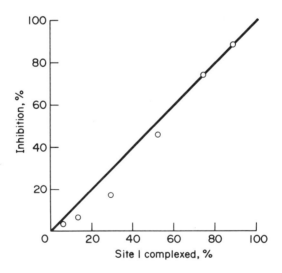

Figure 3.   Relationship between inhibition of ATP synthesis linked to oxidation of succinate and binding of triethyltin to a site in rat liver mitochondria.   The medium was chloride free and contained 100mM K isethionate.

one effect on mitochondria is incorrect.   Selwyn and co-workers have shown that the high sensitivity of mitochondrial functions is attributed to the presence of chloride in the medium used for the experiments (*19, 20*).   Triethyltin hydroxide and chloride are both lipophilic and can pass through membranes.   Mitochondria are able to maintain a hydroxyl ion gradient between an intramitochondrial compartment and the outside medium.   Thus, in the presence of trialkyltins, the mitochondrial membrane which normally maintains a hydroxyl–chloride ion gradient is unable to do so.   The details of how this was established will be dealt with by Selwyn (Chapter 15).   If measurements of mitochondrial functions are made in a medium containing nitrate or isethionate instead of chloride, an approximately one-to-one relationship between binding and inhibition is found (Figure 3).   The presence of chloride is necessary for the high sensitivity of oxidative phosphorylation linked to pyruvate oxidation, and the lower sensitivity is caused by the binding of triethyltin to a component of the energy conservation system.   Triethyltin therefore brings about two effects which are mechanistically completely different, the latter being similar to the effects produced by oligomycin (*20, 21*).

Inhibition of Mitochondrial Functions in the Presence of Chloride

When triethyltin is added to mitochondria, ATP hydrolysis is increased but to a much smaller extent than with classical uncouplers such as 2,4-dinitrophenol (*13*).   The concentrations which cause an increase in ATP hydrolysis also cause an increase in oxygen uptake, limited swelling (*13*), and an uptake of chloride into the mitochondria (*22*).   Inhibitors of energy utilization either derived from the oxidation of substrates or from ATP prevent the uptake of chloride (*22*).   All these effects are associated, and the hydrolysis of ATP and the oxygen uptake are measures of the utilization of energy to generate the hydroxyl ion gradient so that the exchange of hydroxyl and chloride across a mitochondrial membrane may take place (*23*).   The question remains of how this process leads to the effects listed in Table V.   The movement of substrates into the mitochondria is impeded presumably because they are replaced by the chloride anion; in some circumstances this may become an important factor.   In addition, the internal pH of the mitochondrion becomes more acid and may in certain circumstances be such that the dehydrogenases are operating at well below their pH optima (*24*).   However the demonstration that ATP synthesis linked to the oxidation of β-hydroxybutyrate is insensitive to triethyltin even in a chloride-containing medium indicates that many of the reactions necessary for oxidative phosphorylation cannot be affected by the changes in the internal pH of the mitochondrion (*25*).   There are several lines of evidence suggesting that β-hydroxybutyrate dehydrogenase of rat liver mitochondria is in a unique position and may be in contact with the external rather than the internal environment (*43*).   It seems likely that some of the effects in a chloride-containing medium may be best explained if

Table V.   Effects of Triethyltin on Mitochondrial Functions

Systems Showing High Sensitivity in a Chloride-Containing Medium
ATP synthesis linked to oxidation of tricarboxylic acid substrates
ATP hydrolysis stimulated by 2,4-dinitrophenol
Oxygen uptake stimulated by 2,4-dinitrophenol
Increase in ATP hydrolysis
Increase in oxygen uptake (state 4)
Limited swelling in an ATP- and KCl-containing medium
Gross swelling in an NaCl- and NH_4Cl-containing medium

Systems Showing Lower Sensitivity in a Chloride-Containing Medium
ATP synthesis linked to oxidation of reduced cytochrome c or
   β-hydroxybutyrate
Inhibition of ATP hydrolysis
Inhibition of oxygen uptake

Systems Showing Lower Sensitivity in a Chloride-Free Medium
ATP hydrolysis stimulated by 2,4-dinitrophenol
ATP synthesis linked to oxidation of tricarboxylic acids

Systems Insensitive
Oxygen uptake stimulated by 2,4-dinitrophenol in a chloride-free medium
Succinate oxidase of disrupted mitochondria
Various NAD-requiring dehydrogenases

there are regions in the mitochondrion of high hydroxyl and high proton concentrations.   The reduction of the gradient between these regions may well reduce the ability of 2,4-dinitrophenol and other uncouplers to stimulate ATP hydrolysis and oxygen uptake (*23*).   Such experiments have only served to indicate how complicated the situation is.   What is rate limiting under one particular experimental condition will not be the same in another (*25*).   The situation will be further influenced by the external and internal concentration of hydroxyl, chloride, and phosphate ions (*25*), all of which compete for the triethyltin.   Although it seems to be generally accepted that a proton gradient across membranes is involved in the synthesis of ATP (*26, 27*), we are much hampered in interpreting the effects brought about by trialkyltins in a halide-containing medium by our ignorance of the chemical compartmentation of the reactants and the chemistry of the process (*28*).

Chemical Nature of Binding Site

As mentioned earlier, the effects of triethyltin in a chloride-free medium resemble those produced by oligomycin.   The fact that there is an approximately one-to-one relationship between inhibition and binding to a site in the mitochondrion indicates that this site is either directly or indirectly involved in the synthesis of ATP from ADP and P_i.   It is clearly important to find out what is the chemical nature of this group.   Unfortunately we have, as yet, been unable to attack this question directly.   The binding site is on an insoluble fraction from the mitochondrion, and all our attempts to solubilize and purify it have led to the loss of binding properties (*17*).   Since triethyltin binds to rather few proteins, the examination of the nature of binding to these proteins may provide relevant information.   So far the only proteins to which triethyltin binds with an affinity of 10^5–$10^6 M^{-1}$ is rat hemoglobin (*29*), cat hemoglobin (*30*), liver supernatant from several species (*39*), and myelin from rat brain (*31*).   Figure 4 shows the

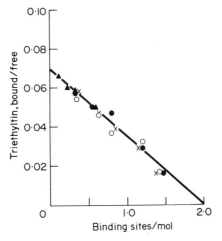

Figure 4.   Binding of triethyltin to cat hemoglobin.   K = 3.5 × $10^4 M^{-1}$ in 0.1M phosphate buffer pH 7.25 (30)

Scatchard binding data for cat hemoglobin.   Only one class of binding sites is found, the affinity is $3.5 \times 10^4 M^{-1}$, and one molecule of hemoglobin binds two molecules of triethyltin.   Previous work has shown that after photooxidation of rat hemoglobin (*32*) or the proteins from guinea pig liver (*33*), only the loss of histidine may be correlated with the loss of binding.   In each case the rate of loss of binding was always twice the loss of histidine (*32, 33*) leading to the general hypothesis that when triethyltin binds to protein with an affinity of 10^5–$10^6 M^{-1}$, it does so by pentacoordination between two histidines.   Such a view would be in accord with other information derived from the formation of triethyltin–imidazole polymers in solvents (*34, 35, 36*).   However, recent experiments in my laboratory have raised some doubts about this general hypothesis.   Using diethylpyrocarbonate as a rather specific reagent for histidine (*37*), it has been shown that after extensive reaction with both cat and rat hemoglobin only one of the binding sites can be removed (Figure 5).   Thus there is confirmatory evidence that one of the two binding sites in rat and cat hemoglobin involve histidine.   However the other binding site appears to be completely insensitive to

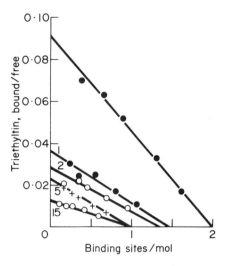

Figure 5. Influence of diethylpy-rocarbonate on binding of triethyltin to cat hemoglobin. Numbers on the lines indicate the times of incubation in min. at 20–22°C (30).

treatment with diethylpyrocarbonate, and binding of one molecule of triethyltin still exists when most of the histidines of rat hemoglobin have been modified (30). Thus we must seriously consider the possibility that in cat and rat hemoglobin there are two quite different binding sites for triethyltin, both of identical affinity. Extension of this work to other proteins is now in progress. Recent work (38) has provided evidence that tri-n-butyltin reacts with the proton-conducting system of the energy-conservation system, and it is clearly very important to increase our knowledge of the kind of chemical groupings for which trialkyltins have affinity.

Table VI.   Concentrations of Triorganotins Causing Different Effects
on Mitochondria (39)

| Compounds | pI_{50}[a] | | |
	Chloride-Dependent Effects	Chloride-Independent Effects	Gross Swelling
Trimethyltin	5.7	3.2[b]	2.1[b]
Triethyltin	7.0	5.25	3.7
Tri-n-propyltin	6.7[b]	6.0[b]	4.6
Tri-n-butyltin	6.3[b]	5.9[b]	5.15
Triphenyltin	5.8[b]	5.6[b]	5.4[b]

[a] pI_{50} = −log molar concentrations for 50% inhibition.
[b] The range of effective concentrations overlap.

In addition to the two different mechanisms whereby trialkyltins may affect mitochondrial function, there is one which may become more important for the higher molecular weight and more lipophilic trialkyltins.   They cause gross swelling (*13, 39*), and it seems possible from the concentrations bringing this about that it may be correlated with the binding to the low affinity sites (*39, 40*).   The concentrations of trialkyltins bringing about changes in mitochondria by three different mechanisms are shown in Table VI.   Each has its own structure–activity relationship.

In conclusion, it is clear that we are far from understanding the biocidal activities of trialkyltins.   In mammals, triethyltin causes a rapid fall in body temperature (*41*) and an increase in water in the brain, particularly associated with nerve tracts (*42*).   In neither case do we know whether these effects are initiated by an attack on mitochondrial function.   If it is assumed that their general biocidal activity may be caused by their efficiency in inhibiting energy-conservation processes which have many similarities in many organisms, then we do not know which of the three mechanisms are involved.

Literature Cited

1. Barnes, J. M., Stoner, H. B., *Pharmacol. Rev.* (1959) **11**, 211–231.
2. Stoner, H. B., Barnes, J. M., Duff, J. I., *Brit. J. Pharmacol.* (1955) **10**, 16–25.
3. Aldridge, W. N., Cremer, J. E., *Biochem. J.* (1955) **61**, 406–418.
4. Cremer, J. E., *Biochem. J.* (1958) **68**, 685–692.
5. Ibid. (1957) **67**, 87–96.
6. Caujolle, F., Lesbre, M., Meynier, D. *CR. Acad. Sci. Paris* (1954) **239**, 556–558.
7. Ibid. (1954) **239**, 1091–1093.
8. Caujolle, F., Lesbre, M., Meynier, D. *Ann. Pharm. Fr.* (1956) **14**, 88–97.
9. Gunsalus, I. C., *J. Cell. Comp. Physiol.* (1953) **41** (Suppl. 1), 113–136.
10. Aldridge, W. N., *Biochem. J.* (1952) **67**, 423–431.
11. Massey, V., Veeger, C. *Biochim. Biophys. Acta* (1961) **48**, 33–47.
12. Ibid. (1960) **40**, 184–185.
13. Aldridge, W. N., Street, B. W., *Biochem. J.* (1964) **91**, 287–297.
14. Aldridge, W. N., *Biochem. J.* (1958) **59**, 367–376.
15. Aldridge, W. N., Street, B. W., *Biochem. J.* (1971) **124**, 221–234.
16. Aldridge, W. N., Threlfall, C. J., *Biochem. J.* (1961) **79**, 214–219.
17. Aldridge, W. N., Street, B. W., *Biochem. J.* (1970) **118**, 171–179.
18. Aldridge, W. N., Rose, M. S., *FEBS Lett.* (1969) **4**, 61–69.
19. Selwyn, M. J., Dawson, A. P., Stockdale, M., Gains, N., *Eur. J. Biochem.* (1970) **14**, 120–126.
20. Stockdale, M., Dawson, A. P., Selwyn, M. J., *Eur. J. Biochem.* (1970) **15**, 342–351.
21. Rose, M. S., Aldridge, W. N., *Biochem. J.* (1972) **127**, 51–59.
22. Skilleter, D. N., *Biochem. J.* (1976) **154**, 271–276.
23. Aldridge, W. N., Street, B. W., Skilleter, D. N., unpublished data.
24. Dawson, A. P., Selwyn, M. J., *Biochem. J.* (1974) **138**, 348–357.
25. Aldridge, W. N., Street, B. W., unpublished data.
26. Mitchell, P., Moyle, J., *Nature (London)* (1965) **208**, 147–151.
27. Mitchell, P., "The Molecular Basis of Membrane Functions," D. C. Tosteson, Ed., p. 483–518, Prentice-Hall, New Jersey 1969.
28. Earnster, L., *Symp. Membranes Electron Transport*, FEBS, Paris (1975) Abstract No. **70**.
29. Rose, M. S., Aldridge, W. N., *Biochem. J.* (1968) **106**, 821–828.
30. Elliott, B., Aldridge, W. N., (1976) unpublished data.

31. Lock, E. A., Aldridge, W. N., *J. Neurochem.* (1976) **25**, 871–876.
32. Rose, M. S., *Biochem. J.* (1969) **111**, 129–137.
33. Rose, M. S., Lock, E. A., *Biochem. J.* (1970) **120**, 151–157.
34. Luijten, J. G. A., Janssen, M. J., van der Kerk, G. J. M., *Rev. Trav. Chim. Pays-Bas* (1962) **81**, 202–205.
35. van der Kerk, G. J. M., Luijten, J. G. A., Janssen, M. J., *Chimia* (1962) **16**, 10–15.
36. Janssen, M. J., Luijten, J. G. A., van der Kerk, G. J. M., *J. Organometal. Chem.* (1964) **1**, 286–291.
37. Holbrook, J. J., Ingram, V. A., *Biochem. J.* (1973) **131**, 729–738.
38. Dawson, A. P., Selwyn, M. J., *Biochem. J.* (1975) **152**, 333–339.
39. Aldridge, W. N., Street, B. W., unpublished data.
40. Wulf, R. G., Byington, K. H., *Arch. Biochem. Biophys.* (1975) **167**, 176–192.
41. Stoner, H. B., *Brit. J. Exp. Path.* (1963) **44**, 384–396.
42. Magee, P. N., Stoner, H. B., Barnes, J. M., *J. Path. Bact.* (1957) **73**, 107–124.
43. Harris, E. J., Manger, J. R., *Biochem. J.* (1969) **113**, 617–628.

RECEIVED May 3, 1976.

14

Bioorganotin Chemistry: Biological Oxidation of Organotin Compounds

RICHARD H. FISH, ELLA C. KIMMEL, and JOHN E. CASIDA

Pesticide Chemistry and Toxicology Laboratory, Department of Entomological Sciences, College of Natural Resources, University of California, Berkeley, Calif. 94720

Previous reports on the biological oxidation of organotin compounds concluded that destannylation (carbon–tin bond cleavage—e.g., tetraalkyltin to tri and dialkyltin derivatives) was the overall consequence of this reaction. Recent studies using tributyltin acetate show that the primary biological oxidation reaction is in fact hydroxylation of carbon–hydrogen bonds that are α, β, γ, and δ to the tin atom. The lability of the major metabolites, the α and β carbon-hydroxylated products, was the reason previous workers concluded that destannylation was the major biological oxidation reaction. A free radical process is invoked to explain the predominance of β and the significant amount of α-carbon-hydroxylation in n-alkyltin compounds. Triphenyltin derivatives are not hydroxylated or destannylated in the monooxygenase enzyme system although in vivo studies with rats reveal destannylation products.

The increased use of organotin compounds has stimulated renewed interest in their biological fate. A perusal of the literature, however, reveals numerous reports on their biological properties but no definitive studies on their metabolic reactions (*1, 2, 3, 4, 5*). This review considers the progress and prospects in understanding a particularly important metabolic reaction of alkyltin compounds—their biological oxidation.

Monooxygenase Enzyme Systems

A widely studied metabolic reaction of organic compounds is their biological oxidation, where the in vitro investigations are usually carried out with monooxygenases of mammalian liver [e.g., rat liver microsomal preparations with reduced nicotinamide adenine dinucleotide phosphate (NADPH) as the essential cofactor]. These are generally cytochrome P-450-dependent monooxygenases which convert carbon–hydrogen to carbon–hydroxyl bonds (6). This monooxygenase system activates oxygen in a two-electron reduction with an iron–porphyrin–oxygen complex playing a major role (Scheme I).

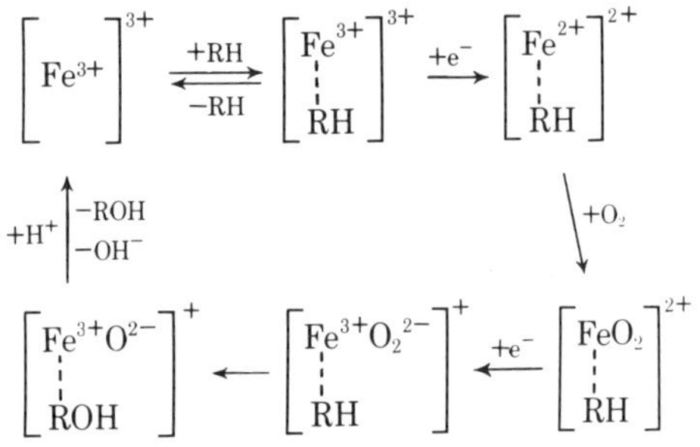

$$\left[Fe^{3+} \right]^{3+} \xrightleftharpoons[-RH]{+RH} \left[\begin{array}{c} Fe^{3+} \\ RH \end{array} \right]^{3+} \xrightarrow{+e^-} \left[\begin{array}{c} Fe^{2+} \\ RH \end{array} \right]^{2+}$$

$$+H^+ \Big\uparrow \begin{array}{c} -ROH \\ -OH^- \end{array} \qquad\qquad \Big\downarrow +O_2$$

$$\left[\begin{array}{c} Fe^{3+}O^{2-} \\ ROH \end{array} \right]^+ \longleftarrow \left[\begin{array}{c} Fe^{3+}O_2^{2-} \\ RH \end{array} \right]^+ \xleftarrow{+e^-} \left[\begin{array}{c} FeO_2 \\ RH \end{array} \right]^{2+}$$

Scheme I

Unlike most compounds metabolized by the microsomal monooxygenases, the organotin derivatives have a propensity to coordinate nonspecifically with various available heteroatoms (7). This works against the specificity needed for binding of the substrate molecule (RH) in proximity to the active iron–oxygen complex (6) (Scheme II).

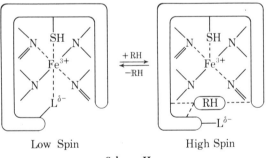

Low Spin   High Spin

Scheme II

It is not surprising, therefore, to find that trialkyltin derivatives give low metabolite yields (<10%) in monooxygenase systems necessitating the use of [^{14}C]-organotin compounds to quantify and to identify the products from the biological oxidation reaction.

In general it is important to compare the in vitro monooxygenase reactions with those occurring in vivo in order to define the products in each situation. If the same products are formed both in vitro and in vivo (using several animal systems), it is likely that these monooxygenase enzyme reactions are important in man.

In Vitro and In Vivo Metabolism of Organotin Compounds

***n*-Alkyltin Compounds.**   The first attempt to understand the metabolic reactions of organotin compounds was the pioneering work of Cremer (8).   It was shown both with a rat liver microsomal enzyme system and with rats (in vivo) that tetraethyltin undergoes destannylation (carbon–tin bond cleavage) to give a triethyltin derivative (Reaction 1).

$$(CH_3CH_2)_4Sn \xrightarrow[\substack{\text{or rats (in vivo)} \\ X = \text{anion}}]{\substack{\text{Monooxygenase system} \\ \text{(in vitro)}}} (CH_3CH_2)_3SnX + [CH_2{=}CH_2 + CH_3CH_2OH] \quad (1)$$

While Cremer (8) only observed tetraethyltin going to a triethyltin derivative, later work by Casida et al. (9) showed that the triethyltin derivatives are biologically oxidized in vitro to diethyltin derivatives.   Furthermore, Bridges et al. (10) established that diethyltin derivatives are converted in vivo to monoethyltin derivatives.   The results (9) with other trialkyltin derivatives (i.e., tripropyltin, tributyltin, tripentyltin, and trihexyltin chlorides) were similar to those with the triethyltin compounds.   Additionally, it was found that thin-layer chromatography (TLC), used in conjunction with selective chromogenic reagents, was more suitable for separating and detecting possible metabolites than the colorimetric analyses used in earlier studies (8, 9).

Casida et al. (9) speculated on the possible formation of carbon-hydroxylated metabolites.   However, in that initial study none were identified, partly because of the low total metabolite yield (<10%), the lability of some metabolites in the acidic cosolvents needed in the TLC separations, and the lack of authentic compounds for comparison with the unknown metabolites.

Recently, Fish et al. (11, 12) utilized tributyltin derivatives and the monooxygenase system as a model for the complete elucidation of the major biological oxidation reaction—carbon-hydroxylation on the butyl group.   In these studies with rat liver microsomes, it was ascertained by two criteria that indeed a cytochrome P-450 monooxygenase enzyme system and not a lipid peroxidase system was responsible for the oxidation reaction.   Thus the biological oxidation of tributyltin acetate was totally inhibited by carbon monoxide and to a large extent by 4(5)-α-naphthylimidazole while this reaction was not affected by ethylenediamine-tetraacetic acid or manganese (2^+).

It is important in this type of study to have available authentically synthesized metabolites and a knowledge of their chemistry, especially in the analytical system used to quantify and to identify them.   Thus, all the possible carbon-hydroxylated metabolites that could be readily synthesized were prepared, and their chromatographic (TLC) and chemical properties were compared with those of the metabolites formed in the monooxygenase enzyme system.   By utilizing [1-^{14}C]-tributyltin acetate, it was feasible to quantify and to identify the following metabolites (11, 12) (Reaction 2):

$$[CH_3(CH_2)_3]_3Sn\text{---}X \xrightarrow[\substack{\text{rat liver micro-}\\\text{somes with NADPH}}]{(37°,\ 1\ hr)} [CH_3(CH_2)_3]_2\underset{\underset{X}{|}}{Sn}CH_2CH_2CH_2CH_2OH +$$

(8%)

$$[CH_3(CH_2)_3]_2\underset{\underset{X}{|}}{Sn}CH_2CH_2\underset{\underset{OH}{|}}{CH}CH_3 + [CH_3(CH_2)_3]_2\underset{\underset{X}{|}}{Sn}CH_2CH_2\underset{\overset{\|}{O}}{C}CH_3 +$$

(14%)   (4%)

$$[CH_3(CH_2)_3]_2\underset{\underset{X}{|}}{Sn}CH_2\underset{\underset{OH}{|}}{CH}CH_2CH_3 + [CH_3(CH_2)_3]_2\underset{\underset{X\ OH}{|\ \ |}}{Sn}CHCH_2CH_2CH_3 \quad (2)$$

(50%)   (24%)

An in vivo study (*13*) with [1-^{14}C]-tributyltin acetate established the significance of the findings in the monooxygenase system by showing the occurrence of similar reactions in mice.

A similar approach was attempted with [1-^{14}C]-tetrabutyltin; however, an in vitro study showed that rapid destannylation of the primary metabolites to a tributyltin derivative greatly complicated and confused the results (*13*). Thus, the tributyltin derivative generated reacts more rapidly than the tetrabutyltin so that the enzyme system gives essentially all the products noted above (Reaction 2). The identified tetraorganotin metabolites of [1-^{14}C]-tetrabutyltin are as follows (*13*) (Reaction 3):

$$[CH_3(CH_2)_3]_4Sn \xrightarrow[\substack{\text{rat liver microsomes}\\\text{with NADPH}}]{(37°,\ 1\ hr)}$$

$$[CH_3(CH_2)_3]_3Sn CH_2\underset{\underset{OH}{|}}{CH}CH_2CH_3 + [CH_3(CH_2)_3]_3Sn CH_2CH_2\underset{\underset{OH}{|}}{CH}CH_3 \quad (3)$$

(77%)   (23%)

Cyclohexyltin Compounds. Studies on the metabolic fate of cyclohexyltin compounds have emphasized tricyclohexyltin hydroxide because of its economic importance. Although not reported in detail, the in vivo findings with this compound indicate successive destannylation reactions ultimately yielding inorganic tin (*14*) (Reaction 4).

$$\left(\left\langle\bigcirc\right\rangle\right)_3 SnOH \longrightarrow \left(\left\langle\bigcirc\right\rangle\right)_2 SnO \longrightarrow \left\langle\bigcirc\right\rangle\text{---}SnO_2H \longrightarrow Sn^{4+} \quad (4)$$

In contrast to this study, work in our laboratory (*13*) on the in vitro metabolism of tricyclohexyltin hydroxide with a cytochrome P-450-dependent monooxygenase enzyme system indicates that carbon-hydroxylation of the cyclohexyl group is a major metabolic reaction. Additionally, with model systems such as triphenylcyclohexyltin or diphenylcyclohexyltin acetate, we tentatively identified the 2-, 3-, and 4-hydroxycyclohexyltin derivatives as metabolites (*15*), and by analogy it is also likely that these metabolites are formed with tricyclohexyltin hydroxide. Consequently, we suggest that the metabolic fate of tricyclohexyltin derivatives is not derived from simple oxidative cleavage of the tin-carbon bond but occurs via carbon-hydroxylation of the cyclohexyl group and the subsequent lability of these hydroxycyclohexyltin metabolites. The chemistry, stereochemistry, and toxicological properties of these metabolites are currently under investigation (*15*).

Triphenyltin Compounds. When [^{113}Sn]-triphenyltin acetate was subjected to in vitro biological oxidation conditions (*13*), it was surprising to find that the phenyl group was not metabolized; only insignificant traces of compounds that could conceivably be hydroxyl metabolites were found (TLC) along with starting material. However, quite different results were obtained in vivo studies (*13*) since rats were found to metabolize triphenyltin acetate, yielding substantial amounts of diphenyl and monophenyltin derivatives. The latter finding indicates that destannylation might occur by a nonenzymatic mechanism or that it might involve enzyme systems other than the cytochrome P-450-dependent monooxygenases.

Mechanistic Aspects of the Monooxygenase Reactions with Organotin Compounds

Although the early studies pointed out that destannylation—i.e., tetraethyltin to triethyltin and then to di and monoethyltin derivatives—was a major metabolic reaction, the mechanism of carbon–tin bond cleavage was completely unknown. It was speculated that ethane was the organic fragment liberated on metabolism of diethyltin compounds (*10*) while 1-butene was identified as the product formed upon acidifying the monooxygenase reaction mixture of a tributyltin derivative (*9*). We can now unequivocally describe the primary reaction occurring with a model tributyltin substrate in the cytochrome P-450-dependent monooxygenase enzyme system as a carbon-hydroxylation occurring primarily at the α- and β-carbons and to a smaller extent at the γ- and δ-carbons to the tin atom (*11, 12*). More importantly, the α-hydroxyl and to a lesser extent the β-hydroxyl metabolites (Reaction 2) are unstable and undergo destannylation reactions. Thus, the α-hydroxyl metabolite undergoes a cleavage reaction at pH 7.4 to give 1-butanol and a dibutyltin derivative while the β-hydroxyl metabolite rapidly undergoes a β-elimination reaction in acidic media to form 1-butene and a dibutyltin derivative (Reaction 5).

$$[CH_3(CH_2)_3]_2Sn\underset{\underset{X}{|}\ \underset{OH}{|}}{CHCH_2CH_2CH_3} \xrightarrow{pH\ 7.4} \begin{cases} CH_3CH_2CH_2CH_2OH \\ \\ [CH_3(CH_2)_3]_2SnX_2 \end{cases} \quad (5)$$

$$[CH_3(CH_2)_3]_2Sn\underset{\underset{X}{|}\ \underset{OH}{|}}{CH_2CHCH_2CH_3} \xrightarrow{H^+} \begin{cases} \\ CH_3CH_2CH{=}CH_2 \end{cases}$$

We suggest that previous workers (8, 9, 10) found overall carbon–tin bond cleavage because of the formation and subsequent decomposition of α- and β-carbon-hydroxylated derivatives. Additionally, recent preliminary results (15) on triphenylcyclohexyltin and diphenylcyclohexyltin acetate verify that carbon-hydroxylation is a general biological reaction for alkyltin compounds.

The finding that the β-carbon–hydrogen bond is a predominant site for hydroxylation strongly suggests that the tin–carbon sigma (σ) electrons play a role in directing, to some extent, the site of hydroxylation. While other studies on the cytochrome P-450-dependent monooxygenases provide some evidence for an oxenoid reaction mechanism (6)—i.e., insertion of an electrophilic oxygen species into a carbon–hydrogen bond—we prefer, for several reasons, to invoke a free radical mechanism for our results (11, 12). First, in attempts to use chemical mimics of the monooxygenases (e.g., peracids, Fe^{2+}/H_2O_2, etc.), only destannylation products (e.g., tributyltin acetate to dibutyltin diacetate) and no carbon-hydroxylated products were obtained. This is consistent with the reactions of electrophilic reagents, which cleave tin–carbon bonds (16). Secondly, we see minimal cleavage of the carbon–tin bonds in the [1-^{14}C]-tributyltin acetate experiments using the monooxygenase enzyme system. Finally, the carbon-hydroxylation pattern we obtain (an α, β, γ, and δ normalized ratio of 4.5:9.4:3.4:1 with [1-^{14}C]-tributyltin acetate) is more consistent with a free radical than with an oxenoid mechanism.

The fact that the α- and β-carbon–hydrogen bonds are the prime sites for hydroxylation is also consistent with the recent electron spin resonance experiments of Kochi (17, 18, 19) and Symons (20, 21, 22) and their co-workers. Their studies show a preferred conformation of the tin–carbon σ electrons in relation to a radical on a β-carbon atom as in 1. They also observed the ready abstraction of hydrogen from carbon atoms α and β to the tin atom. Both of these findings support our suggestion that organotin compounds undergo carbon-hydroxylation by a free radical pathway.

1

Conclusions

The first steps have been taken in defining the mechanism of biological oxidation of organotin compounds. Additional studies on cyclohexyltin derivatives should provide needed information on the stereochemistry of this important reaction. It is also desirable to find a suitable model for the P-450-dependent monooxygenase system that will allow an in depth study of the role of the tin atom in these reactions.

Literature Cited

1. Rose, M. S., "Pesticide Terminal Residues," A. S. Tahori, Ed., p. 281, Butterworths, London, 1971.
2. Luijten, J. G. A., "Organotin Compounds," A. K. Sawyer, Ed., Vol. 3, p. 931, Marcel Dekker, New York, 1972.
3. Piver, W. T., *Environ. Hlth. Perspect., Expt. Issue* (1973) 4, 61.
4. Thayer, J. S., *J. Organometal. Chem.* (1974) 76, 265.
5. Smith, P., Smith, L., *Chem. Britain* (1975) 11, 208.
6. Ullrich, V., *Angew. Chem., Internat. Edit.* (1972) 11, 701.
7. Okawara, R., Wada, M., *Adv. Organometal. Chem.* (1967) 5, 137.
8. Cremer, J. E., *Biochem. J.* (1958) 68, 685.
9. Casida, J. E., Kimmel, E. C., Holm, B., Widmark, G., *Acta Chem. Scand.* (1971) 25, 1497.
10. Bridges, J. W., Davies, D. S., Williams, R. T., *Biochem. J.* (1967) 105, 1261.
11. Fish, R. H., Kimmel, E. C., Casida, J. E., *J. Organometal. Chem.* (1975) 93, C1.
12. Fish, R. H., Kimmel, E. C., Casida, J. E., *J. Organometal. Chem.* in press.
13. Kimmel, E. C., Fish, R. H., Casida, J. E., *J. Agr. Food Chem.*, in press.
14. Blair, E. H., *Environ. Qual. Saf. Suppl.* (1975) 3, 406.
15. Fish, R. H., Kimmel, E. C., Casida, J. E., unpublished results.
16. Gielen, M., Nasielski, J., "Organotin Compounds," A. K. Sawyer, Ed., Vol. 3, p. 625, Marcel Dekker, New York, 1972.
17. Krusic, P. J., Kochi, J. K., *J. Amer. Chem. Soc.* (1971) 93, 846.
18. Kawamura, T., Kochi, J. K., *J. Amer. Chem. Soc.* (1972) 94, 648.
19. Kawamura, T., Meakin, P., Kochi, J. K., *J. Amer. Chem. Soc.* (1972) 94, 8065.
20. Lyons, A. R., Symons, M. C. R., *J. Chem. Soc., Chem. Comm.* (1971) 1068.
21. Lyons, A. R., Symons, M. C. R., *J. Chem. Soc., Faraday II* (1972) 68, 622.
22. Symons, M. C. R., *Tetrahedron Lett.* (1975) 793.

RECEIVED May 3, 1976. Work supported by the National Institute of Environmental Health Sciences (NIH grant 2 P01 ES00049) and the Rockefeller Foundation.

15

Triorganotin Compounds as Ionophores and Inhibitors of Ion Translocating ATPases

M. J. SELWYN

School of Biological Sciences, University of East Anglia, Norwich, NR4 7TJ, England

Triorganotin compounds, especially the more hydrophobic ones such as triphenyltin and tributyltin, inhibit electron transport phosphorylation in mitochondria, chloroplasts, and bacteria. The inhibitory site is on the membrane-bound components of the coupling ATPases, and the action of the triorganotins is inhibition of proton flow through these components. Triorganotins also inhibit the Na^+-K^+ ATPase of cell membranes and the Ca^{2+}-translocating ATPase of sarcoplasmic reticulum. Triorganotins are also ionophores which mediate an exchange of Cl^-, Br^-, SCN^-, I^-, or F^- ions for OH^- ions across biological and artificial phospholipid membranes. In media which contain chloride, the Cl^-/OH^- exchange causes clamping of pH differences across membranes with consequent shifts in the apparent pH optima of enzymes, uncoupling of electron transport phosphorylation, and enhanced ion movements.

Triorganotin compounds have several biological applications, for example, as fungicides, molluscicides, acaricides, and in marine antifouling paints. Tin compounds have the advantage over the very effective mercury compounds in that they are degraded to relatively harmless inorganic tin compounds. However, as discovered from the unintentional but disastrous use of triethyltin in a patent medicine, at least some of the triorganotin compounds are very highly toxic to mammals including man (*1*). The LD_{50} values for some of these compounds are very low, for example, about 5 mg/kg body weight for triethyltin which gives an average concentration in the body fluids of about 3 $\times 10^{-5}M$, and in-vitro effects on metabolic processes and enzymes should be manifested at these or lower triorganotin concentrations.

Electron Transfer Phosphorylation

A major advance was made by Aldridge and his collaborators who established that mitochondrial oxidative phosphorylation is extremely sensitive to inhibition by triorganotin compounds (*2, 3, 4, 5, 6*). This not only provided an explanation in biochemical terms for the very high toxicity of these compounds but also prompted their use as probes for investigating the chemistry of oxidative phosphorylation. Later work by J. S. Kahn (*7, 8*) and A. S. Watling-Payne (*9*) in our laboratory has shown that the photosynthetic phosphorylation system in chloroplasts is also very sensitive to inhibition by some triorganotin compounds. These two processes—mitochondrial and photosynthetic phosphorylation—are of fundamental importance in the energetics of living organisms and share a number of common features, not least of which is our ignorance of the chemical details of the reactions in which energy is converted from one chemical form to another (*10, 11*). In both processes the phosphorylation of ADP to form ATP is coupled to the passage of electrons along a chain of carriers, and these coupling reactions

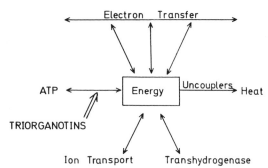

Figure 1. Schematic representation of electron transport phosphorylation and associated reactions

take place in or on the lipid membranes of vesicular organelles—mitochondria and chloroplasts. As shown in Figure 1, a variety of other processes are linked by the energy conversion system; ion transport is the most significant, and it is now clear that ion transport and uncoupling are both secondary effects of proton pumping across the coupling membranes. This proton pumping can be driven either by electron transfer or by hydrolysis of ATP, and according to Mitchell's hypothesis of chemiosmotic coupling (*12, 13*), these proton transfers are the means of coupling the two processes. The importance of proton pumping is now generally accepted although hopes still linger in some quarters that the phosphorylation and electron transfer processes may be found to be more directly coupled; little is yet known about the details of how the chemical reactions of electron transfer or ATP hydrolysis actually result in movement of protons across the membrane. One of the difficulties associated with studying electron transfer phosphorylation is that the process requires the lipid membrane in a relatively undamaged state and probably forming the enclosing surface of a vesicle. One of the approaches of classical biochemistry, the step-by-step isolation of individual enzymes catalyzing steps in the process, has not been altogether successful, partly because of the need for the vesicle membrane but also because some of the components do not have a function in terms of a chemical reaction. Consequently a great deal of use has been made of another classical biochemical approach—inhibitors or other modifiers of the biochemical processes. Before describing these it is worth noting a few features of the coupling system outlined in Figure 1. Once the energy is in the black box, it can be used for a variety of functions irrespective of its source. For example, electrons can be driven backward in an energy-requiring process, and the energy for this can be provided by hydrolysis of ATP or by energy-producing electron transfer in another region of the electron transfer chain. Even in the presence of electron donors and acceptors, the rate of electron transfer is normally limited by the rate at which phosphorylation of ADP can take place. Thus in the absence of ADP or inorganic orthophosphate, the rate of electron transfer is slow but becomes rapid when ADP and orthophosphate both are added. This dependence of the rate of electron transfer on the presence of ADP and orthophosphate is known as respiratory control (in mitochondria) or photosynthetic control (in chloroplasts). The other energy using processes such as ion transport or the transhydrogenase reaction can also allow electron transfer to proceed at a rapid rate. Uncoupling, which dissipates the energy, also allows rapid electron transfer in the absence of ADP and phosphate, and this effect is described as release of respiratory (or photosynthetic) control. Figure 1 shows a general scheme which also applies to bacterial oxidative and photosynthetic phosphorylation (14) although some features, for example, the transhydrogenase reaction, may not always be present.

Several types of action on these processes can be defined:

(a) Electron transfer inhibitors: these act directly on the components of the electron transfer chain.

(b) Inhibitors of the phosphorylation process: otherwise known as energy transfer inhibitors.

(c) Uncouplers: these release control of the electron transfer and can also allow the phosphorylation system to run freely in the hydrolytic direction, i.e., an uncoupler-induced ATPase activity.

(d) Some of the processes may require transport of substrates for oxidation or reduction or of ADP and inorganic orthophosphate across the membrane by specific carriers, and inhibitors of transport processes are also known. These transport inhibitors may appear superficially to act as inhibitors of the electron transfer or phosphorylation systems.

The coupling membranes are generally impermeable to ions including protons and hydroxide ions. Many uncouplers act by rendering the membrane permeable to protons (15) thus releasing control of electron transfer or ATP hydrolysis. A result of the electron transfer control is that inhibitors of phosphorylation will inhibit electron transfer which is accompanied by phosphorylation. They do not, however, inhibit electron transfer released by uncouplers.

All four types of compound, if added to a phosphorylating system, will cause a decrease in the amount of ATP produced. Since the sites and mechanism of actions and associated effects are different, it is important to distinguish between the four types of effect particularly as will be seen when dealing with the trialkyltins. Unfortunately many of the inhibitors which are most specific in their action are natural products often of the antibiotic type and, being complex molecules sometimes used without full knowledge of their structure, they have revealed little about the chemistry of their site of action.

Interaction with Electrolytes

The triorganotins by contrast are relatively simple molecules and offer the possibility of defining the chemistry of their site of action. However, when applied to mitochondrial or photosynthetic phosphorylation, the triorganotins can produce a variety of effects: phosphorylation of ADP is usually decreased, but this appears to be caused not only by inhibition of the phosphorylation step but also by uncoupling, inhibition of electron transfer, and inhibition of substrate transport. Insight has come from the observation in our laboratory that triorganotins mediate the exchange of chloride and hydroxide ions across many natural membranes (16). This was originally reported for mitochondrial and erythrocyte membranes and also artificial phospholipid membranes in the form of vesicles (known as liposomes) by Selwyn et al. (16) and for chloroplasts membranes by Watling and Selwyn (17). In recent experiments with D. Fulton, we have demonstrated this anion exchange across bacterial membranes (E. coli). The hydroxide transporting action of triethyltin compounds was suggested by Mitchell (18) in 1963, but the significance was not realized at the time partly because neither the importance of the proton in phosphorylation processes nor the significance of different modes of transport were generally understood (19). A carrier may transport a substance across a membrane and return empty to collect another molecule—this is termed uniport. Alternatively it may *have* to exchange one molecule for another which is called antiport; or it may *have* to take two or more molecules across together, and this is called symport. The emphasis on *have* in these definitions is important because these definitions imply *molecular* coupling of the transport of the different molecules. A uniporter which can carry two or more different molecules can effect an exchange, and if the molecules are ions or weak electrolytes, the process may appear to be tightly coupled on account of the macroscopic requirement for preserving electrical neutrality and absence of change in pH differences across the membrane.

These two macroscopic requirements can be used in investigating the nature of electrolyte transport processes. We have made particular use of two techniques, passive osmotic responses of vesicular structures (20) and pulse or step titrations also on vesicular structures.

The lipid membranes are readily permeable to water, and the external medium must contain solutes to balance the osmotic pressure of material, such as proteins and potassium ions, contained within the vesicles. If it does not, water will cross the membrane causing the vesicles to swell and possibly burst. If the solute in the external medium can also pass across the vesicle membrane, then although it will provide osmotic balance initially, it will contribute to the internal osmotic pressure as it crosses the membrane, water will follow to maintain osmotic balance, and the vesicles will swell. If the solute enters at a rate which is slow compared with the rate at which water can enter, the initial rate of swelling can be used as a measure of the rate of solute entry. Volume changes in these vesicles can be readily monitored by recording changes in light scattering by their suspensions—the light scattering decreasing as the vesicles swell.

When the solute is an electrolyte, failure to observe swelling does not mean that all the ions present are unable to cross the membrane. To observe extensive swelling, the two conditions mentioned above must be met. First, the overall transport processes must be electrically neutral as, owing to the minute electrical capacity of the vesicle membrane, a small charge imbalance produced by transport of an ion produces a large potential difference across the membrane and this prevents further entry of ions carrying charge of the same sign. Secondly, where weak acids or bases form part of the osmotic activity of the medium, the transport processes must produce no net movement of protons or hydroxyl

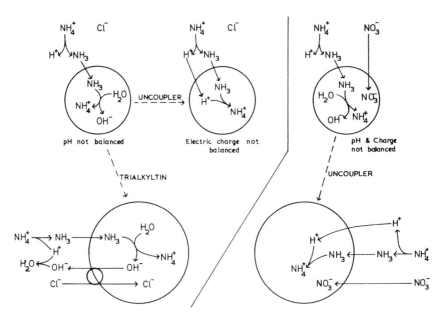

Figure 2. Requirement for electrical charge and pH balance in swelling of vesicles in electrolyte solutions. Conditions which allow extensive swelling are indicated by the larger circles.

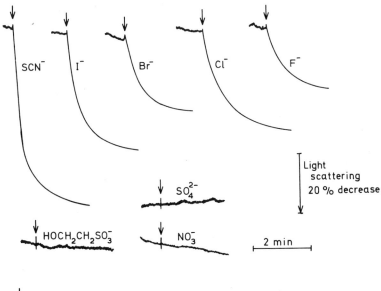

↓ addition of   tributyltin chloride   (µM)

Figure 3. Specificity of anion–hydroxide exchange mediated by tributyltin chloride acting on chloroplasts suspended in various ammonium salts. Chloroplasts from pea plants were suspended at 5 µg chlorophyll/ml in 100mM solutions of ammonium salts, adjusted to pH 7.7. Narrow angle scattering was measured as discribed by Watling and Selwyn (17). A decrease in light scattering corresponds to an increase in chloroplast volume.

ions; if such movement does occur, a pH difference develops across the membrane, and the different degree of ionization of weak electroyltes on the two sides of the membrane causes cessation of transport. These limitations are diagramed in Figure 2.

With electrolytes the rate of swelling may indicate the rate of entry of the slower ion (or weak acid or weak base), or the rate of proton (or hydroxyl ion) conduction across the membrane. When two or more of the permeant substances have similar rates of entry, the rate of swelling will be a function of both rates. Using different salts and different ionophores of known function enables one to determine the permeability properties of particular ions or weak electrolytes and the mode of action of another ionophore.

Using this technique with ammonium salts, we have shown that the triorganotins mediate an exchange of SCN^-, I^-, Cl^-, Br^-, or F^- ions for hydroxide ions across natural membranes or phospholipid liposome membranes. Figure 3 shows the relative effectiveness of tributyltin cations' acting on chloroplasts with these anions, and it also shows that NO_3^-, SO_4^{2-}, and isethionate (hydroxyethyl sulfonate) do not participate in this exchange.

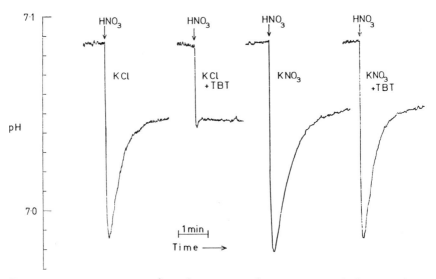

Figure 4. Step titration of erythrocytes in the presence and absence of tri-butyltin chloride in KCl and KNO₃ media. pH recording of a suspension of erythrocytes, .05 ml packed cells, in 4.0 ml of 0.1M KNO₃ or KCl containing 0.1mM Diamox (to inhibit carbonic anhydrase). A small change in pH was produced by rapid injection of 50 μmole HNO₃. Tributyltin when added was 5μM. Other conditions as described in Selwyn et al. (16).

Using a similar technique with liposomes containing a KCl solution we have shown that both valinomycin and an uncoupler (proton uniporter) are required as well as a triorganotin to allow electroneutral movement of KCl across the membrane (*16*). This shows that exchange of anions is compulsory, and the trialkyltin compounds cannot act as either a chloride or hydroxide uniporter.

The pulse- or step titration (*16, 21*) in which a small finite quantity of acid or alkali is added to a suspension of vesicles also demonstrates the need for an appropriate anion for the net movement of acid across the membrane. In the absence of trialkyltin compounds, added acid initially titrates only the external buffer, and the pH drops to a low value almost instantaneously (Figure 4); subsequently the pH rises as acid passes across the membrane, and the internal buffer capacity becomes effective. In a KCl medium tributyltin cations allow very rapid titration of the internal buffer, and very little overshoot is observed, but in a KNO₃ medium tributyltin cations have virtually no effect (Figure 4).

Effect on Proton-Translocating ATPases

With a knowledge of the role of chloride ions in the triorganotin-mediated movement of hydroxide ions, the experimental conditions can be adapted to eliminate this activity. Chloride ions are normally used to maintain osmotic balance and prevent damage to vesicular organelles since in many cases they constitute a large part of the natural osmotic balance; they can be replaced with nitrate, sulfate, or isethionate ions, or the osmotic balance can be maintained by a nonelectrolyte such as sucrose.   Buffers and components can also be chosen to eliminate chloride or other active ions.

When this is done, the effect of low concentrations of triorganotin compounds on the mitochondrial (22, 23, 24, 25) or chloroplast (9) systems is a simple inhibition of the phosphorylation process.   Figure 5 shows this action as observed by the inhibition of controlled electron transport during photophosphorylation by isolated chloroplasts.   Incorporation of radioactively labelled inorganic phosphate into ATP shows a similar decrease in synthesis of ATP.   Figure 6 shows that when chloride is carefully excluded, triethyltin cations have little or no effect on the rate of electron flow in chloroplasts in the absence of ADP and P_i.   When a low concentration of Cl^- (6.6 mM) is present, a marked enhancement of the rate is observed, and this release of photosynthetic control is caused by an uncoupling action that triethyltin cations have in the presence of Cl^- ions.   Another method for observing the inhibition by triorganotins is the uncoupler-stimulated ATPase activity.   Figure 7 shows the effect of trimethyltin chloride on the

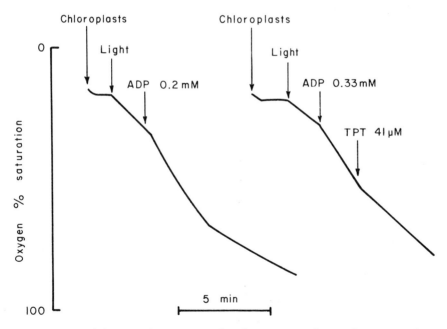

Figure 5.   Inhibition of ADP-stimulated oxygen evolution by tripropyltin chloride (TPT).   Broken chloroplasts at a concentration of 60 μg of chlorophyll/ml were suspended in 300mM sorbitol, 25mM Hepes adjusted to pH 7.6 with NaOH, 2.2mM $K_3Fe(CN)_6$, 3.3mM $MgCl_2$, 10mM KH_2PO_4/KOH (pH 7.6) at 18°C in a total of 4.5 ml.   (Left:) Photosynthetic control of the rate of evolved O_2; (Right:) inhibition by tripropyltin of ADP-stimulated O_2 evolution. (Adapted from Biochemical Journal, Ref. 9.)

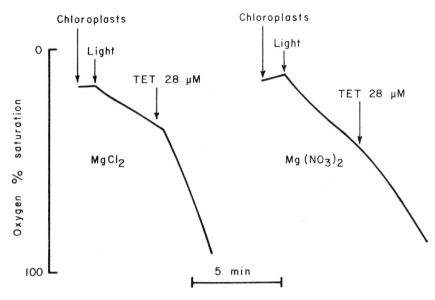

Figure 6.   Effect of anions on the stimulation of basal oxygen evolution by triethyltin sulfate (TET). Broken chloroplasts at a concentration of 60 μg of chlorophyll/ml were suspended in 300mM sorbitol, 25mM Hepes adjusted to pH 7.6 with NaOH, 4.4mM $K_3Fe(CN)_6$ at 18°C in a total of 4.5 ml. (Left:) plus 3.3mM $MgCl_2$ (Right:) 3.3mM $Mg(NO_3)_2$. (Adapted from Biochemical Journal, Ref. 9.)

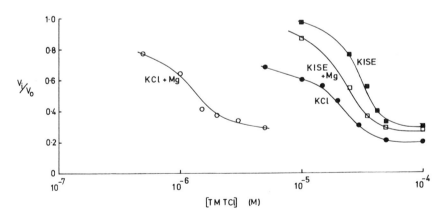

Figure 7.   Inhibition of uncoupler-stimulated ATPase of rat liver mitochondria by trimethyltin chloride in potassium chloride and potassium isethionate media, in the presence and absence of added Mg^{2+} ions. Rat liver mitochondria, 7–12 mg protein, were suspended in 10 ml 0.1M KCl or potassium isethionate containing 2 μg rotenone, 2.25mM ATP, and 0.1mM 2,4-dinitrophenol (uncoupler). $MgSO_4$ where added was 1.25mM. ATPase was estimated by measuring acid production with a Radiometer pH Stat holding the pH at 6.8.

2,4-dinitrophenol-stimulated ATPase of whole mitochondria (*26*). In a potassium isethionate medium, high concentrations of trimethyltin compounds are required, and this corresponds to the poor ability of the trimethyltin species to inhibit the phosphorylation process directly; the very low concentrations of trimethyltin compounds required to inhibit the ATPase activity in the presence of Mg^{2+} ions in a KCl medium should be noted.   This is thought to be an effect of the collapse of the pH gradient across the mitochondrial membrane and the requirement of the adenine nucleotide carrier for free ATP (not the ATP–Mg^{2+} complex).   The

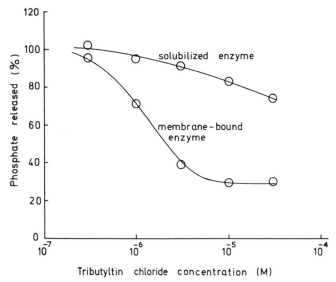

Figure 8.   Inhibition of the Mg^{2+}-activated membrane ATPase from E. coli by tributyltin chloride.   Membrane-bound ATPase was prepared by disrupting E. coli in a French press and collecting the membrane fraction by differential centrifugation.   Soluble ATPase was prepared by treating the membrane-bound ATPase with 2mM EDTA (pH 7.3) and separating the membranes by centrifugation.   ATPase was estimated by liberation of inorganic phosphate at 30°C in the following medium: 50mM tris-HCl (pH 7.8), 3.5mM ATP, and 1.75mM $MgSO_4$.

effect of triorganotins on the ATPase activity of sub-mitochondrial particles in which the ATPase component is exposed to the suspending medium is affected little by the presence or absence of chloride ions.   When the ATPase is dissociated from the membrane, it is no longer sensitive to inhibition by the triorganotins (*26, 27, 28*).

There are similar ATPases in chloroplasts (*29*) and bacteria (*30*) although the mitochondrial complex has been studied most.   Both appear to be multi-component enzyme systems comprising a surface component, F_1 in mitochondria, CF_1 in chloroplasts, and BF_1 in bacteria which can be detached from other components, some of which are very firmly imbedded in the membrane. Figure 8 shows that the bacterial membrane ATPase is inhibited by low concentrations of tributyltin compounds only when the ATPase is bound to the membrane. A similar dependence on the membrane component has been shown for the chloroplast ATPase by Gould (31). Information about the role of the intrinsic membrane components has come from work on sub-mitochondrial particles from which the F_1 component of the ATPase has been removed. Such particles have very poor P/O ratios and exhibit little or no respiratory control, but their performance is improved by the addition of oligomycin, an energy-transfer inhibitor which, like the trialkyltins, inhibits the membrane-bound but not the solubilized mitochondrial ATPase. An explanation for these effects has been given by Mitchell and Moyle (32) who showed that when the F_1 component is removed, the membrane components of the ATPase act as proton conductors through the membrane and that oligomycin blocks this proton conduction. Similar results indicating blockage of proton movement through the membrane-bound components of the ATPase have been reported for the mitochondrial system (33) and

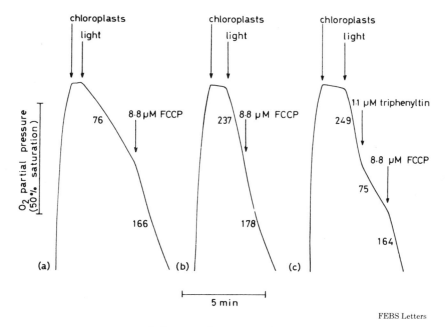

FEBS Letters

Figure 9.   Restoration of photosynthetic control to CF_1-deficient chloroplasts by triphenyltin chloride (36). CF_1 was removed by treating with EDTA. Oxygen evolution from EDTA-uncoupled and non-treated chloroplasts. Chloroplasts at a concentration of 60 µg chlorophyll/ml were added to a medium, deoxygenated by bubbling with N_2, containing 300mM sorbitol, 12.5mM Hepes adjusted to pH 7.6 with NaOH, 4.4mM $K_3Fe(CN)_6$, and 3.3mM $MgCl_2$ at 18°C in a total of 4.5 ml. (a) Non-treated chloroplasts (b) and (c) EDTA-uncoupled chloroplasts. Numbers along the traces refer to rates of O_2 evolution expressed as µg-atoms of oxygen/hr per mg chlorophyll.

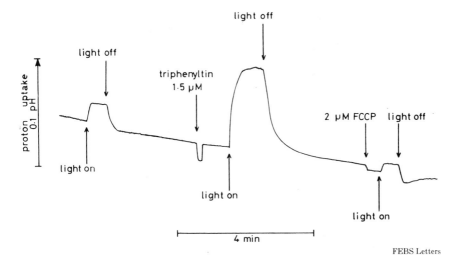

Figure 10.   Proton uptake in EDTA-uncoupled chloroplasts during periods of actinic illumination (36).   Chloroplasts were suspended at 60 μg chlorophyll/ml in 150mM sorbitol, 5mM KNO_3, 2mM $Mg(NO_3)_2$, and 20μM phenazine methosulfate at 17°C in a total of 5.0 ml.   The pH of the suspending medium was adjusted to pH 7.5 with dilute HNO_3.

chloroplasts (34) using the inhibitor dicyclohexylcarbodiimide.   Dawson and Selwyn (35) have shown that tributyltin compounds also block the proton permeability of F_1 ATPase component-deficient, sub-mitochondrial particles. Watling-Payne and Selwyn (36) have shown that triphenyltin chloride has a similar effect on chloroplasts which have been treated with EDTA to detach the CF_1 ATPase component.   Figure 9 shows the restoration by triphenyltin chloride of a pseudo-control of photosynthetic electron transfer and its release by the proton-conducting uncoupler carbonyl cyanide *p*-trifluoromethoxyphenylhydrazone (FCCP).   Another mode of demonstrating this is to observe the pH imbalance across the chloroplast membrane produced by illumination in the absence of ADP and P_i.   In this system, EDTA-treated chloroplasts show a very poor response and a very rapid decay of proton uptake when the light is switched off.   In the presence of triphenyltin chloride, the steady-state level of H^+ uptake is much greater, and the rate of decay is slower (Figure 10) corresponding to the reduced H^+ permeability of the membrane.

The action of triorganotins on the proton-translocating ATPases is illustrated schematically in Figure 11.   A precise significance should not be placed on the number and arrangement of the components, and the proton movement through the membrane components should not necessarily be interpreted to mean the existence of a channel or well such as that described by Mitchell and Moyle (32). Alternative modes of proton conduction are some form of rotating or shuttle-type carrier or even a chain of bound protons which can hop from one acceptor group to another in a Grotthuss type of mechanism.   When the F_1 ATPase and pro-

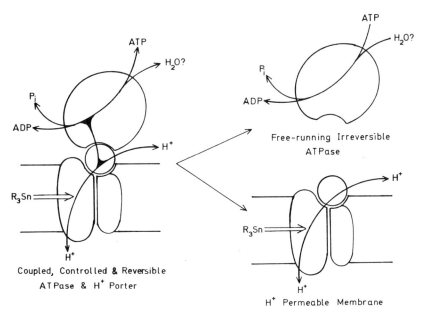

Figure 11.   Caricature of the functional and structural relationship between ATPase and H^+-porter components of H^+-translocating ATPases

ton-conducting components are joined together, the two functions are tightly coupled; ATP hydrolysis does not occur unless protons can be pumped through the membrane and proton movement is restricted unless it is driving or being driven by the ATPase reaction.   The precise mode of coupling is as yet unknown. It is possible that the protons moving via the membrane components actually take part in the ATPase reaction, and when the F_1 ATPase component is detached from the membrane, the hydrolytic site is freely accessible to protons from the surrounding water.   Alternatively the two processes may be coupled by protein interactions and conformation changes.   Boyer (37) has proposed that conformation changes occur in the F_1 ATPase reaction and that these could well be transmitted to or reversed by conformation changes in the proton conducting components.   The triorganotins are of particular significance in emphasizing the similarity of these proton-translocating ATPases since they are effective in all three types.

     In the absence of chloride and other anions which are active in the triorganotin-mediated anion-hydroxide exchange, the effects of low concentrations of the triorganotins on mitochondrial and chloroplast metabolism are simple.   At higher concentrations, direct inhibition of, for example, electron transfer in chloroplasts (7, 9) or the adenine nucleotide transporter in mitochondria (38) may occur, but these effects are probably not very important in the toxicity of the triorganotins and can often be avoided in in vitro experiments.

pH Differential

In media containing appreciable concentrations of chloride ions, such as blood plasma, sea water, Ringer's solution, or the simple KCl and NaCl solutions used with isolated mitochondria and chloroplasts, the chloride–hydroxide exchange has the effect of acting as a pH clamp across biological membranes. To a first approximation it can be assumed that it will remove and prevent the generation of a pH difference across the membrane, but two effects can result in the existence of a pH difference. Under equilibrium conditions the chloride–hydroxide exchange will result in the following equality:

$$[Cl^-]_R[OH^-]_L = [Cl^-]_L[OH^-]_R$$

or

$$[Cl^-]_R[H^+]_R = [Cl^-]_L[H^+]_L$$

(where the subscripts R and L refer to the two sides of the membrane). Although this relationship is algebraically similar to that given by the Donnan equilibrium, there is an important difference. The Donnan equilibrium assumes that the ions move by independent uniport processes which are linked by the electrical potential difference across the membrane, and it assumes the need for macroscopic charge balance. The concentrations of both ions are, therefore, related to this electrical potential difference in the Donnan equilibrium, but in the case of an electrically neutral, molecularly coupled antiport, the ion concentrations are independent of the electrical potential difference across the membrane. Much of the anionic charge inside natural vesicular organelles is provided by organic groups, and there may well be a lower concentration of chloride inside the vesicle than outside. Under such conditions the chloride–hydroxide exchange will tend to clamp the pH difference not at zero but at a fixed difference with a lower pH inside the vesicle. A static equilibrium of this type can be produced by extremely low concentrations of the triorganotins.

The other situation in which a pH differential can exist is when there is another process actively pumping protons or hydroxide ions across the membrane. Under these conditions a steady-state pH difference will be set up, the magnitude of which will depend on the relative rate constants for the pumping process and the triorganotin Cl^-/OH^- exchange. Since the latter is proportional to the organotin concentration, reduction of this type of pH difference can be brought about by increasing the organotin concentration.

One result of the clamping of the pH difference across the mitochondrial membrane is the apparent direct inhibition of electron transfer, i.e., of uncoupler-stimulated respiration, by low concentrations of trialkyltins in chloride media (3, 39). Coleman and Palmer (23) found that this was apparent only below pH 7 and Dawson and Selwyn (25) showed that the effect is caused by a shift in the apparent pH optimum (apparent because the pH is measured in the external medium not inside the vesicle) of the intramitochondrial respiratory system, and

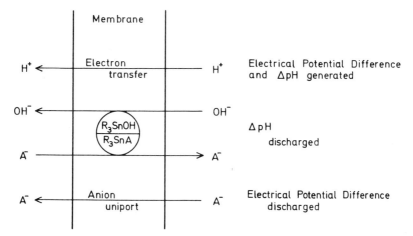

Figure 12. Mechanism of uncoupling by triorganotins showing the two-step nature of the process: neutralization of pH difference by anion–hydroxide exchange and neutralization of electrical potential difference by uniport of anion.

at high pH values the trialkyltins stimulate uncoupler-released respiration. This effect resembles the action on uncoupler-stimulated ATPase (Figure 7) in being produced by very low concentrations of trialkyltins.

Uncoupling requires higher concentrations of trialkyltins and depends on the natural permeability of the anion (Figure 12). The reasons for this are that the electron transfer is continually pumping protons across the membrane by a uniport type of process, and this rate will depend on the rate at which the chloride–hydroxide exchange can transport hydroxide ions in the same direction. The uniport-type pumping of protons produces not only a pH difference but also an electrical charge imbalance which is manifested as an electrostatic potential difference across the membrane. The chloride–hydroxide exchange being electrically neutral discharges only the pH component and also leads to the accumulation of chloride in mitochondria. Uncoupling of mitochondria by triorganotins in a chloride medium is poor, but if chloride is replaced by thiocyanate, uncoupling is effective because the thiocyanate can cross the membrane by its intrinsic uniport-type permeability thereby discharging the electrical potential difference (Figure 12). Chloroplasts are sufficiently permeable to chloride for the triethyltin species to act as an effective uncoupler in a chloride medium, but with other triorganotins the uncoupling effect is largely masked by their inhibitory effects (9).

When the electrical imbalance can be neutralized by a cation moving in the opposite direction to the pumped protons the anion hydroxide exchange can increase the movement of cations. For example a massive uptake of Ca^{2+} ions by mitochondria in a KCl medium in the presence of tripropyltin chloride was the observation that led to the discovery of the chloride hydroxide exchange (22), and the presence of trialkyltins permits extensive light driven shrinkage of chloroplasts in an NaCl medium (*17*).

Cation-Translocating ATPases

Certain triorganotins have been reported to inhibit the Na^+-K^+ translocating ATPases of cell membranes at low concentrations (*4, 40, 41, 42*), and in our laboratory M. G. Rhodes has extensively studied the inhibition of the Na^+-K^+ ATPases of membranes from human erythrocytes and from ox brain. The inhibition of the Na^+-K^+ ATPase activity of a partially purified preparation from ox brain is shown in Figure 13. The affinities of the triorganotins for the enzyme are generally lower than for the proton translocating ATPases, but they are still very high for the more hydrophobic triorganotin compounds. The relative in-

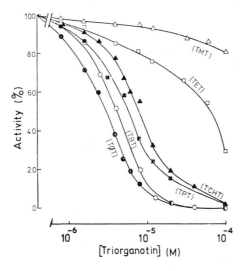

Figure 13. Inhibition of the Na^+-K^+ ATPase from bovine brain by triorganotins. Enzyme was prepared from homogenized brains by differential centrifugation and treatment with NaI. The preparations had specific activities of 70–130 μmoles P_i/mg protein/hr. 20–30 μg enzyme protein were incubated for 10 min following a 10-min pre-incubation with inhibitor in 1 ml of a medium containing: 1mM EGTA, 5mM $MgCl_2$, 75mM tris-HCl (pH 7.5), 20mM KCl, 100mM NaCl, and 2mM ATP. ATPase activity was measured by estimating inorganic orthophosphate.

effectiveness of triethyltin and trimethyltin compounds suggests that the hy-drophobic interaction is very important in this action. This enzyme has a phosphoesterase activity which also is inhibited by triorganotins although under some conditions a small stimulation of p-nitrophenol phosphatase activity by triorganotins at low concentrations has been observed. The inhibitory effect is not instantaneous but develops with a half-time of about 1 min (Figure 14). Measurement of the amount of the triphenyltin moiety bound to the preparation suggests that at virtually complete inhibition about 60 molecules of the tri-phenyltin moiety are bound per active ATPase complex. However, the number actually required to inhibit the ATPase may be much smaller since some may be bound to inactive enzyme, to non-specific sites, or dissolved in the membrane lipid.

The generality of inhibition of membrane-bound ATPase by triorganotins prompted us to test their action on the Ca^{2+}-translocating ATPase of sarcoplasmic reticulum. Figure 15 shows that tributyltin chloride at about $10\mu M$ produces considerable inhibition of the ATPase activity of this enzyme. Parallel experi-

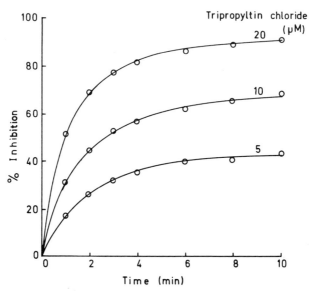

Figure 14. Time-dependence of inhibition of the Na^+-K^+ ATPase from bovine brain by tripropyltin chloride. Experimental conditions were as in Figure 13 with the addition of: 2mM phosphoenolpyruvate, 0.4mM NAOH, 2.0 units/ml pyruvate kinase, and 3.85 units/ml lactate dehydrogenase. ATPase activity was followed continuously by recording oxidation of NADH and the progressive inhibition estimated from these curves.

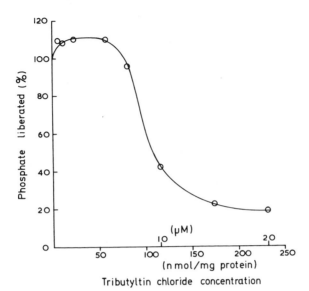

Figure 15.   Inhibition of the Ca^{2+} ion translocating ATPase of sarcoplasmic reticulum by tributyltin chloride.   Sarcoplasmic reticulum vesicles were prepared from rabbit muscle.   86 μg of enzyme protein was incubated for 10 min in 50mM tris-NO_3 (pH 7.2), 100mM KNO_3, 5mM $Mg(NO_3)_2$, 25μM $CaCl_2$, and 5mM ATP, total volume 1.0 ml at 30°C. ATPase was estimated by measuring inorganic phosphate liberated.

ments showed that Ca^{2+} transport was also inhibited.   Triphenyltin and tributyltin compounds are the most effective of the organotins that we have tried, and as with the Na^+-K^+ ATPase, inhibition takes measurable time to develop, the half-time being about 5 min.   Measurements of the binding of the organotin showed many molecules bound per molecule of enzyme.   These characteristics are similar to those of the inhibition of the Na^+-K^+ ATPase, and it is suggested that in both these metallic cation translocating ATPases the triorganotins may not act at specific sites on the protein molecules but rather by substituting for some of the lipid molecules which have a strong interaction with the protein and which are known to be important in the activity of the enzyme (43).

Properties of Anion–Hydroxide Exchange

The anion–hydroxide exchange has considerable interest, not only intrinsically and in relation to its biological effects but also because it constitutes a simple model for natural carriers.   In our study of the properties of the anion exchange we have made extensive use of the artificial phospholipid vesicles known as

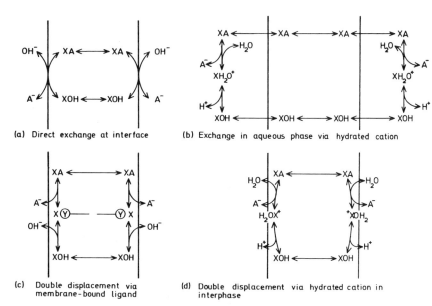

(a) Direct exchange at interface

(b) Exchange in aqueous phase via hydrated cation

(c) Double displacement via membrane-bound ligand

(d) Double displacement via hydrated cation in interphase

Figure 16.   Proposed mechanisms for the anion/hydroxide exchange mediated by triorganotins

smectic mesophases or liposomes (16, 44). These have bilipid layer membranes which resemble natural biological membranes in several respects and, being vesicular, osmotic effects can be observed by changes in light scattering. These preparations have the advantages over natural vesicles in that the composition of the membrane and the internal aqueous phase can be controlled; furthermore, the possibility that the observed effects are caused by natural carriers or modification of natural carriers is eliminated. The activity of the triorganotins in these liposomes therefore shows that it is the triorganotins which are acting as ionophores.

In an aqueous medium we cannot distinguish, on kinetic data, between Cl^-/OH^- antiport and Cl^-/H^+ symport. However, the chemical properties of the triorganotins include the existence of covalent non-polar solvent-soluble chlorides and hydroxides but not complexes with both Cl^- and H^+ (16, 45). Complexes with ammonia are known and may occur in the experiments with ammonium salt solutions, but such complexes are not an essential feature of the anion exchange since the exchange can be demonstrated in ammonia-free media.

Selwyn et al. (16) postulated a shuttle-type carrier, Figure 16(a), in which the triorganotin moiety, represented by X, moves across the membrane, and while remaining in the membrane exchanges anions at the lipid–water interface. We can consider a modification of this scheme in which double displacement reactions occur at the interface; the triorganotin compound binding to a nucleophilic group in the membrane surface, Figure 16(c), or forming an aquo–cation which remains trapped at the interface, Figure 16(d). Although some triorganotins are very highly partitioned into hydrophobic phases, others such as triethyltin and trimethyltin species have appreciable concentrations in the aqueous phases; since the mass of the aqueous phase is always very large compared with the mass of membrane material, the total amount of triorganotin species in the aqueous phase is always appreciable. To allow for this and to take into account the work of Prince (46) who suggested that formation of trialkyltin hydroxides from the halides takes place via the aquo–cation, we propose the scheme shown in Figure 16(b). The rate equations for these mechanisms are complicated but one feature which is clear is that in the mechanisms (a, c, and d) in which the organotin remains in the membrane (if other factors remain unchanged), the effectiveness should increase with increasing partition into the lipid membrane phase. In mechanism b the rate equation confirms what examination of the mechanism suggests; that too high a partition into the lipid phase will reduce the effectiveness (other factors remaining constant) of the ionophore by reducing the mass in the aqueous phases.

Table I. Survey of Biochemical Activities of Triorganotins

Partition into Organic Solvents: (Partition Coefficient)

TCHT	TØT	TBT	TPT	TET	TMT	
—	2000	300	—	10	0.15	chloroform[a]
20000	12000	1300	52	3.7	0.5	n-octanol[b]

Effectiveness in Cl⁻/OH⁻ Exchange: (Relative Concentration)

TBT	TPT	TØT = TET	TCHT	TMT	
1	1.2	10 10	20	51	liposomes in NH_4Cl[a,c]

TBT	TPT	TCHT	TPT	TØT	TET	TMT	
1		5	7	11	100	1500	erythrocytes in NH_4Cl[a,c]

TBT	TPT	TØT	TET	TMT	
1	5	9	115	480	chloroplasts in NH_4Cl[d]

TPT	TBT	TØT	TMT		
1	3	11	540		chloroplasts in NaCl[d] (light driven shrinkage)[d]

TET	TPT	TBT	TØT	TMT	
1	2	4	8	32	mitochondria in NaCl[e]

Effectiveness as Inhibitors: (Relative Concentration)

TBT	TPT	TØT	TET	TMT		
1	1.1	3	4	350		mitochondrial phosphorylation[e]
TBT	TØT	TPT	TMT	TET		
1	2	40	650	2100		chloroplast phosphorylation[d]
TØT	TBT	TPT	TCHT	TET	TMT	
1	1.4	1.9	2.6	~10	~100	Na⁺–K⁺ ATPase bovine brain[f]
TBT	TØT = TPT	TET	TMT			
1	2 2	8	20			Ca²⁺ ATPase of sarcoplasmic reticulum[a,g]

[a] Unpublished data, this laboratory.
[b] Ref. 47.
[c] Ref. 16.
[d] Ref. 48.

[e] Refs. 16, 22, and 25.
[f] From Figure 13.
[g] See Figure 15.

The partition coefficients of triorganotins into chloroform and *n*-octanol
(47) are given in Table I together with data on the effectiveness of the triorganotins in the chloride–hydroxide exchange in liposomes. The data are certainly
in accord with the hypothesis that too high a partition coefficient, as in the triphenyltin and tricyclohexyltin species, reduces the effectiveness, but other ex-
planations may account for the poor performance of these two compounds. Both
have large organic groups which may slow ligand exchange and/or movement
through the membrane; tricylcohexyltin compounds have a low solubility and
triphenyltin hydroxide may have a tendency to dehydrate to the oxide.

It is possible that more than one of the mechanisms shown in Figure 16 may
be operative but the original suggestion, Figure 16(a), now seems oversimplified
and Figure 16(b) is the most plausible mechanism. One feature to be expected
from all these mechanisms is that the exchange should exhibit an optimum pH.
On the low side of the optimum, formation of the triorganotin hydroxide is decreased while on the high side of the optimum formation of the triorganotin halide
is decreased. The presence of an optimum pH is shown in Figure 17, and it can
be seen that the optimum pH is dependent on the anion being exchanged. This
is to be expected since the overall rate depends on the balance between formation
and breakdown of both the triorganotin halide and hydroxide. An anion with
a high affinity for the triorganotin cation will be expected to exhibit a high pH

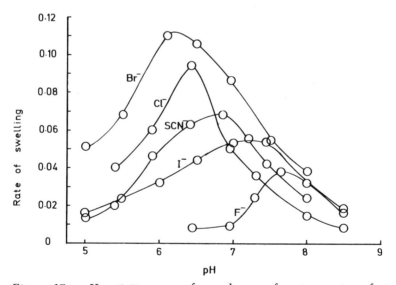

*Figure 17.   pH activity curves for exchange of various anions for
hydroxide across liposome membranes mediated by triethyltin sulfate.
Liposomes were prepared as described in Selwyn et al. (16) and suspended in 100mM ammonium salts of the anions shown on the figure.
Swelling was measured by light scattering changes (16).   Triethyltin
was added to a final concentration of 1.25μM except with NH₄F when
a triethyltin concentration of 3.75μM was used.*

optimum. The low rate even at the optimum pH with fluoride ions (note the higher concentration of triethyltin, Figure 17) suggests an additional factor, presumably a low velocity constant for dissociation of fluoride ions as well as a high equilibrium formation constant indicated by the high pH optimum.

The order of effectiveness of the trioganotin compounds as ionophores depends to a certain extent on which membrane preparation is used, Table I. The different relative effectiveness of tricyclohexyltin chloride under different conditions may be a result of its low solubility, and with all compounds the effectiveness will depend on the pH and lipid composition of the membrane. With chloroplasts and mitochondria the relatively greater effectiveness of triethyltin and tripropyltin compounds is thought to reflect the greater binding of tributyltin compounds to the ATPase complex and its immobilization by this binding.

Table I also shows the relative effectiveness of the triorganotins in the inhibition of ion translocating ATPases and the phsophorylation processes. A hydrophobic site of action is indicated for all these enzymes. Although the absolute concentrations required to inhibit the Na^+-K^+ ATPase and the Ca^{2+} ATPase are higher than those required to inhibit the phosphorylation processes, they are still low enough to be important in the toxicity of these compounds.

Acknowledgment

It is a pleasure to acknowledge the collaboration of colleagues and graduate students in this research, particularly, Alan Dawson, Angela Watling, Michael Rhodes, Derek Fulton, and Barry Farrow.

Literature Cited

1. Barnes, J. M., Magos, L., *Organometal. Chem. Rev.* (1968) 3, 137–150.
2. Aldridge, W. N., Cremer, J. E., *Biochem. J.* (1955) 61, 406–418.
3. Aldridge, W. N., *Biochem. J.* (1958) 69, 367–376.
4. Aldridge, W. N., Street, B. W., *Biochem. J.* (1964) 91, 287–297.
5. Aldridge, W. N., Rose, M. S., *FEBS Lett.* (1969) 4, 61–68.
6. Sone, N., Hagihara, B., *J. Biochem.* (1964) 56, 151–156.
7. Kahn, J. S., *Biochim. Biophys. Acta* (1968) 153, 203–210.
8. Kahn, J. S., *Biochem. J.* (1970) 116, 55–60.
9. Watling-Payne, A. S., Selwyn, M. J., *Biochem. J.* (1974) 142, 65–74.
10. Baltscheffsky, H., Baltscheffsky, M., *Ann. Rev. Biochem.* (1974) 43, 871–897.
11. Dawson, A. P., Selwyn, M. J., "Companion to Biochemistry," A. T. Bull, J. R. Lagnado, J. O. Thomas, K. F. Tipton Eds., pp. 553–586, Longman, London, 1974.
12. Mitchell, P., "Chemiosmotic Coupling and Energy Transduction," Glynn Research Ltd., Bodmin, 1968.
13. Greville, G. D., *Curr. Topics Bioenerget.* (1969) 3, 1–78.
14. Harold, F. M., *Bacteriol. Rev.* (1972) 36, 172–230.
15. Hopfer, U., Lehninger, A. L., Lennarz, W. J., *J. Membrane Biol.* (1970) 3, 142–155.
16. Selwyn, M. J., Dawson, A. P., Stockdale, M., Gains, N., *Eur. J. Biochem.* (1970) 14, 120–126.
17. Watling-Payne, A. S., Selwyn, M. J., *FEBS Lett.* (1970) 10, 139–142.
18. Mitchell, P., *Biochem. Soc. Symp.* (1963) 22, 142–168.

226   ORGANOTIN COMPOUNDS: NEW CHEMISTRY AND APPLICATIONS

19. Mitchell, P., Adv. Enzymol. (1967) 29, 33–87.
20. Chappell, J. B., Haarhoff, K. N., "Biochemistry of Mitochondria," E. C. Slater, Z. Kaniuga, L. Wojtczak, Eds., pp. 75–91, Academic, London and New York, 1967.
21. Mitchell, P., Moyle, J., Biochem. J. (1967) 104, 588–600.
22. Stockdale, M., Dawson, A. P., Selwyn, M. J., Eur. J. Biochem. (1970) 15, 342–351.
23. Coleman, J. O. D., Palmer, J. M., Biochim. Biophys. Acta (1971) 245, 313–320.
24. Rose, M. S., Aldridge, W. N., Biochem. J. (1972) 127, 51–59.
25. Dawson, A. P., Selwyn, M. J., Biochem. J. (1974) 138, 349–357.
26. Selwyn, M. J., Dunnett, S. J., Philo, R. D., Dawson, A. P., Biochem. J. (1972) 127, 66–67P.
27. Kagawa, Y., Racker, E., J. Biol. Chem. (1966) 241, 2461–2466.
28. Byington, K. H., Biochem. Biophys. Res. Commun. (1971) 42, 16–22.
29. Leech, R. M., Biochem. Soc. Spec. Publ. (1974) 4, 19–25.
30. West, I. C., Biochem. Soc. Spec. Publ. (1974) 4, 27–38.
31. Gould, J. M., Eur. J. Biochem. (1976) 62, 567–575.
32. Mitchell, P., Moyle, J., Biochem. Soc. Spec. Publ. (1974) 4, 91–111.
33. Beechey, R. B., Biochem. Soc. Spec. Publ. (1974) 4, 41–62.
34. Uribe, E. G., Biochem. (1972) 11, 4228–4235.
35. Dawson, A. P., Selwyn, M. J., Biochem. J. (1975) 152, 333–339.
36. Watling-Payne, A. S., Selwyn, M. J., FEBS Lett. (1975) 58, 57–61.
37. Boyer, P. D., Wolcott, R. W., Yuthavong, Y., Degani, C., Biochem. Soc. Trans. (1974) 2, 27–30.
38. Harris, E. J., Bangham, J. A., Zukovic, B., FEBS Lett. (1973) 29, 339–344.
39. Manger, J. R., FEBS Lett., (1969) 5, 331–334.
40. Minakami, S., Saito, T., Kume, S., Yoshikawa, H., J. Biochem. (1965) 57, 221–222.
41. Matsui, H., Schwartz, A., Biochim. Biophys. Acta (1966) 128, 380–390.
42. Schwartz, A., Moore, C. A., Am. J. Physiol. (1968) 214, 1163–1167.
43. Warren, G. B., Toon, P. A., Birdsall, N. J. M., Lee, A. G., Metcalfe, J. C., Proc. Nat. Acad. Sci. USA (1974) 71, 622–626.
44. Bangham, A. D., De Gier, J., Greville, G. D., Chem. Phys. Lipids (1967) 1, 225–246.
45. Ingham, R. K., Rosenberg, S. D., Gilman, H., Chem. Rev. (1960) 60, 459–539.
46. Prince, R. H., J. Chem. Soc. (1959) 1783–1790.
47. Wulf, R. G., Byington, K. H., Arch. Biochem. Biophys. (1975) 167, 176–185.
48. Watling, A. S., Doctoral Thesis, University of East Anglia (1972).

RECEIVED May 3, 1976. Work supported by the Science Research Council and the Dow Chemical Co. Ltd.

16

Photoelectron Spectroscopy as a Tool in Organotin Chemistry

Y. LIMOUZIN AND J. C. MAIRE

Universite de Droit, d'Economie et des Sciences, Laboratoire des Organometalliques, Faculte des Sciences et Techniques, Saint-Jerome, 13397 Marseille Cedex 4, France

This paper reviews published work on x-ray and UV–photo-electron spectroscopy as applied to physicochemical problems in organotin chemistry. Chemical shift sensitivity of tin core levels is examined. The application of x-ray–PES to problems of charge distribution, coordination changes, and oxidation states are reviewed, and attention is given to the combination of x-ray–PES and Mössbauer spectroscopy. New results are reported on complexation of amino acids with Bu_3SnCl in relation to the toxicity of organotin compounds. UV–PES has been used to study ring strain in cyclostannanes, and it has been established that the C—Sn—C ring angle is the largest in the six-membered dimethylstannacyclohexane ring. The question of $p\pi$–$d\pi$ bonding in aromatic organotin compounds has been examined using a combination of UV–PES and nuclear quadrupole resonance. A correlation exists between the first ionization potential and the toxicity of tetraalkyltins. All published data on experimental binding energies and UV–photo-toelectron spectroscopy of tin compounds are listed in the Appendix.

The measurement of core-electron binding energies by ESCA spectroscopy can provide information about the nature of the bonds around the atom studied, and particularly its coordination number or oxidation state and the charge distribution in the molecule or the complex to which this atom is attached. Although many different compounds have been studied, the exact relationship between core-electron binding energy and chemical environment is still uncertain, especially for fourth group elements because of difficulties with surface charging, low sensitivity, and the lack of appropriate theoretical methods to handle a heavy atom such as tin.

The problem of sensitivity has been discussed by Grutsch and coworkers (1) who measured 35 tin compounds. The binding energy values span only 2 eV, the smallest value being for $PtCl(SnCl_3)[(C_6H_5)_3As]_2$ and the largest for SnF_4.

About the same time W. E. Swartz et al. (2) in a study of tetramethylammoniumhexahalostannates observed a chemical shift range of only 0.7 eV. More recently, S. C. Avanzino and W. L. Jolly (3) examined 15 tin compounds in the gas phase, and the range of binding energies extended from 491.01 eV for hexamethylditin to 495.36 eV for $Sn(NO_3)_4$. This disappointing behavior was not entirely unexpected since the change in binding energy is approximately proportional to the change in atomic charge, but the proportionality constant is inversely proportional to the atomic radius; therefore, the insensitivity of the tin binding energies to chemical environment is not surprising, and this will not be a straight-forward way to approach chemical problems. This relative insensitivity, compared with other fourth group elements is shown in Figure 1 (4). On the x-axis are the IP values of typical germanium compounds, and on the y-axis are the IP values of homologous compounds of C, Si, Sn, and Pb. The slopes of the straight lines obtained give an estimate of the sensitivity of the different atoms, and if the sensitivity of carbon is chosen as the arbitrary unit, that of tin is only 0.17. Since carbon itself is not highly sensitive, the situation is by no means encouraging. Nevertheless, interesting results have been obtained in four areas: charge distribution, change in coordination, oxidation state, and correlation of IP values with Mössbauer shifts.

The question of charge distribution can be approached in different ways. The simplest is to use the classical charge-shell atomic model which represents the valence electrons by a spherical valence shell of electric charge surrounding the core electrons. If the charge of the valence shell suffers a change of q electronic charges, the potential energy of any inner electrons is changed by q/r. Thus, the chemical shift sensitivity decreases with increasing atomic size. Of course the most suitable value of r is not easily determined, but if we compare tin and carbon, using for r the radius of the $+4$ ion (4), a relative sensitivity of 0.21 is obtained; this compares favorably with the experimental value of 0.17. To estimate the charges, it is possible to use classical electronegativities. In fact Swartz and co-workers (2) reported a fairly good correlation for hexahalostannates: IP $= 483.84 + 1.09 E_p$ between experimentally obtained binding energies and the average Pauling electronegativity of the halide ligands. Thus, the same type of correlation does exist with the estimated atomic charges, δ, on the central tin atom as that obtained from a method of electronegativity equalization: IP $= 486.4 + 0.57\delta$.

More recently S. C. Avanzino and W. L. Jolly (3) reinvestigated this problem using the point-charge potential model and their own electronegativity equalization method called CHELEQ (5). They introduced in the classical equations: IP $= kQ + V + l$, where Q is the charge of the ionized atom, V is the Coulomb potential energy at the site of the ionized atom resulting from other charged atoms in the molecule, k and l are empirical constants, a relaxation energy correction which could be evaluated by a so-called "transition state" method which is as-

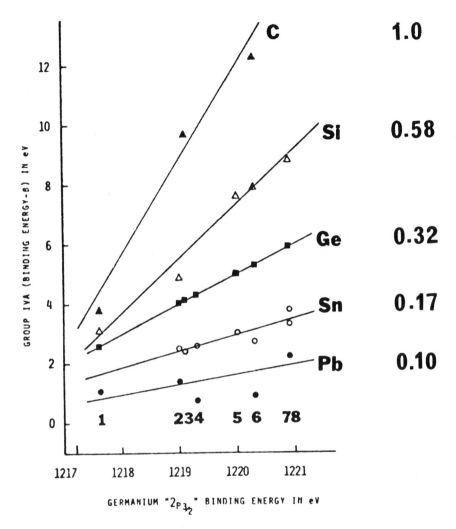

Figure 1. *Relative chemical shift effects in group IV compounds* (4). *Numbers at the bottom indicate compound type:* $1 = M$; $2 = (C_6H_5)_4M$; $3 = Na_2MO_3$; $4 = MS$; $5 = MS_2$; $6 = MO_2$; $7 = MF_2$; $8 = KMF_6$; $M = C$, Si, Ge, Sn, Pb.

sociated with the shift of electron density toward the created core hole. This correction is estimated (5) by considering a "transition state" halfway between the molecule itself and the molecule in which the central atom is ionized. Thus they succeeded in calculating the Sn $3d_{5/2}$ binding energies of 12 compounds and found $k = 14.77$ and $l = 493.57$ with a correlation coefficient of 0.974.

The possibility of determining coordination changes by measuring binding energies has been considered, but the low sensitivity of tin is a problem. In addition, the effect of ligand electronegativity can mask the influence of hybrid-

Table I. Complexation Shift of the Ligand Atom (P $2p$ and N $1s$ Levels) in Some Hexacoordinated Tin(IV) Complexes

Compounds	Shift
$[(C_6H_5)_3PO]SnBr_4$	0.2
$[(C_6H_5)_3PO]SnCl_4$	0.2
$\{[(CH_3)_2N]_2CS\}_2SnCl_4$	0.9
$(\alpha,\alpha'-C_{10}H_8N_2)SnBr_4$	1.7
$(C_5H_5N)_2SnCl_4$	2.0

ization change. However, if the IP shift of the central atom is not significant, the IP shift of the ligand can be. P. A. Grutsch et al. (1) have examined a number of hexacoordinated tin complexes and measured the shift undergone by the ligand atom. Table I presents some of their results.

The shift in binding energy of a given level of an atom is often interpreted in terms of a specific oxidation state, but the oxidation state of a metal atom is often not apparent from the molecular formula of its compounds, especially not in cluster compounds. A pioneering work has been published by G. W. Parshall (6) on platinum–tin cluster compounds such as $[(C_2H_5)_4N]_4[Pt_3Sn_8Cl_{20}]$. The tin $3d_{5/2}$ binding energies were consistent with an oxidation state of $+4$ for the above complex but less convincing for some parent compounds. A more elaborate study has been published jointly by T. J. Marks and J. J. Zuckerman and their co-workers (7), but it combines ESCA and Mössbauer spectroscopy (reviewed below). Correlations between core-level shifts in ESCA and isomeric shifts in Mössbauer spectroscopy are on firm ground. Isomer shifts (IS) in Mössbauer spectroscopy give a measure of the s-electron density at the nucleus; therefore, it is reasonable to expect that in a series of compounds where the stereochemistry and the oxidation number of the central atom are kept constant, a close relationship will exist between the s-electron density and the total electron density in the valence shell and consequently between Mössbauer IS and binding energies as measured by ESCA spectroscopy.

The first paper dealing with this was published by Barber and co-workers. They found a linear relationship between the $4d$ binding energies of tin in a series of Sn(IV) octahedral complexes and the corresponding Mössbauer IS values (8). The compounds examined were a series of dioxinates and SnO_2. According to the valence shell approximation, the same linear relationship should hold for the other electronic levels which are experimentally easier to measure. Grutsch and co-workers (1) examined 35 compounds and found for Sn $4d$ and Sn $3d_{5/2}$ the expected trend of increasing binding energy with decreasing IS. W. E. Swartz and co-workers (2), in their study of the hexahalostannates, observed a good correlation IP $= 486.8 - 0.641$ IS between Sn $3d_{5/2}$ binding energy and Mössbauer shift (correlation coefficient, $r = 0.94$). The good correlation reflects the fact that the compounds met the following criteria: same stereochemistry and same oxidation state for tin. Thus all tin compounds would not be expected to give values lying on the same line, but if the requirements above are fulfilled, a good correlation should be obtained for a homologous series of compounds. We have studied (9) two series of compounds [Et_2SnX_2 (X = F, Cl, Br, I which have a hexacoordinated structure, and $Ph_4Sn_2(OOCR)_2$] where x-ray structure determination (10) showed tin to be pentacoordinated. The results are in Figure 2 which shows a linear relationship between binding energies and isomer shifts for similar compounds. Thus it is possible to decide if a compound, which falls within a given series according to its stoichiometry, also has the required structure.

Marks, Zuckerman, and co-workers (7) have combined Mössbauer spectroscopy and ESCA to determine simultaneously the oxidation state and bonding in dialkyltin pentacarbonylchromium base and tetracarbonyliron complexes. When dialkyltin dihalides are treated in basic media (THF, DMSO) with $Na_2Cr_2(CO)_{10}$ or when basic cleavage of the tin–iron bond in the cyclic $[R_2SnFe(CO)_4]_2$ dimer occurs, compounds of type I: $R_2SnM(CO)_x$ (with M = Fe, x = 4 or M = Cr, x = 5) are obtained. X-ray data show the base, B, bonded to tin. Electronic configurations which could be written for these compounds are shown below. Structures Ia and Ib depict the use of the lone pair available in the three-coordinated tin(II) structure to establish a σ-donor interaction with

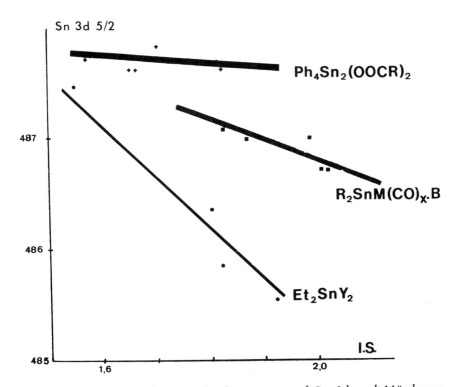

Figure 2. Correlation between binding energy of Sn 3d and Mössbauer shift

M. These two structures can be ruled out on the basis of ESCA and Mössbauer results. The observed isomer shift is 1.82–2.11 when tin(II) compounds give a shift larger than 2.6.

$$
\begin{array}{ccccc}
\underset{R}{\overset{B}{\underset{\diagup}{R\diagdown}}}\!\!\!\underset{}{\overset{\big\downarrow}{Sn}}\!\!\rightarrow M &
\underset{R}{\overset{B}{\underset{\diagup}{R\diagdown}}}\!\!\!\underset{}{\overset{\big\downarrow}{Sn}}\!\!=\!M &
\underset{R}{\overset{B}{\underset{\diagup}{R\diagdown}}}\!\!\!\underset{}{\overset{\big|}{Sn^{+}}}\!\!-\,^{-}M &
\underset{R}{\overset{B}{\underset{\diagup}{R\diagdown}}}\!\!\!\underset{}{\overset{\big|}{Sn}}\!\!-\!M^{-} &
\underset{R}{\overset{B^{+}}{\underset{\diagup}{R\diagdown}}}\!\!\!\underset{}{\overset{\big|}{Sn^{-}}}\!\!=\!M \\[6pt]
\text{Ia} & \text{Ib} & \text{Ic} & \text{Id} & \text{Ie}
\end{array}
$$

The observed isomer shifts correlate very well with Sn $3d_{5/2}$ binding energies which we consider as proof that all the compounds have the same structure. The only tin(II) compound for which tin $3d_{5/2}$ has been measured is $[(C_2H_5)_4N]SnCl_3$ which contains the $SnCl_3^-$ unit: 485.7 eV. This value is outside the range of data observed for type I compounds. The other arguments presented are outside the scope of this paper, but compounds I can be described by contributions of structures Ic to Ie.

Application to Organotin Toxicity

No element can match tin for the various applications of its organoderivatives. However, widespread use of organotins has been hindered by toxicity. Trialkyltin compounds inhibit oxidative phosphorylation and disrupt the basic energy processes in living systems. The blocking mechanism probably involves (11) the formation of pentacoordinated trialkyltin histidine complexes. The same type of bonding operates in the trialkyltin derivatives of some heterocycles (12). The anionic group does not seem to be involved.

For dialkyltin compounds a distinct mechanism has been advanced, involving reaction with thiol groups, disrupting α-ketoacid oxidation (13). Dialkyltin dihalides can react with free thiol groups, but the formation of gunuine Sn–S bonds may be unnecessary.

A pentacoordinated complex between tin and a suitable group in a protein chain could give rise to the observed toxicity. Our attention had been called to the possibility of ESCA spectroscopy as a tool for detecting the formation of complexes by three earlier studies. The first was by D. R. Crist (14) and co-workers who established that in the complexes formed between cations and oxaziridine, the almost exclusive site of complexation was the nitrogen atom. This resulted in a shift of 0.6 eV of the N $1s$ IP to lower energy in the ESCA spectrum no matter what the initial oxaziridine was.

The second encouraging result was observed in our laboratory (15) and concerned the formation of complexes between oximes and oxaziridines and praseodynium salts. We could prove that with oxime ligands the site of complexation was mainly oxygen (the resulting ESCA shift of O $1s$ being between 0.4 and 1 eV to lower energy depending on the oxime) and only to a slight extent

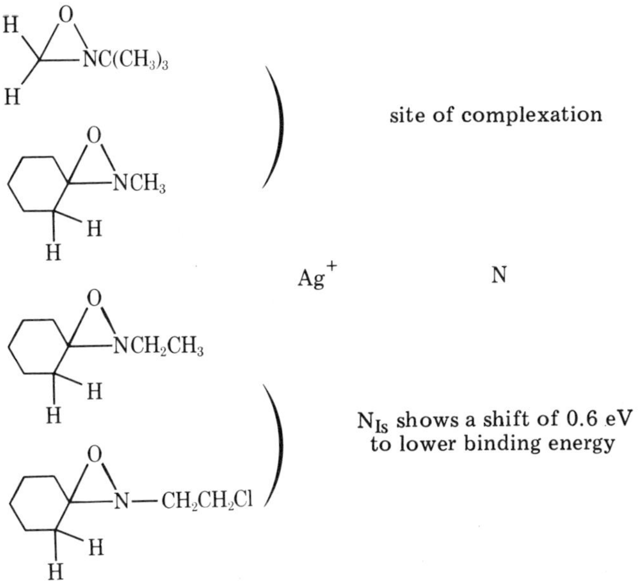

site of complexation

Ag$^+$ N

N_{ls} shows a shift of 0.6 eV
to lower binding energy

nitrogen (but with a wide shift of N 1s IP to higher energy between 1.6 and 2.6
eV). In the oxaziridine series the situation is clearly different. When the α-
positions bear methyl groups, the complexation site is exclusively nitrogen, resulting in a shift to lower energy of 1.1 to 2.4 eV; when there is a CH_2 group next
to the aziridine function, complexation occurs partly on oxygen with a shift of
2.4 eV in addition to the normal nitrogen complex.

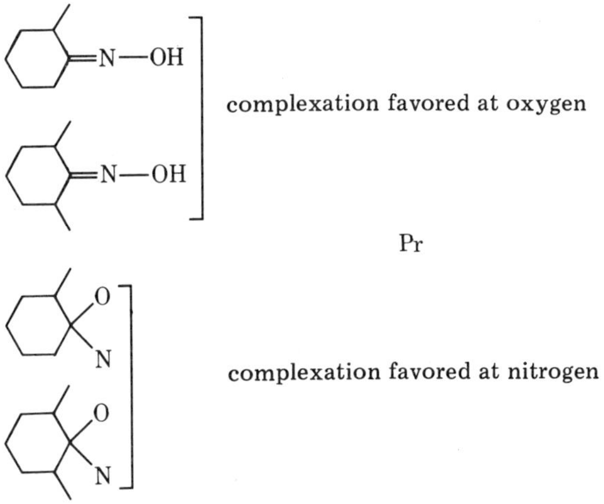

complexation favored at oxygen

Pr

complexation favored at nitrogen

The third set of results which prompted this research was published by G.
Jung (16) who studied metals in proteins. He noted drastic IP differences for

Table II.   Complexation Site of Bu₃SnCl with Nitrogen and Sulfur Ligands

Compound		Binding Energy[a] (eV) of Free Ligand		Binding Energy[a] (eV) of Complex	Complexation Shift
2-Mercapto-benzimidazole (structure with SH)	1	N 1s	400.4	400.7[c]	+0.3
		S $2p_{1/2}$	162.9	164.8	+1.9
		S $2p_{3/2}$	226.7	228.0	+1.3
2,3-Diphenylthiazole (structure)	2	N 1s	399.3	399.9[d]	+0.6
		S $2p_{1/2}$	164.1	165.8	+1.7
		S $2p_{3/2}$	229.2	229.7	+0.5
L (−) methionine (HO₂C—CH(NH₂)—(CH₂)₂S—CH₃)	3	N 1s	400.9	400.9[c]	0.0
		S $2p_{1/2}$	163.8	163.8	0.0
		S $2p_{3/2}$	227.6	227.6	0.0
L (+) cystéine (HO₂C—CH(NH₂)—CH₂—SH)	4	N 1s	400.7	400.7	0.0
		S $2p_{1/2}$	164.4	163.7	−0.7
		S $2p_{3/2}$	227.6	227.5	−0.1
S-methyl-L-cysteine (HO₂C—CH(NH₂)—CH₂—S—CH₃)	5	N 1s	401	400.7	−0.3
		S $2p_{1/2}$	163.8	163.7	−0.1
		S $2p_{3/2}$	227.6	227.6	0.0
Cystine (S—CH₂—CH(NH₂)—CO₂H / S—CH₂—CH(NH₂)—CO₂H)	6	N 1s	400.6	400.8	+0.2
		S $2p_{1/2}$	163.7	163.6	−0.1
		S $2p_{3/2}$	227.6	227.6	0.0
Histidine (imidazole—CH₂—CH(NH₂)—CO₂H)	7	N 1s	399.6	399.6[f]	0.0
(C₂H₅)₂Sn(S—CH₂CH₂—S)	8	S $2p_{1/2}$	163.4		
		S $2p_{3/2}$	227.2		

[a] Reference C 1s: 285.0 eV.
[b] A positive value indicates a shift to higher binding energy.
[c] Sn 3d: 487.3; 495.8; Sn 3d in Bu₃SnCl: 486.7 and 495.3 eV.
[d] Sn 3d: 487.4; 496.0 eV.
[e] Sn 3d: 487.5; 495.2 eV.
[f] Sn 3d: 487.6; 495.9 eV.

transition metals in amino acid complexes and in proteins and inferred that sulfur was not involved in the bonding.

Thus, two facts were clear: the existence of such complexes and the site of complexation. We began our investigations by studying amino acids and model compounds, some bearing additional groups which could function as complexing agents. The results are summarized in Table II and show the following. The (ca.) 165 eV $2p$ band of sulfur is a very good indicator of complexation since differences as large as 1.9 eV can appear between complex and free ligand which proves that sulfur can play an essential part in complex formation. On the other hand, tin is a poor indicator with a shift never exceeding 0.8 eV.

For the details two sets of compounds must be considered: mercapto-2-benzimidazole, **1**, and cystein, **3**; both contain a free SH group but behave in a different ways. For **1** all the binding energies are shifted to higher energy; for **3** only S signals are shifted but to lower energy. Essentially, for complexation through either nitrogen or sulfur a shift of the binding energy of the element acting as donor ligand is expected. However, compare the results with those obtained in the oxime and oxaziridine series, where, by complexation, the N $1s$ electron was shifted to higher energy but O $1s$ to lower energy and where sulfur should behave like oxygen. We are led to conclude that complexation occurs at the SH group of cystein. We have compared **1** with another sulfur–nitrogen aromatic molecule, diphenylthiazole, and the results are similar: S $2p$: + 1.7 N $1s$: + 0.6 Sn $3d$: +0.7. We believe that the aromatic compound acts as bidentate ligand and that a hexacoordinated tin complex is obtained. The possibility of a covalent Sn–S bond must be rejected on the basis of the results obtained for diethyl-1,1-stanna-1-dithia-2,5-cyclopentane **7**.

The other compounds do not contain the SH group. Sulfur atoms are still present but as S–CH$_3$ or S–S groups. Definite evidence is lacking for the formation of a complex, either on nitrogen or on sulfur.

We conclude that complexes involving coordination of tin by Sn–S bonds play a part in the toxicity and other biological activity of organotin compounds and require the presence of free SH groups. In no case did NH$_2$ groups prove basic enough to coordinate to tin. Some reservations must be made in regard to experimental conditions. The amino acid was placed on the sample holder which was then cooled to ca. $-80°$C. Tributyltin chloride was introduced as a vapor and condensed on the amino acid. The spectrum was run at ca. $-20°$C. Of course, these conditions are different from those in living tissues, but the comparison between the functional group candidates for complexation should still hold.

UV–Photoelectron Spectroscopy

Low energy photoelectron spectroscopy gives direct information about valence electrons and therefore about bonding, but it can also reveal some facts on problems as subtle as ring strain.

Cyclostannanes.   This study of the photoelectron spectra of cyclostannanes is a continuation of our earlier interest in the problems of bonding in organotin derivatives.   There has been much interest in the question of d orbital participation, and some important papers have been published ($17, 18$) on tetramethyltin vs. its carbon, silicon, germanium, and lead homologs.   In this series we cannot deal directly with the problem of d orbital participation, but we can try to account for two observations.

The first is a chemical one: under attack by nucleophilic reagents such as bromine the ring compounds open and, for example, dimethyl(5-bromopentyl)tin bromide is obtained from dimethylstannacyclohexane.   The second is related

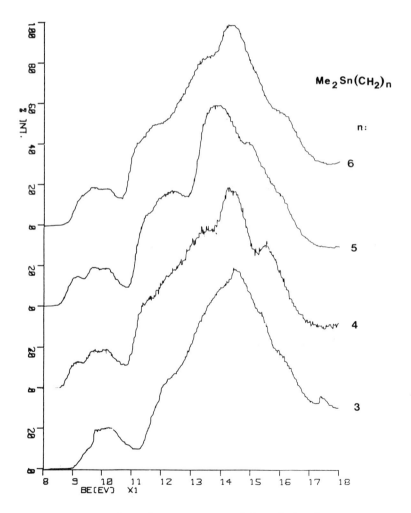

Figure 3. UV-photoelectron spectra of dimethylcyclostannanes

to stereochemistry: is it possible to observe by PES any ring strain effect by measuring the IPs of exo- and intracyclic tin–carbon bond.

In the cyclosilane series C. S. Cundy and co-workers (*19*) demonstrated that the ionization potential of the intracyclic Si–C bond is lower than that of the exocyclic one by about 1 eV. The same trend was expected for tin compounds on the basis of chemical reactivity.

The PE spectra of four dimethylcyclostannanes are given in Figure 3; they are remarkably similar. In each the photoelectron transitions occur in two groups: a single, well defined low ionization energy band with a fine structure and a broad ionization region at higher energy.

The low energy band can be assigned as the ionization of the exocyclic group band on the basis of following arguments. This band has the same contour as all dimethyl compounds, but with diethyl compounds there is no longer any obvious multiplicity. The same band appears in the spectrum of $Sn(CH_3)_4$ that we have seen above, in agreement with the results of A. E. Jonas and co-workers (*17*); thus, there is little doubt that the first ionization peak can be assigned to the exocyclic Sn–C bond which is completely different from what has been observed in the silicon series. However, it is not at variance with chemical reactivity because a more easily ionized bond should be more reactive, whatever the reaction mechanism. Nevertheless, in very simple cases clear correlations do exist between ionization potential and reactivity—e.g., in the basicity of simple alcohols (*20*) or of amines (*21*).

The value observed for the first ionization potential must be discussed as well as its multiplicity. The fact that a single peak is observed with ethyltin compounds demonstrates that the appearance of three peaks is associated with the methyl groups. This is clearly a Jahn–Teller effect. Ionization produces an ion in which the photoelectron can occupy three degenerate orbitals, but this degeneracy can be lifted by Jahn-Teller distortion. The distorting effect and,

Table III. Ionization Potentials of Stannacycloalkanes: $R_2Sn(CH_2)_n$

		R = Me					R = Et		
		1	2-1	2	3-2	3	3-1	Sn–C	Sn	Sn–C
Sn (6)	6	9.40	0.40	9.80	0.50	10.30	0.9	14.6	9.50	14.6
Sn (5)	5	9.15	0.45	9.60	0.65	10.25	1.10	13.8	9.46	13.6
Sn (4)	4	9.20	0.52	9.72	0.48	10.25	1.05	13.4	9.58	13.4
Sn (3)	3	9.65	0.25	9.90	0.45	10.25	0.7	13.8	—	—

therefore, the splitting of the first IP band are expected to vary inversely as the distance between interacting groups—i.e., methyl groups. The splitting diagram is, therefore, not exactly symmetric.

The values of the first ionization bands are listed in Table III, from which the following conclusions can be drawn. The tin–methyl bond is the most easily ionized in the stannacyclohexane (whatever component of the band is chosen for comparison). The methyl groups on the basis of Jahn–Teller effect are closer in stannacyclohexane than in any other compound, which means that the C–Sn–C angle is largest in the series.

These conclusions confirm each other. A change in ring size is probably accompanied by a change in hybridization of the tin atom. A drop in IP means a decrease in s character. If the s character of the extracyclic Sn–C bond decreases, the C–Sn–C angle in the ring will decrease, opening the C–Sn–C extracyclic angle and increasing the distance between the two methyl groups and, therefore, decreasing the Jahn-Teller effect. The largest IP then corresponds to the smallest Jahn-Teller effect. The Jahn-Teller effect is the largest and the IP is the smallest for stannacyclohexane. Thus, the ring strain affecting the tin–carbon bonds is the smallest in stannacyclohexane. Further, IP is close to that of $Sn(CH_3)_4$ (9.73 eV). In the ethyl series the first ionization band is broadened by some unresolved vibrational structure, and no conclusion can be reached.

For diphenyldistannacyclopentane the first band is a benzenoid at 8.8 eV (which is very low with respect to 9.4 eV for benzene itself). Then two peaks appear at 9.58 and 9.94 eV. We could interpret these three peaks as two benzenoid bands, caused by the same $p\pi$-$d\pi$ effect which should be 0.78 eV, the same order of magnitude as the conjugation effect observed in chlorobenzene. We think in this case that the effect is too large and that we have the benzenoid band at 8.8 eV and that the Sn–C peak is split by some Jahn-Teller effect.

Correlation of UV–PES with Nuclear Quadrupole Resonance. The question of $(p\pi \rightarrow d\pi)$ bonding has been the subject of a long and unfinished controversy in which we have participated as defenders. We can assist the recent comeback of $(p\pi \rightarrow d\pi)$ bonding by juxtaposing results of PES and NQR spectroscopies obtained for $Cl–C_6H_4–MMe_3$ with M = C, Si, Ge, and Sn. The ratio of the nuclear quadrupole coupling of a given atom in a molecule to that of the free atom A can be written (22):

$$R = \frac{eqQ}{eqQ_A} = (1 - \alpha^2)(1 - i) - \pi$$

where α is the s character of the bond binding A to the rest of the molecule, M, i is the ionicity of the bond A–M, and π is the percentage of the double bonding of A–M. In this series where we are dealing with a C_{Ar}–Cl bond, α is small with respect to unity (23), and therefore, the above ratio depends on i and π. The value of i can be estimated on the basis of the electronegativities of MMe_3 groups

(24) $(E_C$: 2.60; E_{Si}: 1.90; E_{Ge}: 2.00; E_{Sn}: 1.93). Therefore, R should decrease in the order Si > Sn > Ge > C. The experimental order is completely different, and this discrepancy must be accounted for by a variation in π. To influence the double-bond character of the C_{Ar}–Cl bond, the MMe_3 group has to bring π-electrons or a convenient orbital into play, and in that particular case the only way to do it is to use d-orbitals to create a $p\pi$-$d\pi$ bond. The accumulated effects of π and i should restore agreement between the observed and expected order. Figure 4 shows how this is achieved and shows that the $p\pi$-$d\pi$ effect is largest for the Si atom. This is logical since the Si atomic radius is the closest to that of carbon. This conclusion is confirmed by stronger arguments. Photoelectron spectroscopy was an appropriate way to provide the necessary data and was suited

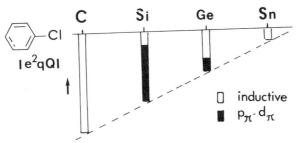

Figure 4. Contribution of the $p\pi$–$d\pi$ *effect to the variation of* $|e^2qQ|$
^{35}Cl *in the series* Cl—C_6H_4—MMe_3, $M = C, Si, Ge, Sn$

	$\nu_{(MHz)}$	$|e^2qQ|_{(MHz)}$
p-*tert*-Bu–C_6H_4–Cl	34,261	−68,522
p-Me_3Si–C_6H_4–Cl	34,428	−68,855
p-Me_3Ge–C_6H_4–Cl	34,465	−88,930
C_6H_5Cl		−69,2
p-Me_3Sn–C_6H_4–Cl	34,663	−69,326

to that series of compounds since it is possible to get many parameters from the spectra—e.g., the ionization potentials of $3p$ chlorine-free doublets of highest occupied orbitals of the aromatic ring and of the metal–methyl bond. The spectra are shown in Figure 5.

The chlorine data show that in all the compounds where a mesomeric effect is possible, the ionization peak of $3p$ Cl is split in two. The high energy component is associated with a conjugated electron pair (stabilized). The difference, δ, in energy of the two components of the doublet is a measure of the mesomeric interaction. Actually, except for M = C, two peaks are observed. The high field peak is essentially fixed in the series, barring a drift to high energy ascribable to the inductive effect of the substituents. However, the low field peak varies according to changes in mesomeric effect. A graph of δ vs. the nature of M (Figure 6) shows that δ rises to a maximum for M = Si. This means that the (−M) in-

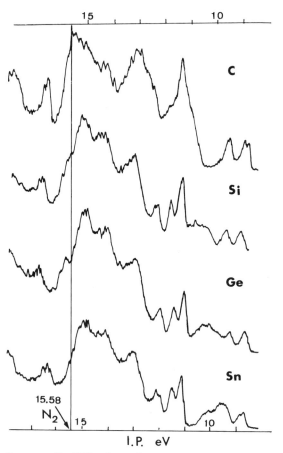

Figure 5. UV-photoelectron spectra of p-$Cl(C_6H_4)MMe_3$ *compounds, M = C, Si, Ge, Sn*

fluence of the $3d$ empty orbitals of the M atom is felt by the chlorine atom in the para position and again that Si is more favorable than tin for $(p\pi \rightarrow d\pi)$ bonding, thus confirming NQR results. The aromatic ring should of course experience this effect. Since the HMO of the ring is affected by what happens to the $3p$ chlorine orbital, the degeneracy of the π highest level is lifted and two ionization energies, π_2 and π_3, are actually measured. The splitting is a measure of conjugation. If most of the π electrons given by chlorine are recaptured by the metal through the $p\pi \rightarrow d\pi$ bond, the electronic structure of the aromatic ring will approximate that of free benzene and the splitting should decrease and the maximum value of δ should correspond to a minimum for the splitting of the π orbitals, which is observed (Figure 7).

Other information can be obtained from the spectra. Ionization of the M–CH$_3$ bond gives rise to a band at ca. 10.5 eV. When M = C, this band is very

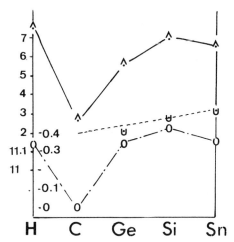

Figure 6. Cl 3p ionization potentials in p-Cl(C$_6$H$_4$)MMe$_3$ compounds. ∧ ∪ are components and ⊙ is splitting of the Cl 3p doublet.

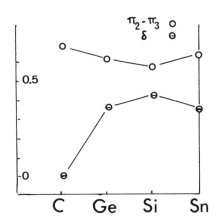

Figure 7. Correlation between the splitting of Cl 3p and the energy difference between the two HMOs of the benzene ring in Cl—C$_6$H$_5$ and p-Cl(C$_6$H$_4$)-MMe$_3$, M = C, Si, Ge, Sn

Table IV.   Ionization Potentials of C—Me Bond in
p-Cl(C$_6$H$_4$)MMe$_3$ and M(Me)$_4$

	p-Cl(C$_6$H$_4$)MMe$_3$	M(Me)$_4$
C	11.35	11.40
Si	10.58	10.6
Ge	10.41	10.20
Sn	10.11	9.62

similar to that of $C(CH_3)_4$, but for all other compounds the IP is larger than that of corresponding $M(CH_3)_4$, thus revealing the electron contribution of the ring (Table IV). When M = Si, the larger electronegativity of the chlorophenyl ring with respect to the methyl group is more than offset by the $(p\pi \rightarrow d\pi)$ effect, making the IP of the Si–C bond close to that observed in $Si(CH_3)_4$. Thus PE spectroscopy can reveal the mystery of some complex bonds and peculiarities of the behavior of elements as similar as these of the fourth group.

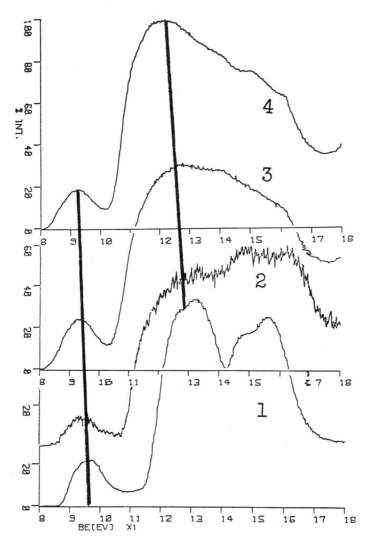

Figure 8. UV–photoelectron spectra of tetraalkyltin compounds

Animal Toxicity of Tetraalkyltin Compounds

The tetraalkyltin compounds are probably not toxic themselves but undergo dealkylation in the liver to form toxic trialkyltin salts (*14*).   If the determining step is the formation of a trialkyltin salt, there could be a correlation between toxicity and the ease of formation of the trialkyltin cation, which could proceed through ionization of one of the tin–carbon bonds.   Figure 8 shows that the PE spectra of some currently available tetraalkyltins are very similar.   The low energy band at ca. 9.5 eV, which does not appear in the spectra of alkanes, is, therefore, attributed to ionization of the tin–carbon bond.   The corresponding ionization potentials vary from 9.20 to 9.65 eV as the alkyl chain lengthens.   At the same time the broad ionization region assignable to the alkyl chain moves to low IP as generally observed.

Table V.   First Ionization Potential and Animal Toxicity
of Tetraalkyltin Compounds

R_4Sn	$IP(eV)$	Lethal Dose, mg/kg
Me$_4$Sn	9.36; 9.65; 9.85	9
Et$_4$Sn	9.65	24
Bu$_4$Sn	9.45	350
Pent$_4$Sn	9.30	450
Oct$_4$Sn	9.20	34000

Table V compares the observed IP with experimental toxicities.   It is obvious that toxicity and the ionization potential vary in the same way—i.e. the easier the Sn–C bond ionizes, the less toxic is the compound.   The reverse trend was expected.   Apparently ionization of a tin–carbon bond and formation of trialkyltin cation are not necessarily the limiting process.   Another possibility is that the formation of the trialkyltin cation is a necessary but not rate-determining step for the toxicity.   Other work on different series of compounds shows them displaying a wider spectrum of activity.

Appendix

Table AI.   Experimental Binding Energies of Tin Compounds

Compound	C 1s Direct	Measured Binding Energy[a], eV 3d$_{3/2}$	3d$_{5/2}$	4d (and Related Data)
1 K$_2$SnF$_6$	286.7	496.3	487.8	
	287.4	496.3	487.8	

Table AI. (Continued)

Compound	C 1s Direct	$3d_{3/2}$	$3d_{5/2}$	4d (and Related Data)
2 SnF$_2$	285.8	495.7	487.3	
	285.7	495.5	487.2	
3 Na$_2$SnO$_3$	285.4		486.4	
4 SnO$_2$	286.9	495.1	486.8	
	286.8	495.1	486.7	
5 SnO	286.3	495.8	487.4	
	286.4	495.4	487.1	
6 SnS$_2$	286.5	495.4	487.0	
7 SnS	284.6	495.1	486.6	
8 (C$_6$H$_5$)$_4$Sn	283.5	494.8	486.5	
	283.4	494.9	486.5	
9 (C$_6$H$_5$)$_3$SnCl	283.6	495.7	487.2	
10 (CH$_2$C$_6$H$_5$)$_3$SnCl	283.6	495.5	487.1	
11 (CH$_2$C$_6$H$_5$)$_2$SnCl$_2$	283.7	496.2	487.8	

Compound				$3d_{5/2}-4d_{5/2}$
12 SnH$_4$		492.74		460.97
13 Sn(CH$_3$)$_4$		491.38		461.03
14 Sn$_2$(CH$_3$)$_6$		491.01		460.96
15 (CH$_3$)$_3$SnCl		492.27		460.85
16 (CH$_3$)$_2$SnCl$_2$		493.21		460.84
17 CH$_3$SnCl$_3$		494.06		460.92
18 SnCl$_4$		494.92		460.85
19 SnBr$_4$		494.46		460.93
20 SnI$_4$		493.75		
21 (CH$_3$)$_3$SnBr		492.23		460.78
22 (CH$_3$)$_3$SnI		492.02		460.80
23 (C$_4$H$_9$)$_3$SnOCH$_3$		491.31		⌡1
24 (CH$_3$)$_3$SnMn(CO)$_5$		491.16		461.01
25 Sn(NO$_3$)$_4$		495.36		
26 Sn{N[Si(CH$_3$)$_3$]$_2$}$_2$		491.93		

Compound				4d
27 SnF$_2$			486.4	26.0
28 SnCl$_2$			485.8	25.2
29 SnCl$_2 \cdot$2H$_2$O			486.8	26.5
30 SnBr$_2$			486.2	26.0
31 SnI$_2$			486.6	26.2
32 SnO			485.9	25.9
33 Sn(C$_4$H$_9$)$_2$(NCS)$_2$ a (o-C$_{12}$B$_8$N$_2$)			485.7	25.9
34 SnBr$_4$(C$_4$H$_8$S)$_2$			486.1	25.6

Table AI. (Continued)

| | | Measured Binding Energy[a], eV | | |
| | C 1s | | | 4d (and |
Compound	Direct	$3d_{3/2}$	$3d_{5/2}$	Related Data)
35 $SnCl_4[(C_4H_9)_3P]_2$			485.4	25.2
36 $SnBr_4(\alpha,\alpha'\text{-}C_{10}H_8N_2)$			486.3	26.3
37 $Sn(C_6H_5)_2(C_9H_6NO)_2$			485.7	25.5
38 $SnCl_4\{[(CH_3)_2N]_2CS\}_2$			486.0	25.8
39 $SnBr_4[(CH_3)_2SO]_2$			486.3	26.0
40 $SnBr_4[(C_6H_5)_3PO]_2$			486.4	25.8
41 $SnCl_4(C_5H_5N)_2$			485.8	25.9
42 $(NH_4)_2SnCl_6$			486.0	26.1
43 $SnCl_4[(CH_3)_2SO]_2$			486.2	26.1
44 $SnCl_4\{[(CH_3)_2N]_2CO\}_2$			486.2	26.1
45 $SnCl_4\{[(CH_3)_2N]_3PO\}_2$			486.5	26.4
46 SnO_2			486.0	26.0
47 SnF_4			487.2	27.0
48 $Pt(SnCl_3)_2[(CH_3)_2N]_2CS\}_2$			485.4	25.4
49 $Pd(SnCl_3)_2\{[(CH_3)_2N]_2CS\}_2$			485.6	25.4
50 $PtCl(SnCl_3)[(C_6H_5)_2\text{-}P(CH_2)_2P(C_6H_5)_2]$			485.7	25.8
51 $PdCl(SnCl_3)[(C_6H_5)_2\text{-}P(CH_2)_2P(C_6H_5)_2]$			485.6	25.6
52 $Pt(SnCl_3)_2[C_5H_{10}N)_3PS]_2$			486.0	25.6
53 $Pd(SnCl_3)_2[(C_5H_{10}N)_3PS]_2$			485.8	25.7
54 $PtCl(SnCl_3)[(C_6H_5)_3Sb]_2$			485.6	25.3
55 $PtCl(SnCl_3)[(C_6H_5)_3As]_2$			485.3	25.4
56 $PtCl(SnCl_3)[(C_6H_5)_3P]_2$			485.8	25.5
57 $Au(SnCl_3)[(C_6H_5)_3P]_3$			485.8	25.0
58 $Ag(SnCl_3)[(C_6H_5)_3P]_3$			485.8	25.2
59 $Cu(SnCl_3)[(C_6H_5)_3P]_3$			485.6	25.3
60 $(CH_3)_4NSnCl_3$			485.7	25.6
61 $KSnF_3$			486.0	26.2
62 $[(C_2H_5)_4N]_4[Pt_3Sn_8Cl_{20}]$			487.3	
63 $(C_8H_{12})_3Pt_3SnCl_6 \cdot CH_3NO_2$			487.3	
64 $[(C_2H_5)_4N]_3[Pt(SnCl_3)_5]$			487.1	
65 $[(C_2H_5)_4N]_2[SnCl_6]$			486.3	
66 $[(C_2H_5)_4N][SnCl_3]$			485.7	
67 $[(CH_3)_2SnFe(CO_4)]_2$			486.9	
68 $[(C_6H_5)_2SnFe(CO)_4]_2$			486.8	
69 $[(tert\text{-}C_4H_9)_2SnFe(CO)_4]_2$			487.1	
70 $(tert\text{-}C_4H_9)_2SnFe(CO_4) \cdot DMSO$			487.0	
71 $(tert\text{-}C_4H_9)_2SnFe(CO)_4 \cdot py$			487.1	
72 $(tert\text{-}C_4H_9)_2SnCr(CO)_5 \cdot THF$			486.7	
73 $(tert\text{-}C_4H_9)_2SnCr(CO)_5 \cdot DMSO$			487.0	
74 $(tert\text{-}C_4H_9)_2SnCr(CO)_5 \cdot py$			486.7	

Table AI. (Continued)

		Measured Binding Energy[a], eV		
Compound	C 1s Direct	$3d_{3/2}$	$3d_{5/2}$	4d (and Related Data)
75 $Et_2Sn(OX)_2$				22.3
76 $Ph_2Sn(OX)_2$				22.9
77 $SnBr_4 \cdot 2OX$ H				23.3
78 $SnI_2(OX)_2$				23.5
79 $SnBr_2(OX)_2$				23.8
80 $SnCl_2(OX)_2$				24.3
81 SnO_2				26.3
82 $trans$-$Fe(isocy)_4(SnCl_3)_2$		only Fe bands reported[d]		
83 $[Fe(SnCl_3)(isocy)_5]ClO_4$				
84 ⟨S⟩SNEt₂		495.1	486.6	
85 K_2SnF_6			488.3	
86 $(Et_4N)_2SnCl_4F_2$			487.3	
87 $(Et_4N)_2Br_4F_2$			487.2	
88 $(Et_4N)_2SnCl_6$			487.1	
89 $(Et_4N)_2SnCl_4Br_2$			487.2	
90 $(Et_4N)_2SnCl_4I_2$			487.4	
91 $(Et_4N)_2SnBr_6$			487.2	
92 $(Et_4N)_2SnBr_4Cl_2$			487.1	
93 $(Et_4N)_2SnBr_4I_2$			487.0	
94 $(Et_4N)_2SnI_4Cl_2$			486.9	
95 $(Et_4N)_2SnI_4Br_2$			486.8	
96 $(Et_4N)_2SnI_6$			486.7	
97 K_2SnBr_6			487.6	
98 $Sn(\beta)$			486.9	
99 $Et_2 SnF_2$			487.4	
100 Et_2SnCl_2			486.3	
101 Et_2SnBr_2			485.8	
102 Et_2SnI_2			485.5	
103 $Ph_4Sn(OOCCH_3)_2$			487.6	
104 $Ph_4Sn(OOCCH_2Cl)_2$			487.7	
105 $Ph_4Sn(OOCCCl_3)_2$			487.3	
106 $Ph_4Sn(OOCCF_3)$			486.6	
107 $Ph_4Sn(OOCC_6H_5)_2$			487.6	

[a] Compounds 1 to 11: Measurements referenced to hydrocarbon-contaminant C 1s line, assuming a value of 285.0 eV (1).
Compounds 12 to 26: Measurements referenced against Ar $2p_{3/2}$ line (248.45 eV) (2). The energy of the Sn $3d_{5/2}$ line is not given.
Compounds 27 to 61: ionization potentials based on Au $4f_{5/2}$: 87.0 eV (3).
Compounds 62 to 73: calibrated to e C_{1s} binding energy of 285.0 eV (4,5).
Compounds 74 to 80, no standardization reported (6).
Compounds 84 referenced against C_{1s}: 285.0 eV (3).
Compounds 85 to 98: MeO_3 as internal standard (232.5 eV) (3).
Compounds 99 to 107: referenced against O_{1s}: 532.6 eV (10).

Table AII. UV–PES Data

$R\ Sn(CH_3)_3$ R	PI		
1 n-Bu	9.52[a,c]	11.31	
2 iso-Bu	9.33[c]	11.06	
3 Allyd	8.70[c]	9.76	10.87
4 Cyclopropyl–CH_2–	8.35[c]	10.09	10.45
5 4-Butenyl	9.71[a,c]	11.74	
6 5-Pentenyl	9.72[a,c]	11.40	
7 Me_4Sn	9.65[d]	12.9	
8 Et_4Sn	9.65	13.0	
9 Bu_4Sn	9.45	12.7	
10 Am_4Sn	9.30	12.5	
11 Oct_4Sn	9.20	11.8	
12 $Me_2Sn(CH_2)_3$	10.2[t]		
13 $Me_2Sn(CH_2)_4$	9.80[t]		
14 $Me_2Sn(CH_2)_5$	9.65[t]		
15 $Me_2Sn(CH_2)_6$	9.60[t]		
16 $Et_2Sn(CH_2)_4$	9.60		
17 $Et_2Sn(CH_2)_5$	9.40		
18 $Et_2Sn(CH_2)_6$	9.50		
19 $Ph_2Sn(CH_2)_5$	8.8[a] 9.58	9.94	
20 SnF_4	[b]		
21 $SnCl_4$	[b]		
22 SnH_4	[b]		
23 $ClC_6H_4SnMe_3$	[b]		

[a] Weak shoulder at 9.1.
[b] See original paper for details on the spectrum.
[c] Compounds 1 to 6, *see* Ref. 25.
[d] See Refs. *17, 18, 26.*

Literature Cited

1. Grutsch, P. A., Zeller, M. V., Fehlner, T. P., *Inorg. Chem.* (1973) **12**, 1432.
2. Swartz, W. E., Watts, P. H., Lippincott, E. R., Watts, J. C., Huheey, J. E., *Inorg. Chem.* (1972) **11**, 2632.
3. Avanzino, S. C., Jolly, W. L., *J. Electron. Spectrosc.* (1976) **8**, 15.
4. Morgan, W. E., Van Wazer, J. R., *J. Phys. Chem.* (1973) **77**, 964.
5. Jolly, W. L., Perry, W. B., *Inorg. Chem.* (1974) **13**, 2686.
6. Parshall, G. W., *Inorg. Chem.* (1974) **11**, 433.
7. Grynkewich, G. W., Ho, B. Y. K., Marks, T. J., Tomaja, D. L., Zuckerman, J. J., *Inorg. Chem.* (1973) **12**, 2522.
8. Barber, M., Swift, P., Cunningham, D., Frazer, M. J., *J.C.S., Chem. Comm.* (1970) 1338.
9. Limouzin, Y., Maire, J. C., *J. Organometal. Chem.* (1974) **82**, 99.
10. Bandoli, G., Clemente, D. A., Panatoni, C., *J.C.S., Chem. Comm.* (1971) 311.
11. Rose, M. S., Mock, E. A., *Biochem. J.* (1970) **120**, 151.
12. Gassend, R., Thesis, Marseille, 1976.
13. Barnes, J. M., Magos, L., *Organometal. Chem. Rev.* (1968) **3**, 137.
14. Crist, D. R., Jordan, G. J., Hashmall, J. A., *J. Amer. Chem. Soc.* (1974) **96**, 4923.
15. Limouzin, Y., Rivière, M., unpublished work.

16. Jung, G., Ottnad, M., Hartmann, H. J., Rupp, H., Weser, U., Z. Anal. Chem. (1973) 263, 282.
17. Jonas, A. E., Schweitzer, G. K., Grimm, F. A., Carlson, T. A., J. Electron Spectrosc. (1972/73) 1, 29.
18. Boschi, R., Lappert, M. F., Pedley, J. B., Schmidt, W., Wilkins, B. T., J. Organometal. Chem. (1973) 50, 69.
19. Cundy, C. S., Lappert, M. F., Pedley, J. B., Schmidt, W., Wilkins, B. T., J. Organometal. Chem. (1973) 51, 99.
20. Robin, M. B., Kuebler, N. A., J. Electron Spectrosc. (1972/73) 1, 13.
21. Yoshikawa, K., Hashimoto, M., Morishima, E., J. Amer. Chem. Soc. (1974) 96, 288.
22. Hownes, C. H., Dailey, B. P., J. Chem. Phys. (1949) 17, 782.
23. Heel, H., Zeil, W., Z. Elektrochem. Ber. Bunsen Gesel. Phys. Chem. (1960) 64, 692.
24. Allred, A., Rochow, E. G., J. Inorg. Nucl. Chem. (1958) 5, 269.
25. Brown, R. S., Eaton, D. F., Hosomi, A., Traylor, T. G., Wright, J. M., J. Organometal. Chem. (1974) 66, 249.
26. Evans, S., Green, J. C., Joachim, P. J., Orchard, A. F., Turner, T. W., Maier, J. P., J. Chem. Soc., Trans. (1972) 2, 903.

RECEIVED May 3, 1976.

The Optical Stability of Organotin Compounds

MARCEL GIELEN, CORNELIS HOOGZAND, SERGE SIMON, YVES TONDEUR,
IVAN VAN DEN EYNDE, and MICHEL VAN DE STEEN

Free University of Brussels, Pleinlaan, 2, A.O.SC.-T.W., B-1050 Brussels, Belgium

Eight different types of triorganotin compounds RR′R″SnY are configurationally stable within the NMR time scale. Three of them are even optically stable for several weeks. Thus, the synthesis of optically active RR′R″SnY compounds and the study of the stereochemistry of substitution reactions at the tin atom of these compounds are both feasible.

Bimolecular nucleophilic substitutions at a saturated carbon atom have been studied extensively, and their stereochemistry, the well known Walden inversion, seems to be quite general (*1*) and can be easily interpreted (*2*). Organic halides RR′R″CX and analogous derivatives which undergo S_N2 reactions are optically stable.

Bimolecular electrophilic substitutions at a saturated carbon atom have been less extensively studied, and their stereochemistry is not so clear cut. Sometimes the relative configuration is retained, sometimes it is inverted and the factors influencing their stereochemistry are not yet well defined (*3, 4, 5, 6*). Organomercury-, organotin-, organoboron-, and organolithium compounds RR′-R″C—MX$_n$, for which the stereochemistry of these S_E2 reactions has generally been studied, are optically stable at carbon.

Bimolecular nucleophilic substitutions at tetrahedrally substituted silicon or germanium atoms have been studied during the past 20 yr (*7*). Here also, the reactions are often stereospecific and may occur either with retention or with inversion. Now we can begin to understand their stereochemistry (*5*).

Nothing seems to be known about the stereochemistry of S_E2 reactions at silicon or germanium. Many optically stable triorgano RR′R″Si*Y or diorgano RR′Si*XY silicon compounds or triorganogermanium compounds RR′R″Ge*X are available.

In organotin chemistry, the situation is somewhat different. Ten years ago, racemic triorganotin halides RR′R″SnX and tetraorganotin RR′R″SnR‴ compounds were not readily available. This is not the case any more (*10–16*). Tetraorganotin compounds RR′R″SnR‴ are optically stable. Unfortunately, triorganotin halides RR′R″SnX are not configurationally stable even within the NMR time scale, especially in the presence of nucleophiles ($17, 18, 19, 20, 21$). This optical instability prevents the study of the stereochemistry at tin of the halodemetallation reaction. The kinetics of this cleavage reaction of the carbon–tin bond has been studied the most ($10, 11, 12$).

To study the stereochemistry of substitution reactions at tin, it is necessary to use two optically stable compounds RR′R″SnY and RR′R″SnZ, which can be interconverted either by:

$$RR′R″SnY + ZU = RR′R″SnZ + YU$$

or by

$$RR′R″SnZ + YU = RR′R″SnY + ZU.$$

This is why we have proceeded in two different directions:

(1) To determine how to explain the configurational instability of the triorganotin halides in order to find out how to increase their optical stability

(2) To find optically stable triorganotin compounds RR′R″SnY, RR′R″SnZ, . . . which will be used as substrates for the study of the stereochemistry of substitution reactions at tin.

Configurational Instability of Triorganotin Halides

It has been shown by NMR at 60, 100, and 270 MHz that the racemization of triorganotin halides is a third-order process: first order with respect to RR′R″SnX and second order with respect to the nucleophile N (22), as for triorganosilicon halides ($23, 24, 25$).

The rate r_{coal} at which diastereotopic groups are permuted at the coalescence of the neophylic methyl groups in methylneophyl-*tert*-butyltin bromide (**I**) or

$$C_6H_5-\underset{\underset{CH_3}{|}}{\overset{\overset{CH_3}{|}}{C}}-CH_2-\underset{\underset{R}{|}}{\overset{\overset{X}{|}}{Sn}}-CH_3$$

I R = *tert*-Bu, X = Br
II R = C$_6$H$_5$, X = Cl
III R = C(C$_6$H$_5$)$_3$, X = Br

in methylneophylphenyltin chloride (**II**) as a consequence of a change of the absolute configuration of the tin atom, can be expressed by ($26, 27$):

$$r_{coal} = (\pi/\sqrt{2})\,\Delta\nu_\infty(\text{Herz}) \times [RR′R″SnX]$$

or

$$r_{coal} = (\pi/\sqrt{2})\,\Delta\nu_\infty(\text{ppm}) \times \nu_0(\text{MHz}) \times [RR′R″SnX]$$

On the other hand, one can write:

$$r_{coal} = k_2[RR′R″SnX]\,[N]_{coal} + k_3[RR′R″SnX][N^2]_{coal}$$

so that, at a given temperature and with a constant concentration of RR′R″SnX, one has:

$$\frac{\Delta\sigma_\infty \times \nu_{0,i}}{[N]_{coal,i}} = (\sqrt{2}/\pi)\, k_2 + k_3[N^2]_{coal,i}$$

where $[N]_{coal,i}$ is the concentration of the nucleophile which causes the coalescence at that temperature and concentration of RR′R″SnX at a given field $\nu_{0,i}$. The experimental results are shown in Figure 1.

From the experimental results shown in Figure 1, it is clear that the racemization of these triorganotin halides is second order in N. The most reasonable mechanism for the racemization consistent with this observation is:

$$\underset{\diagdown}{\diagup}Sn-X + N \rightleftharpoons N^{(+)}\overset{|}{\underset{\diagup}{-}}Sn\overset{(-)}{\underset{\diagdown}{-}}X \rightleftharpoons N^{(+)}\overset{\diagup}{\underset{\diagdown}{-}}Sn + X^{(-)}$$

$$N^{(+)}\overset{\diagup}{\underset{\diagdown}{-}}Sn + N \xrightarrow{\text{rate determining}} N^{(+)}\overset{|}{\underset{\diagup}{-}}Sn\overset{(-)}{\underset{\diagdown}{-}}N^{(+)}$$

Bulky groups on tin inhibit such a complexation (28). This is probably why **I** racemizes more slowly than **II**, as shown by the slopes of the lines in Figure 1.

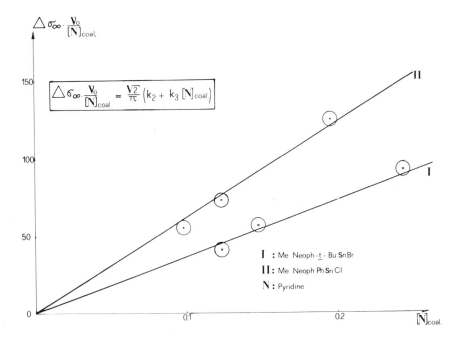

*Figure 1. Determination of reaction order with respect to pyridine (N) for the racemization of 0.262M methylneophyl-tert-butyltin bromide (**I**) and of 0.332M methylneophylphenyltin chloride (**II**) at 22°C*

Enhanced Optical Stability of Sterically Hindered Triorganotin Halides

One might thus expect an enhanced optical stability for sterically hindered triorganotin halides, for instance for methylneophyltrityltin bromide (**III**) (29). The addition of DMSO (up to 0.5M) to a CCl_4 solution of **III** does not cause the diastereotopic neophylic methyl groups to coalesce, whereas smaller quantities of DMSO (less than 0.05M) induce the coalescence of the diastereotopic groups of **I**. Addition of pyridine (up to 3 M in $CDCl_3$) or an increase in temperature (up to 150°C) does not cause the diastereotopic methyl peaks to broaden (29).

Optical Stability of Tetraorganotin Compounds

The optical stability of tetraorganotin compounds has been studied by NMR. Even in the presence of strong nucleophiles, diastereotopic groups do not coalesce (8–21, 30–35).

It was also possible to synthesize an optically pure chiral tetraorganotin compound in which the tin atom is the only chiral center (12, 36, 37, 38) starting from tetra-p-anisyltin, An_4Sn:

$$An_4Sn \xrightarrow{SnCl_4} An_2SnCl_2 \xrightarrow[NpMgBr]{MeMgI} MeAn_2SnNp \xrightarrow{HCl} MeAnNpSnCl \xrightarrow{LiAlH_4}$$

$$MeAnNpSnH \xrightarrow{CH_2=CH-COO(-)Menthyl} MeAnNpSnCH_2CH_2COO(-)Menthyl$$

Four steps are needed to get methylnaphthyl-p-anisyltin hydride MeAnNpSnH which is added to (−)menthylacrylate to yield two diastereoisomers which can be distinguished by NMR and separated. One of these with $[\alpha]_D$ − 24°, reacted with methylmagnesium iodide to give the corresponding chiral alcohol $[\alpha]_D$ + 9°, which is optically stable for years. ($[\alpha]_D$ does not change significantly after three years).

Lequan has recently synthesized another optically active tetraorganotin compound by a similar procedure (34). Taddei (35) has also prepared an optically active, but probably not optically pure, benzylmethylphenylisopropyltin, $[\alpha]_D$ + 4.6°, by an asymmetric synthesis:

$$MePh\text{-}iso\text{-}PrSn\text{---}O(-)Menthyl \xrightarrow{PhCH_2MgX} MePh\text{-}iso\text{-}PrSnCH_2Ph$$

Configurational Stability of Triorganostannylamines, -phosphines, and -arsines

Triorganotin halides are generally not configurationally stable even within the NMR time scale. Organotin compounds with a tin–oxygen bond RR'-R''SnOR''' are configurationally stable within the NMR time scale for some R''' groups and configurationally unstable for other R''' groups (*40*). Thus it seems that the configurational stability of RR'R''SnY increases when replacing Y by an element Z which is found at the left of Y in the periodic table. This is the reason why we studied the configurational stability of triorganostannylamines, -phosphines, and -arsines **IV** and **V** (Y = NEt₂, PPh₂, and AsPh₂).

We have prepared dimethyl(2-phenylpropyl)stannyl and methylphenyl(2-phenylpropyl)stannyl compounds of nitrogen, phosphorus, and arsenic (*41*) and showed by NMR that they are configurationally stable within the NMR time scale even in the presence of up to 2.5*M* HMPT.

Three Other Classes of Configurationally Stable Organotin Compounds

Triorganotin Hydrides. Tetraorganotin compounds are optically stable. We expected triorganotin hydrides to be also configurationally stable within the NMR time scale. This has been tested on **VI** (Y = H).

VI (Y = H) had been already prepared by Peddle and Redl (*8, 9*), but they did not see any anisochronism of the diastereotopic groups at 100 MHz. However, we observed anisochronous neophylic methyl groups in C₆D₆ at 270 MHz (Δδ = 0.016 ppm) and an AB pattern for the diastereotopic methylenic protons when the H on tin is decoupled ($\Delta\delta$ = 0.059 ppm, $|J_{AB}|$ = 12 Hz) (21). These aniso-chronisms remain unchanged upon addition of nucleophiles such as diphenyl-phosphine, bipyridyl, and HMPT and even when VI (Y = H) is dissolved in ac-etonitrile (+LiBr) or in DMSO-d_6(+LiBr). The configurational stability of triorganotin hydrides in the presence of nucleophiles has also analogously been checked IV (Y = H) and V (Y = H).

Tin–Germanium Compounds. We expected organotin compounds with a tin–germanium bond also to be configurationally stable and tested for this on VII (42)

$$
\begin{array}{ccc}
\text{Me} & & \text{Me} \\
| & & | \\
\text{Ph}-\text{Sn}- & & \text{Ge}-\text{Ph} \\
| & & | \\
\text{Me} & & \text{Np}
\end{array}
$$

VII

The anisochronism $\Delta\delta$ of the two diastereotopic methyl groups is observed even in nucleophilic solvents like DMSO ($\Delta\delta$ = 0.043 ppm) or HMPT ($\Delta\delta$ = 0.058 ppm).

Tin–Transition Metal Compounds. Organotin compounds with a tin–transition metal bond have also been studied (42). VI [Y = Mo(η-C$_5$H$_5$)(CO)$_3$] shows two diastereotopic neophylic methyl groups separated by about 0.16 ppm in a CD$_2$Cl$_2$ solution containing up to 6M of DMSO, phenol, aniline, or pyridine. One also observes an AB pattern for the diastereotopic methylene protons ($\Delta\delta$ ~ 0.18 ppm; $|J_{AB}|$ = 12.6 Hz).

Organotin compounds with a tin–iron bond behave analogously (43): VI [Y = Fe(η-C$_5$H$_5$)(CO)$_2$] shows two diastereotopic neophylic methyl groups, even in nucleophilic solvents like acetone or in the presence of very strong nucleophiles like DMSO.

In pyridine the anisochronisms of the diastereotopic groups of VI [Y = Fe(η-C$_5$H$_5$)(CO)$_2$] remain practically unaffected up to 80°C. Its 22.63-MHz ^{13}CMR spectrum in the same solvent shows two diastereotopic signals for the neophylic methyl carbons ($\Delta\delta_{AB}$ = 77 Hz) showing that this compound is confi-gurationally stable about the tin atom within the NMR time scale.

The anisochronisms of the methyl groups of VIII in the PMR and ^{13}CMR spectra show that VIII is configurationally stable around the iron atom as well as around the tin atom, within the NMR time scale.

$$
\begin{array}{ccc}
\text{Me} & & \text{CO} \\
| & & | \\
\text{Ph}-\text{Sn}- & & \text{Fe}\,(\eta-\text{C}_5\text{H}_5) \\
| & & | \\
\text{Me} & & \text{PPh}_3
\end{array}
$$

VIII

The Optical Stability of Tin–Germanium Compounds

Two optically active, but probably not optically pure, compounds containing a tin–germanium bond have been synthesized, the first having an asymmetric tin atom and the second having an asymmetric germanium atom.

Methylphenylisopropyltin chloride was converted into the corresponding (−)-menthoxy compound, which was allowed to react with triphenylgermyllithium. The product obtained, **IX,** is optically active: $[\alpha]^{20°C}_{\lambda=435.8\ nm} = +0.66 \pm 0.05°$ (C = 6.33).

$$\text{PhMe-iso-PrSnCl} \xrightarrow[\text{2) Ph}_3\text{GeLi}]{\text{1) (−) Menthyl OLi}} \underset{\textbf{(IX)}}{\text{PhMe-iso-PrSnGePh}_3}$$

The 70-eV mass spectrum of **IX** shows a tin-containing fragment at $m/e = 515$ (Ph$_4$MeGeSn$^{(+)}$), another at $m/e = 481$ (Ph$_3$MeisoPrGeSn$^{(+)}$), another at $m/e = 439$ (Ph$_3$MeGeSnH$^{(+)}$) with a metastable peak at about $m/e = 400$ (481 → 439), and another tin-containing ion at $m/e = 423$ (Ph$_3$GeSn$^{(+)}$). The NMR spectrum of **IX** is compatible with the proposed structure. $J(^{119}\text{Sn—C}^1\text{H}_3) = 52$ Hz as expected.

The optical activity remained practically unchanged for several weeks, showing that tin–germanium compounds are optically stable at tin. They are also optically stable at germanium as shown by the optical activity obtained for (dimethylphenylstannyl)methyl-1-naphthylphenylgermanium **X** which also remained unaltered during several weeks.

$$(+)\text{–PhMeNpGeH} \xrightarrow{\text{PhMe}_2\text{SnNEt}_2} \underset{\textbf{X}}{\text{PhMeNpGeSnMe}_2\text{Ph}}$$

(optical purity: 30%)

$$\text{in CCl}_4 \quad : [\alpha]_D^{20} -6.91° \pm 0.05° \ (C = 10.68)$$
$$\text{in HMPT} : [\alpha]_D^{20} -6.27° \pm 0.05° \ (C = 12.75)$$

Optical Stability of Tin–Iron Compounds

The optical stability of tin–iron compounds has been shown by the synthesis of two different mixtures of diastereoisomers and by the fact that these mixtures remain different after several weeks, showing that tin retains its configuration for long periods. The proportions of the two diastereoisomers were determined by integrating the two diastereotopic methyl–tin signals which appear at 0.489 and 0.543 ppm *vs.* TMS in pyridine-d_5 and by that of the diastereotopic cyclopentadienyl protons. The syntheses of these mixtures of diastereoisomers can be described as follows:

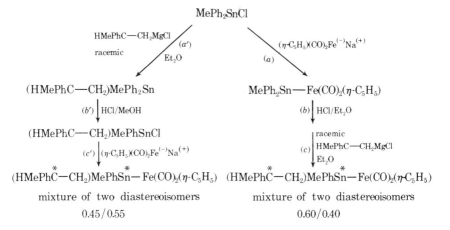

The yields are 70% for (a) and 50% for (bc) (35% for the first route), 70% for (a'), 90% for (b'), and 70% for (c') (thus 44% for the other one).

Conclusion

Since optically stable triorganotin compounds are available, the study of the stereochemistry of substitution reactions at tetracoordinated tin atoms should be feasible. Optically active triorganotin compounds must, therefore, be synthesized, and we will focus all our efforts to reach that goal in the very near future.

Acknowledgments

The authors wish to thank F. Resseler, M. Desmedt, R. Polain, and C. Moulard for recording the spectra. They also thank the "Fonds voor Kollektief Fundamenteel Onderzoek F.K.F.O." and the "Nationale Raad voor Wetenschapsbeleid" for a grant. They are grateful to the "Institut pour l'Encouragement de la Recherche Scientifique dans l'Industrie et l'Agriculture I.R.S.I.A." (Y.T., M. V.D.S.). Financial help from the "Nationaal Fonds voor Wetenschappelijk Onderzoek," and from the "Ministerie van Nationale Opvoeding" is also gratefully acknowledged.

Literature Cited

1. Maryanoff, C. A., Ogura, F., Mislow, K., *Tetrahedron Lett.* (1975) 4095.
2. Bunton, C. A., "Nucleophilic Substitution at a Saturated Carbon Atom," in "Reaction Mechanisms in Organic Chemistry," C. Eaborn, Ed., Elsevier, 1966.
3. Jensen, F. R., Rickborn, B., "Electrophilic Substitution of Organomercurials," McGraw-Hill, New York, 1968.
4. Abraham, M. H., "Electrophilic Substitutions at a Saturated Carbon Atom," in "Comprehensive Chemical Kinetics," C. H. Bamford and C. F. H. Tipper, Eds., Vol. 12, Elsevier, 1973.

5. Matteson, D. S., "Organometallic Reaction Mechanisms," Academic, New York, 1974.
6. Reutov, O. A., Beletskaya, I. P., "Reaction Mechanisms of Organometallic Compounds," North Holland, Amsterdam, 1968.
7. Sommer, L. H., "Stereochemistry, Mechanisms and Silicon," McGraw-Hill, 1965.
8. Gielen, M., Dehouck, C., Mokhtar-Jamai, H., Topart, J., *Rev. Silicon, Germanium, Tin, Lead Compd.* (1974) 1, 9.
9. Gielen, M., "Stéréochimie Dynamique," Freund, Tel Aviv, 1974.
10. Gielen, M., *Acc. Chem. Res.* (1973) 6, 198.
11. Gielen, M., Nasielski, J., *Organotin Compd.* (1972) 3, 625–825.
12. Gielen, M., *Ind. Chim. Belge.* (1973) 38, 20, 138.
13. Boué, S., Gielen, M., Nasielski, J., *Tetrahedron Lett.* (1968) 1047.
14. Boué, S., Gielen, M., Nasielski, J., Lieutenant, J. P., Spielmann, R., *Bull. Soc. Chim. Belg.* (1969) 78, 135.
15. Gielen, M., De Poorter, B., Sciot, M. T., Topart, J., *Bull. Soc. Chim. Belg.* (1973) 82, 271.
16. Gielen, M., Nasielski, J., Topart, J., *Recl. Trav. Chim. Pays-Bas* (1968) 87, 1051.
17. Peddle, G. J. D., Redl, G., *J. Am. Chem. Soc.* (1970) 92, 365.
18. Stynes, D. V., Allred, A. L., *J. Am. Chem. Soc.* (1971) 93, 2666.
19. Boer, F. P., Doorakian, G. A., Freedman, H. H., McKinley, S. V., *J. Am. Chem. Soc.* (1970) 92, 1225.
20. Holloway, C. E., Kandil, S. A., Walker, I. M., *J. Am. Chem. Soc.* (1972) 94, 4027.
21. Gielen, M., De Clercq, M., Mayence, G., Nasielski, J., Topart, J., Vanwuytswinkel, H., *Recl. Trav. Chim. Pays-Bas* (1969) 88, 1137.
22. Gielen, M., Mokhtar-Jamai, H., *J. Organomet. Chem.* (1975) 91, C33.
23. Corriu, R., Henner-Leard, M., *J. Organomet. Chem.* (1974) 64, 351.
24. *Ibid.* (1974) 65, C39.
25. *Ibid.* (1974) 74, 1.
26. Pople, J. A., Schneider, W. G., Bernstein, H. J., "High Resolution Nuclear Magnetic Resonance," p. 223, McGraw-Hill, New York, 1959.
27. Kost, D., Carlson, E. H., Raban, M., *Chem. Comm.* (1971) 656.
28. Gielen, M., De Clercq, M., *J. Organomet. Chem.* (1973) 47, 351.
29. Gielen, M., Mokhtar-Jamai, H., *Bull. Soc. Chim. Belg.* (1975) 84, 1037.
30. Peddle, G. J. D., Redl, G., *J. Am. Chem. Soc.* (1964) 86, 2628.
31. Morris, D. R., Rockett, B. W., *J. Organomet. Chem.* (1972) 40, C21.
32. Gielen, M., Barthels, M. R., De Clercq, M., Dehouck, C., Mayence, G., *J. Organomet. Chem.* (1972) 34, 315.
33. Jean, A., Lequan, M., *J. Organomet. Chem.* (1972) 36, C9.
34. Lequan, M., Meganem, F., *J. Organomet. Chem.* (1975) 94, C1.
35. Folli, U., Iarossi, D., Taddei, F., *J. Chem. Soc. Perkin II* (1973) 638.
36. Mokhtar-Jamai, H., Gielen, M., *Bull. Soc. Chim. Fr.* (1972) 9B, 32.
37. Gielen, M., Mokhtar-Jamai, H., *Ann. N. Y. Acad. Sci.* (1974) 239, 208.
38. Gielen, M., Boué, S., De Clercq, M., De Poorter, B., *Rev. Silicon, Germanium, Tin, Lead Compd.* (1974) 1, 97.
39. Gielen, M., Mokhtar-Jamai, H., *Bull. Soc. Chim. Belg.* (1975) 84, 197.
40. Folli, U., Iarossi, D., Taddei, F., *J. Chem. Soc. Perkin II* (1973) 1284.
41. Gielen, M., Tondeur, Y., *Bull. Soc. Chim. Belg.* (1975) 84, 933.
42. Gielen, M., Simon, S., Tondeur, Y., Van De Steen, M., Hoogzand, C., *Bull. Soc. Chim. Belg.* (1974) 83, 337.
43. Gielen, M., Hoogzand, C., Van Den Eynde, I., *Bull. Soc. Chim. Belg.* (1975) 84, 939.

RECEIVED May 3, 1976.

18

The Structural Chemistry of Some Organotin Oxygen-Bonded Compounds

PHILIP G. HARRISON

Department of Chemistry, University of Nottingham, University Park, Nottingham, England

Crystallographic structural data for organotin compounds of composition R_3SnOE and $R_2Sn(OE)_2$ (E = an organic residue) were examined. Preferred coordination numbers for these species are five and six, respectively. The stereochemistry at tin in six-coordinated dimethyltin compounds depends not only on the nature of the electronegative ligands but also on the structure of the crystal lattice. The magnitude of the tin-119m Mössbauer quadrupole splitting is characteristic of the stereochemistry at tin in both tri- and diorganotin derivatives. In addition, the temperature dependence of the Mössbauer recoil free fraction of several compounds of known structure has been used to illustrate how this parameter can be used to distinguish different lattice types.

Determination of molecular and lattice structure are problems fundamental to chemists in general. Of the "direct" methods which may be used, x-ray crystallography is now at that stage where, for most materials, procurement of the single crystal is virtually the only obstacle to complete structural identification. Organotin chemists, however, now have a vast array of "indirect" physicochemical techniques, which, when used together, can provide just as potent (though only qualitative) a source of structural information. Tin 119m Mössbauer spectroscopy in particular is perhaps the most rewarding single technique available. Careful interpretation of the parameter data that it yields can furnish structural information for those materials which frustrate the crystallographer—amorphous powders which resist crystallization. The validity of any such interpretation relies on confirmation from model compounds for which both Mössbauer and x-ray structural data are available. This article describes some major aspects of the structural chemistry of di- and triorganotin(IV) oxygen-bonded compounds and the applicability of the Mössbauer quadrupole splitting and recoil-free fraction in determining molecular and lattice structure.

X-ray Crystallographic Results

Crystallographic data for simple four-coordinated compounds of the types R_3SnOE and $R_2Sn(OE)_2$ are, surprisingly, lacking. To date, the structures of only two rather sophisticated examples, $Et_3SnOC_6Cl_4OSnEt_3$ (*1*) (by two-dimensional methods) and $Ph_3SnOC_5Ph_4[Mn(CO)_3]$ (*2*), both containing slightly distorted tetrahedral configurations, have been determined. Tetrahedral geometries are also undoubtedly possessed by derivatives of more bulky –OE ligands, and it seems likely that the bis(triorganotin) oxides as well as Ph_3SnO-$GePh_3$ and $Ph_3SnOSiPh_3$ have angular structures similar to that of $Ph_3PbOSiPh_3$ (*3*).

A reduction in the steric bulk of R and E in R_3SnOE derivatives permits the oxygen atom to function as a bridging ligand which links adjacent planar R_3Sn units and raises the coordination number of the tin to five. Me_3SnOH (*4*), Me_3SnOMe (*5*), and $Me_3SnON{=}C_6H_{10}$ (*6*) have structures of this type in the crystal, with the E groups pendant from the oxygen atoms of the parallel infinite $-\!(Me_3SnO)\!-$ chains. The resultant trigonal bipyramidal geometry of the tin in the repeating unit of $Me_3SnON{=}C_6H_{10}$ and projections of the unit cell illustrating the infinite chain structure of the solid are shown in Figures 1 and 2, respectively.

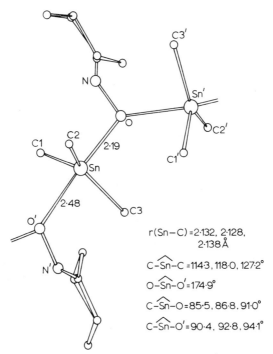

$r(Sn-C) = 2.132, 2.128, 2.138\,\text{Å}$

$C-\widehat{Sn}-C = 114.3, 118.0, 127.2°$

$O-\widehat{Sn}-O' = 174.9°$

$C-\widehat{Sn}-O = 85.5, 86.8, 91.0°$

$C-\widehat{Sn}-O' = 90.4, 92.8, 94.1°$

Figure 1.   Repeating unit of $Me_3SnON{=}C_6H_{10}$

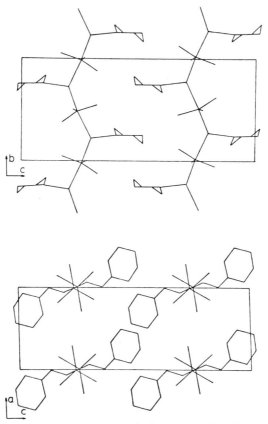

Figure 2. Projection of the unit cell of Me₃-SnON=C₆H₁₀ onto (a) the bc plane and (b) the ac plane illustrating the infinite chain structure

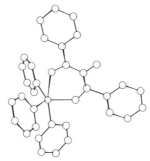

Figure 3. Molecular geometry in Ph₃Sn-(dibenzoylmethanate)

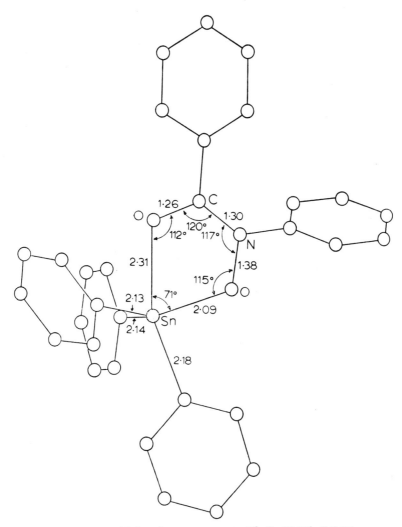

Figure 4.   Molecular geometry in $Ph_3SnONPh \cdot CO \cdot Ph$

*Figure 5.   Re-
peating unit in
$Me_3SnO_2CCF_3$*

The inclusion of a second potential donor site in the E residue allows the –OE ligand to function in three possible ways: (a) as a unidentate group, (b) as a chelating ligand, or (c) as a bridging group giving rise to chain structures similar to those of the Me_3SnOE (E = H, Me, N=C_6H_{10}) compounds mentioned earlier. Only when the organic groups attached to tin are so bulky as to preclude chelation or bridging does coordination in a unidentate fashion persist, and the only example of this type as yet characterized is $(C_6H_{11})_3SnOAc$ (7) which has a distorted tetrahedral geometry. The second tin–oxygen bond distance, however, is only 2.95 Å, and might represent some degree of bonding interaction. Two geometries are possible when the ligand functions as a chelating group: the cis and meridional trigonal bipyramidal configurations I and II, respectively.

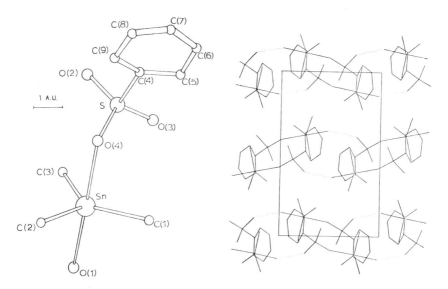

The meridional geometry has not been conclusively identified although it has been proposed for the trimethyltin derivatives of β-ketoenolates (8) and tricyclohexyltin glycylglycinate. The cis configuration, on the other hand, has been distinguished for the triphenyltin derivatives of the anions of 1,3-diphenylpro-

Figure 6. (a) *Environment about tin in* $Me_3SnO_3SPh \cdot H_2O$. (b) *Projection of the unit cell of* $Me_3SnO_3SPh \cdot H_2O$ *onto the* bc *plane illustrating the hydrogen-bonded chain structure. Hydrogen bonds are broken lines.*

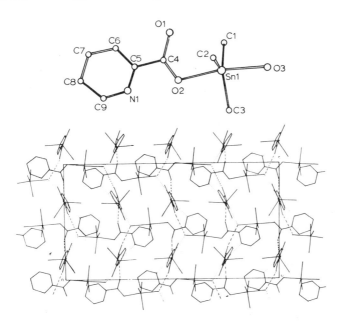

Figure 7.   (a) Environment about tin in Me₃Sn(picolinate)·H₂O.   (b) Projection of the unit cell of the Me₃Sn(picolinate)·H₂O onto the ac plane.   Hydrogen bonds are dotted and broken lines.

pane-1,3-dione and N-phenyl-N-benzoylhydroxylamine (9, 10). These derivatives have very similar geometries, though both are somewhat distorted from ideality (Figures 3 and 4).

The trans-trigonal bipyramidal geometry, III, is exhibited by the three triorganotin carboxylates, Me_3SnOAc (11), $Me_3SnO_2CCF_3$ (11), and $(PhCH_2)_3SnOAc$ (12), the carboxylate group bridging planar C_3Sn moieties in each case.   Figure 5 shows the repeating unit in $Me_3SnO_2CCF_3$.   Analogous chain structures have also been proposed for triorganotin arylsulfonates (13) and nitrates (14).

Trimethyltin nitrate (15), phenylsulfonate (13), and picolinate (2-pyridylcarboxylate) (16) all form stable monohydrates.   The trans trigonal bipyramidal stereochemistry is preserved in all three cases, but now a water molecule occupies the second axial coordination site (Figures 6a and 7a).   The lattice structures of $Me_3SnO_3SPh·H_2O$ and $Me_3SnNO_3·H_2O$ are similar, and both consist of infinite chains of $[Me_3SnL·H_2O]$ units linked together by hydrogen bonds (Figure 6b).   That of $Me_3Sn(picolinate)$ is more complex, with both carboxyl oxygen atoms and the pyridyl nitrogen atom taking part in hydrogen bonding with the coordinated water molecule, thus affording a more rigid lattice (Figure 7b).   Even though the picolinate anion is potentially terdentate, the preferred coordination number at tin is still five, and no six-coordinated trimethyltin derivatives have been characterized crystallographically.

The introduction of a second electronegative substituent increases the effective nuclear charge of the tin and hence the tendency for tin to seek coordination numbers greater than four. Crystal data for simple diorganotin derivatives of monofunctional oxygen ligands are unavailable, and the same situation exists for diorganotin biscarboxylates. However, the structures of the dimethyltin derivatives of several other bifunctional ligands have been determined, and all contain hexacoordinated tin. Observed structures encompass nearly the whole range of CSnC valence bond angles (θ) from almost ideal cis-(IV) ($\theta = 90°$) to the trans geometry (VI) ($\theta = 180°$), via many intermediate cases (V) where $90° < \theta < 180°$.

Examples of both bridging and chelating bifunctional ligands are also known. The structure of dimethyltin bis(fluorosulfonate) (17) consists of infinite sheets in which fluorosulfonate groups bridge linear dimethyltin units although $(Me_2Sn)_3(PO_4)_2 \cdot 8H_2O^{18}$ has a similar structure in which phosphate groups bridge both linear and nonlinear ($\theta = 147°$, $150°$) dimethyltin units to give a lattice consisting of infinite ribbons. Water molecules also coordinate to the nonlinear dimethyltin species.

β-Ketoenolate, N-acylhydroxylamino, nitrate, dithiocarbamate, and oxinate all function as chelating ligands towards the dimethyltin residue. Table I gives the available CSnC bond angle data for $Me_2Sn(chelate)_2$ derivatives as well as the angle subtended by the ligands at tin (the ligand "bite" angle) in each case.

Table I. Valence Bond Angle and Tin 119m Mössbauer Data for Six-Coordinated Dimethyltin Compounds

Compound	CSnC Angle[a]	Ligand "bite"[a]	Quadrupole Splitting[b]
$Me_2Sn(acac)_2$	180	86	4.02
$Me_2Sn(salen)$	160.0	80.5	3.46
$Me_2Sn(ONH \cdot CO \cdot Me)_2 \cdot H_2O$	156.8	71.5	3.42
$Me_2Sn(ONMe \cdot CO \cdot Me)_2$	145.8	71.3	3.31
$Me_2Sn(NO_3)_2$	144	54.6	4.20
$Me_2Sn(S_2CNMe_2)_2$	136	64.3	3.14
$Me_2Sn(oxin)_2$	110.7	73.6	2.02
$Me_2Sn(ONH \cdot CO \cdot Me)_2$	109.1	81.5	1.99

[a] Degrees.
[b] mm sec^{-1}.

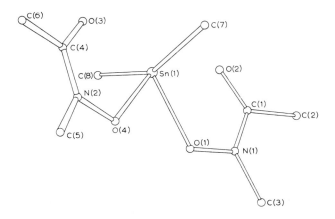

Figure 8.   Geometry about tin in Me₂Sn(ONMe·CO·Me)₂.

The factors governing the particular stereochemistry adopted seem quite complex. The value of the ligand "bite" angle seems relatively unimportant since the values for complexes with both low and high CSnC bond angles are similar. However, for compounds with lattices composed of essentially non-interacting molecules, crystal-packing effects will be minimized and the geometry at tin will be determined largely by the particular demands (steric, electronic, etc.) of each individual ligand, as exemplified by the structures of $Me_2Sn(acac)_2$ *(19)*, $Me_2Sn(salen)$ *(20)*, $Me_2Sn(ONMe·CO·Me)_2$ *(21)*, $Me_2Sn(NO_3)_2$ *(22)*, $Me_2Sn(dtc)_2$ *(23)*, and $Me_2Sn(oxin)_2$ *(24)*. Lattice structure can be important in determining molecular geometry as shown by comparing structural data for the three similar compounds $Me_2Sn(ONMe·CO·Me)_2$ *(21)*, $Me_2Sn(ONH·CO·Me)_2$ *(25)*, and its monohydrate $Me_2Sn(ONH·CO·Me)_2·H_2O$ *(25)*. Crystals of the former compound consist of independent, non-interacting $Me_2Sn(ONMe·CO·Me)_2$ molecules possessing an intermediate, type V, geometry ($\theta = 145.8°$) (Figure 8). The value of the ligand "bite" angle (71.3°) compares well with that of $Ph_3SnONPh·CO·Ph$ *(10)* (71.5°). Here the molecules are well separated in the crystal, affording some corroboration that the N-acylhydroxylamino ligands have indeed assumed an equilibrium molecular geometry about tin. Replacement of the N-methyl group on the hydroxylamino ligand by a hydrogen dramatically changes the stereochemistry to the cis geometry (IV) (Figure 9a). The CSnC angle closes to 109.1°, the ligand "bite" angle increases to 81.5°, and adjacent molecules are now held in the crystal lattice by four NH···O=C hydrogen bonds. The lattice is thus composed of parallel infinite chains although no significant interaction takes place between adjacent chains (Figure 9b). The inclusion of a molecule of water into the lattice of $Me_2Sn(ONH·CO·Me)_2$ has an equally drastic effect on the stereochemistry at tin, even though there is no direct interaction between the solvating water molecules and the tin atoms [*cf.* $Me_3SnO_3SPh·H_2O$ *(13)*, $Me_3SnNO_3·H_2O$ *(15)*, $Me_3Sn(picolinate)$ *(16)*, and $(Me_2Sn)_3(PO_4)_2·8H_2O$ *(18)*].

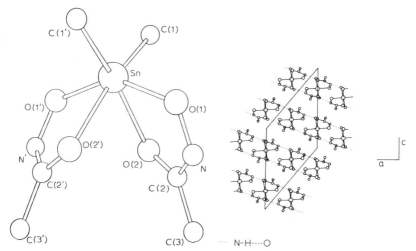

Figure 9. (a) *Geometry about tin in* Me₂Sn(ONH·CO·Me)₂. (b) *Unit cell projection of* Me₂Sn(ONH·CO·Me)₂ *onto the* ac *plane showing* NH···OSn *hydrogen bonds* (*dotted lines*).

The coordination at tin in the two crystallographically distinct Me₂Sn(ONH·CO·Me)₂ molecules of the monohydrate, Me₂Sn(ONH·CO·Me)₂·H₂O, is quite different from that in the anhydrous material but resembles that in Me₂Sn(ONMe·CO·Me)₂. The CSnC bond angle is increased to 156.8°, and the ligand "bite" angle closes to a value almost identical to those of Me₂Sn(ONMe·CO·Me)₂ and Ph₃SnONPh·CO·Ph (Figure 10). Crystals of the monohydrate are composed of alternate double layers of Me₂Sn(ONH·CO·Me)₂ molecules and water molecules connected by a complex system of hydrogen bonds (Figure 11). Within each Me₂Sn(ONH·CO·Me)₂ double layer, the molecules are of the same crystallographic type, with layers of each type alternating. Adjacent Me₂Sn-(ONH·CO·Me)₂ molecules within each layer are connected via NH···O- Sn hydrogen bonds (note the change of hydrogen bond type from the anhydrous material), while the water molecules hold adjacent Me₂Sn(ONH·CO·Me)₂ double layers together. Each water molecule is hydrogen bonded to its two nearest neighbor water molecules, and each forms two additional hydrogen bonds with hydroxylamino oxygen or nitrogen atoms of Me₂Sn(ONH·CO·Me)₂ molecules in the nearest Me₂Sn(ONH·CO·Me)₂ double layer. The lattice is thus quite rigid.

Why are there such drastic changes in stereochemistry at tin from seemingly minor changes in constitution? It is not unreasonable to expect the –ONMe·CO·Me and –ONH·CO·Me ligands to have similar steric and electronic requirements. However such a supposition would be true only for isolated molecules in, for example, dilute solution or the gas phase. In the solid phase, interactions between adjacent molecules must also be considered, and it appears that, for these three examples at least, the dominant factor controlling the immediate geometry adopted at tin is the structure of the lattice as a whole. In crystals of Me$_2$Sn(ONMe·CO·Me)$_2$, adjacent molecules are non-interacting and therefore free to adopt an equilibrium geometry resulting from a balance of the different requirements of the methyl and N-acylhydroxylamino ligands. In both Me$_2$Sn(ONH·CO·Me)$_2$ and its monohydrate, crystals are held together by systems of hydrogen bonds. The crystal of the monohydrate is exceptionally complex, and the immediate ligand geometry at tin is modified in each case to allow the most stable lattice structure to be adopted. Comparison of the intra-ring bond distances of the Sn·O·N·C:O heterocyclic rings of the three compounds reveals that some electronic redistribution within the ligands also occurs (Table II), but this is an expected effect of the change in stereochemistry rather than a cause. Moreover, some perturbation of the ring-bonding parameters is expected from the participation of some of the constituent atoms of the heterocyclic rings in the two N-proto-N-acetylhydroxylamino derivatives in hydrogen bonding.

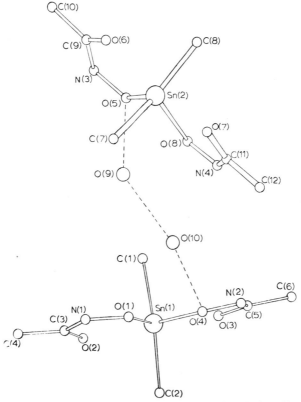

Figure 10. Geometry about the two crystallographically distinct Me$_2$Sn(ONH·CO·Me)$_2$ molecules in Me$_2$Sn(ONH·CO·Me)$_2$·H$_2$O

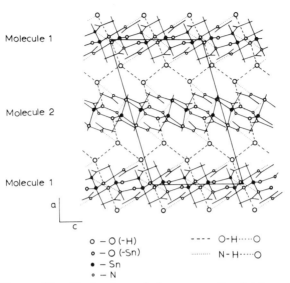

Figure 11. Projection of the unit cell of Me$_2$-Sn(ONH·CO·Me)$_2$·H$_2$O onto the ac plane showing the system of hydrogen bonds (broken and dotted lines)

Table II. Comparison of Bond Distance and Angle Data for Organotin(IV) Derivatives of N-Acylhydroxylamines

Compound	C–Sn–C(°)	Sn–C (Å)	Sn–O (Å)	Sn···O (Å)
Me$_2$Sn(ONH·CO·Me)$_2$	109.1(4)	2.144 (6)	2.106(4)	2.228(4)
Me$_2$Sn(ONH·CO·Me)$_2$·				
H$_2$O top molecule	156.7(7)	2.15(1)	2.16(2)	2.35(1), 2.45 (1)
bottom molecule	156.9(8)	2.15(2)	2.16(2)	2.36(2), 2.43(2)
Me$_2$Sn(ONMe·CO· Me)$_2$	145.8(3)	2.11(1)	2.116(4)	2.377(5)
Ph$_3$SnONPh·CO·Ph	—	eq. 2.14(1)	2.091(5)	2.308(4)
		ax. 2.18 (1)		

Compound	O–Sn–O(°)	N–O (Å)	C–N (Å)	C=O (Å)
Me$_2$Sn(ONH·CO·Me)$_2$	81.5(1)	1.362(5)	1.307(5)	1.277(6)
Me$_2$Sn(ONH·CO·Me)$_2$				
H$_2$O top molecule	71.4(4)	1.40(2)	1.32(3)	1.25(3)
bottom molecule	71.6(5)	1.37(3)	1.34(3)	1.25(4)
Me$_2$Sn(ONMe·CO· Me)$_2$	71.3(1)	1.38(1)	1.31(1)	1.25(1)
Ph$_3$SnONPh·CO·Ph	71.3(2)	1.38(1)	1.30(1)	1.26(1)

Implications of the Structural Data to Tin 119m Mössbauer Spectroscopy

Mössbauer spectroscopy can give information on electronic distribution and stereochemistry about tin through isomer shift and quadrupole splitting data. The use of the technique in this respect has been amply reviewed (at least annually), and the point charge approximation has been used to assign different isomeric stereochemistries of compounds of the same coordination number (*26*, *27*). Available structural data corroborate the predictions of such treatment for the five-coordinate R_3SnL_2 and six-coordinate R_2SnL_4 (L = electronegative ligands such as O, N, halogen, etc.). In the former case, the point charge treatment predicts values of ca. 1.75 mm sec^{-1} (R = Me) and ca. 1.65 mm sec^{-1} (R = Ph) for the cis geometry I but much larger values for the *trans*-(III) [ca. 3.09 mm sec^{-1} (R = Me) and ca. 2.85 mm sec^{-1} (R = Ph)] and *meridional*-(II) [ca. 3.55 mm sec^{-1} (R = Me) and ca. 3.28 mm sec^{-1} (R = Ph)] geometries (*28*). Observed quadrupole splitting values for five-coordinate compounds of known structure are collected in Table III and show that splitting values can be used to distinguish the cis from trans and meridional geometries but are equivocal for the two latter types. However, since $(C_6H_{11})_3SnOAc$ exhibits a splitting of 3.27 mm sec^{-1} caution is necessary, and severe distortion from regular tetrahedral geometry can increase the quadrupole splitting considerably.

Using the point charge approximation Bancroft (*29*) has calculated the expected variation of the quadrupole splitting with the CSnC bond angle for dimethyltin compounds. Predicted values increase smoothly from $\gtrsim 2$ mm sec^{-1} for *cis*-(IV) to $\gtrsim 4$ mm sec^{-1} for the *trans*-(VI). The available structural data again support the model, and observed quadrupole splittings for six-coordinate dimethyltin compounds are related in a fairly simple way to the value of the CSnC bond angle (Table I, Figure 12). The converse deduction—i.e., an estimate of the CSnC bond angle from the value of the quadrupole splitting for such species—therefore has reasonable foundation. Dimethyltin dinitrate is obviously exceptional, and the high observed quadrupole splitting for this compound is probably a result of the high electronegativity of the nitrate group.

Table III. Tin 119*m* Mössbauer Data for Five-Coordinate Triorganotin Derivatives of Known Geometry (mm sec^{-1})

Compound	Isomer Shift	Quadrupole Splitting	Geometry
$Ph_3SnONPh \cdot CO \cdot Ph$ [a]	1.26	1.94	distorted cis
$Ph_3SnO_2C_3HPh_2$ [b]	1.23	2.25	distorted cis
Me_3SnOAc [c]	1.31–1.35	3.43–3.68	trans
$Me_3SnO_2CCF_3$ [c]	1.38	4.22	trans
Me_3SnOH [c]	1.08–1.19	2.71–2.95	trans
$Me_3SnON{=}C_6H_{10}$ [d]	1.43	2.96	trans

[a] Ref. *30*.
[b] Ref. *28*.
[c] Ref. *31*.
[d] Ref. *32*.

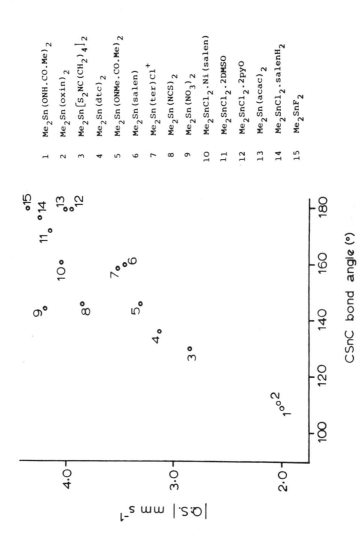

1 $Me_2Sn(ONH.CO.Me)_2$
2 $Me_2Sn(oxin)_2$
3 $Me_2Sn[S_2NC(CH_2)_4]_2$
4 $Me_2Sn(dtc)_2$
5 $Me_2Sn(ONMe.CO.Me)_2$
6 $Me_2Sn(salen)$
7 $Me_2Sn(ter)Cl^+$
8 $Me_2Sn(NCS)_2$
9 $Me_2Sn(NO_3)_2$
10 $Me_2SnCl_2.Ni(salen)$
11 $Me_2SnCl_2.2DMSO$
12 $Me_2SnCl_2.2pyO$
13 $Me_2Sn(acac)_2$
14 $Me_2SnCl_2.salenH_2$
15 Me_2SnF_2

Figure 12. Plot of quadrupole splitting vs. CSnC bond angle for six-coordinate dimethyltin compounds. Except for compounds 1 and 5, the data are taken from Ref. 29.

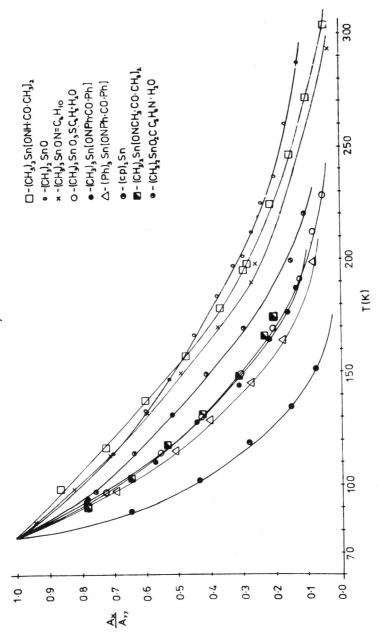

Figure 13. Plot of the relative percentage resonance effect, A_T/A_{77}, vs. T

272 ORGANOTIN COMPOUNDS: NEW CHEMISTRY AND APPLICATIONS

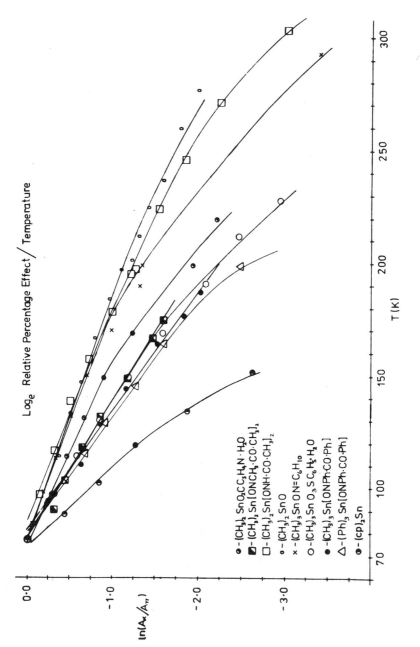

Figure 14. Plot of ln(relative percentage resonance effect), $ln(A_T/A_{77})$, vs. T

The Mössbauer recoil-free fraction, f, is a much less used parameter, but it can indicate binding strength of the lattice as a whole since it is a function of the mean square displacement $\langle x^2 \rangle$ of the tin atom from its equilibrium position:

$$f = \exp\left[-E\gamma^2 \frac{\langle x^2 \rangle}{(hc)^2} \right]$$

where E_γ is the energy of the γ-radiation. For thin absorbers, assuming the Debye model, the recoil-free fraction is related to the percentage resonance effect, A_T, and its temperature dependence is given by:

$$A_T \propto f = \exp\left[-\frac{6E_r}{k\theta^2} \right]_T \quad \text{for } T \geqslant \theta/2$$

where E_r is the Mössbauer recoil energy, and θ is the Debye temperature of the material. Thus plots of $\ln(A_T)$ vs. T should be linear. Moreover, the more tightly bound the tin atoms are in the lattice, the slower will be the decrease of f and hence A_T as the temperature is increased.

Plots of the relative percentage effect, A_T/A_{77}, and $\ln$ (relative percentage effect), $\ln(A_T/A_{77})$, vs. temperature at $77°$–$300°$ K for nine compounds are shown in Figures 13 and 14, respectively. The data have been normalized to $77°$ K to allow direct comparison. The three compounds which show the slowest decrease in resonance effect are dimethyltin oxide (O), whose structure is actually unknown but is accepted to be polymeric in nature, $Me_2Sn(ONH\cdot CO\cdot Me)_2$ (□), which has a hydrogen-bonded lattice, and $Me_3SnON=C_6H_{10}$ (X), which consists of one-dimensional polymeric chains. In contrast, the materials known to consist of non-interacting molecules in the crystal, $Ph_3SnONPh\cdot CO\cdot Ph$ (△) and $Me_2Sn(ONMe\cdot CO\cdot Me)_2$ (■), show a much more rapid decay. The compound $Me_3SnONPh\cdot CO\cdot Ph$ (●), which is considered to have a structure similar to $Ph_3SnONPh\cdot CO\cdot Ph$ (30), behaves identically as does $Me_3SnO_3SPh\cdot H_2O$ (O) although it, like $Me_3SnON=C_6H_{10}$, consists of infinite one-dimensional chains held together by single hydrogen bonds. Me_3Sn(picolinate), in contrast, has a much more extensive hydrogen-bonding system, which results in a significant lowering of $d/dT[\ln(A_T/A_{77})]$. The most rapid rate of resonance decay is exhibited by dicyclopentadienyltin(II); for this compound the infinite cyclopentadienyl-bridged structure, determined crystallographically for the lead analog, has been proposed. The observed behavior, however, would argue against such a proposition and favor an essentially monomeric structure similar to that characterized for cyclopentadienyltin(II) chloride (31).

Literature Cited

1. Wheatley, P. J., *J. Chem. Soc.* (1961) 5027.
2. Bryan, R. F., Weber, H. P., *J. Chem. Soc.* (A) (1967) 843.
3. Harrison, P. G., King, T. J., Phillips, R. C., Richards, J. A., *J. Organometal. Chem.*, in press.

4. Kasa, N., Yasuda, K., Okawara, R., *J. Organometal. Chem.* (1965) 3, 172.
5. Domingos, A. M., Sheldrick, G. M., *Acta Cryst.* (1974) B30, 519.
6. Ewings, P. F. R., Harrison, P. G., King, T. J., Phillips, R. C., Richards, J. A., *J. C. S. Dalton,* (1975) 1950.
7. Alcock, N. W., Timms, R. E., *J. Chem. Soc.* (A) (1968) 1876.
8. Bancroft, G. M., Davies, B. W., Payne, N. C., Sham, T. K., *J. C. S. Dalton* (1975) 973.
9. Harrison, P. G., King, T. J., *J. C. S. Chem. Comm.* (19o) 815.
10. Harrison, P. G., King, T. J., *J. C. S. Dalton* (1974) 2298.
11. Chih, H., Penfold, B. R., *J. Cryst. Mol. Struct.* (1973) 3, 285.
12. Alcock, N. W., Timms, R. E., *J. Chem. Soc.* (A) (1968) 1873.
13. Harrison, P. G., Phillips, R. C., Richards, J. A., *J. Organometal. Chem.*, in press.
14. Potts, D., Sharma, H. D., Carty, A. J., Walker, A., *Inorg. Chem.* (1974) 13, 1205.
15. Drew, R. E., Einstein, F. W. B., *Acta Cryst.* (1972) B28, 345.
16. Harrison, P. G., Phillips, R. C., to be published.
17. Allen, F. H., Lerbscher, J. A., Trotter, J., *J. Chem. Soc.* (A) (1971) 2507.
18. Ashmore, J. P., Chivers, T., Kerr, K. A., Van Roode, J. H. G., *J. C. S. Chem. Comm.* (1974) 653.
19. Miller, G. A., Schlemper, E. O., *Inorg. Chem.*, (1973) 12, 677.
20. Calligaris, M., Nardin, G., Randaccio, L., *J. C. S. Dalton* (1972) 2003.
21. Harrison, P. G., King, T. J., Richards, J. A., *J. C. S. Dalton* (1975) 826.
22. Hilton, J., Nunn, E. K., Wallwork, S. C., *J. C. S. Dalton* (1973) 173.
23. Kimura, T., Yasuoka, N., Kasai, N., Kakudo, M., *Bull. Chem. Soc. Japan* (1972) 45, 1649.
24. Schlemper, E. O., *Inorg. Chem.* (1967) 6, 2012.
25. Harrison, P. G., King, T. J., Phillips, R. C., *J. C. S. Dalton,* in press.
26. Bancroft, G. M., "Mössbauer Spectroscopy," McGraw-Hill, London, 1973.
27. Bancroft, G. M., Platt, R. H., *Adv. Inorg. Radiochem.* (1972) 15, 59.
28. Ho, B. Y. K., Zuckerman, J. J., *Inorg. Chem.* (1973) 12, 1552.
29. Sham, T. K., Bancroft, G. M., *IFORG/ Chem.* (1975) 14, 2281.
30. Harrison, P. G., *Inorg. Chem.* (1973) 12, 1545.
31. Bos, K. D., Bulten, E. J., Noltes, J. G., Spek, A. L., *J. Organometal. Chem.* (1975) 99, 71.
32. Smith, P. J., *Organometal. Chem. Revs.* (1970) 5, 373.
33. Harrison, P. G., Zuckerman, J. J., *Inorg. Chem.* (1970), 9, 177.

RECEIVED May 3, 1976.

Chiral Pentacoordinate Triorganotin Halides

G. VAN KOTEN and J. G. NOLTES

Institute for Organic Chemistry, TNO, Utrecht, The Netherlands

*The optical instability of triorganotin halides has been ex-
plained in terms of conversion of the tetrahedral structure of
these compounds into stereochemically nonrigid penta- or hex-
acoordinate complexes. We have found that chiral
triorganotin halides with remarkable optical stability may be
obtained by blocking these stereoisomerization pathways. A
variety of chiral triorganotin halides of the type RMePhSnBr
have been synthesized in high yields via the 1/1 reaction of
MePhSnBr₂ with organolithium RLi or organocopper compounds RCu, in which R represents a ligand-carrying organo
group such as 2-Me₂NCH₂C₆H₄, 4-MeO-8-Me₂NCH₂-naph-
thyl, or 2-Me₂NCH₂C₆H₄CH₂-. The tin atom in these
monomeric compounds is pentacoordinate as a result of intramolecular Sn–N coordination. Variable temperature ¹H
NMR studies (in particular of prochiral groupings present in
these molecules) have provided information concerning Sn–N
coordination, the optical stability of these organotin compounds, and the conformational changes of the intramolecular
chelate ring Sn-C.*
$$\uparrow$$
$$N$$

Interest in chiral organotin compounds dates back over a long period.
As early as 1900 Pope and Peachey (1) claimed the synthesis of optically active
(+)-MeEtPrSnI, but this claim was later shown to be in error. This early work
was carried out with triorganotin halides, the class of organotin compounds which
in recent years has been shown to be optically very unstable (2,3). In contrast,
tetraorganotin compounds have been shown to be optically stable (4,5,6).

The stereochemical instability of triorganotin halides arises from the fact
that, contrary to tetraorganotin compounds, triorganotin compounds can readily
expand their coordination number. In the presence of nucleophiles the trior-

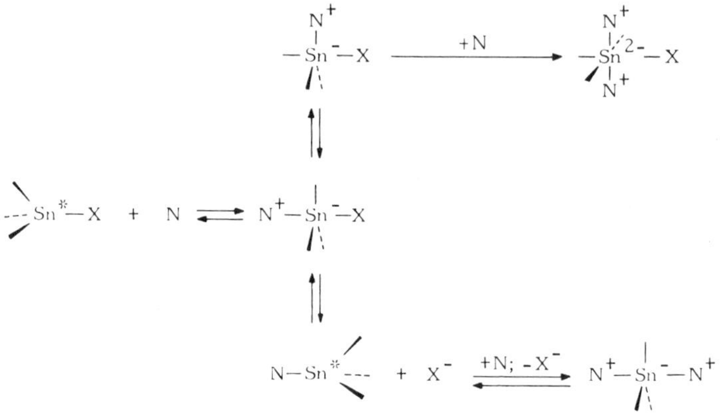

Figure 1. Possible mechanisms for the racemization of triorganotin halides (7).

ganotin enantiomers isomerize either via formation of penta- or hexacoordinated intermediates which are stereochemically nonrigid, or via an S_N 2-type mechanism which leads to direct inversion of configuration (7). As a consequence, enhanced optical stability can be expected if the first step in the mechanism is hindered or even completely blocked (*see* Figure 1).

Recently Gielen and Jamaï have shown that this goal can be reached by introducing organic groups with very high steric requirements (8). We have observed that a second, more general approach for the synthesis of optically stable triorganotin halides involves the introduction of an organic group (R–L) which is potentially bidentate so that the tin atom can become pentacoordinate by intramolecular coordination.

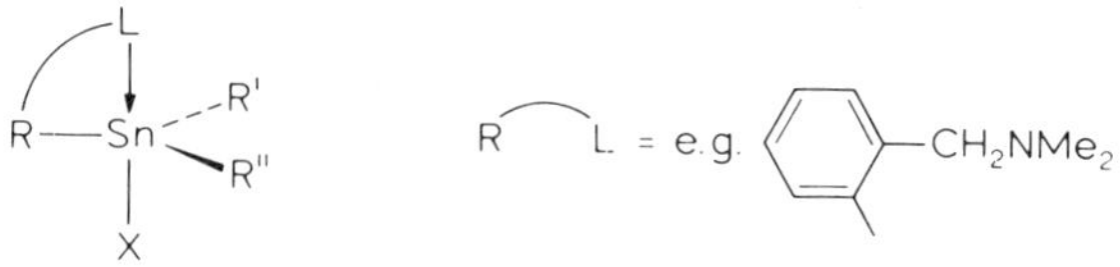

A few functionally substituted triorganotin halides of this type have been prepared recently, e.g. $Ph_2ClSn-(CH_2)_3-C(=NOH)Me$ *(9,10)* and $Me_2ClSn-(CH_2)_n-C(O)R$ *(10)*. However, the synthesis of such compounds in which the tin atom constitutes a chiral center has not yet been reported.

In the course of our investigations on the reactivity of organometal IB compounds in organometallic synthesis we found that arylcopper compounds display a surprising selectivity in the transmetallation reaction with SnX_4 or organotin halides *(12)*. Substitution of aryl groups for halogen atoms takes place up to the triorganotin halide stage (*see* Ref. *12*), e.g.:

$$3PhCu + SnX_4 \rightarrow \frac{Ph_3SnX + 3CuX}{(no\ Ph_4Sn\ formed)}$$

$$PhCu + Me_2SnBr_2 \xrightarrow{-CuBr} \underset{(70\%)}{PhMe_2SnBr} + \underset{(10\%)}{Ph_2Me_2Sn}$$

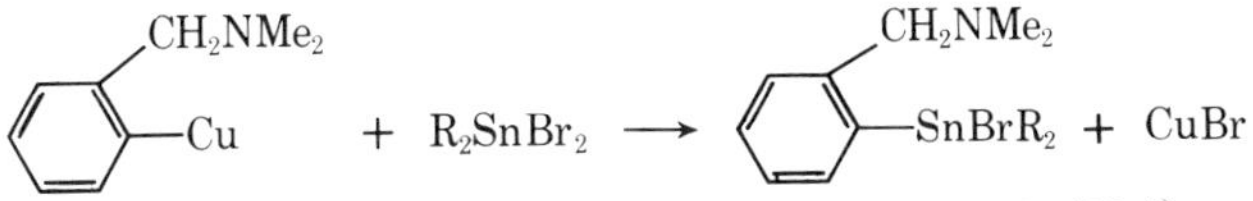

2, R = Ph (80%)
3, R = Me (81%)

The latter two examples show that the organocopper route provides a facile, one-step synthesis for mixed triorganotin halides. This result contrasts with that obtained in the corresponding reaction of PhLi and of 2-Me$_2$NCH$_2$C$_6$H$_4$Li with diorganotin halides, e.g.:

$$PhLi + Me_2SnBr_2 \xrightarrow{-LiBr} \underset{trace}{PhMe_2SnBr} + \underset{(88\%)}{Ph_2Me_2Sn}$$

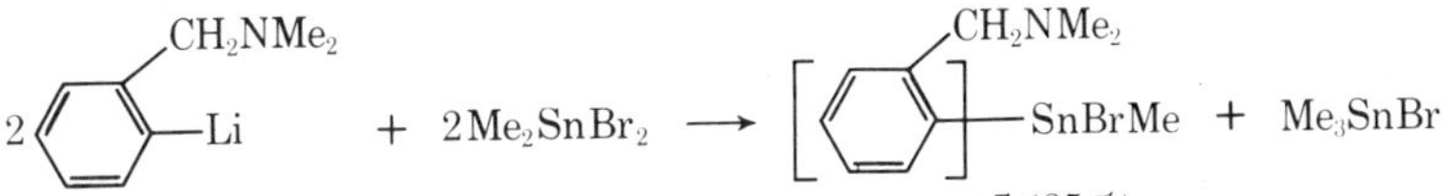

The first reaction yields, irrespective of the order of addition of the reagents, nearly exclusively the tetraorganotin derivative instead of PhMe$_2$SnBr. However, when Me$_2$SnBr$_2$ is added to **4** the first step of the reaction is again the formation of the tetraorganotin derivative, but this step is followed by a rapid and complete redistribution reaction with unreacted Me$_2$SnBr$_2$ resulting in the formation of **5**. In contrast, addition of **4** to Me$_2$SnBr$_2$ proved to be a second method for the synthesis of **3**. This selectivity, which is surprising for an organolithium compound, has been ascribed to the presence of a built-in ligand in **4** (*12*).

With the mixed diorganotin halide PhMeSnBr$_2$ as a starting material, these novel synthetic routes were applied to the synthesis of a variety of chiral triorganotin halides (**6–8**). The 2-[(dimethylamino)methyl]aryl group was selected as a ligand-carrying organo group because, in addition to coordinating properties, this group has unique advantages from an NMR point of view.

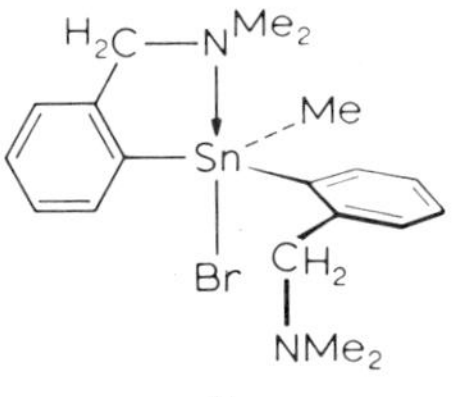

5

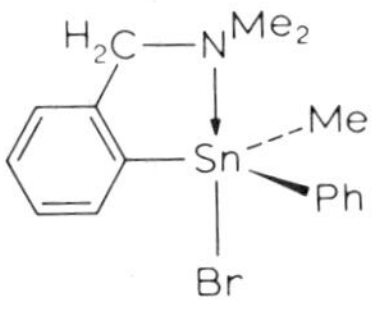

6

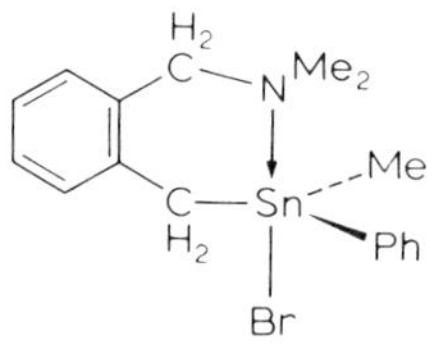

7

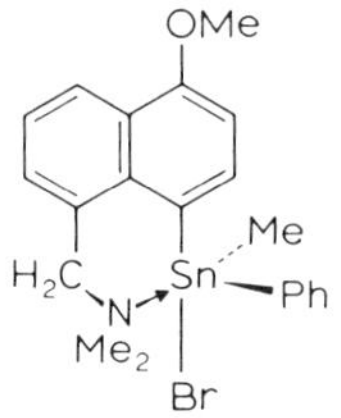

8

The —CH_2NMe_2 group is an excellent probe for the detection by NMR spectroscopy of dissymmetry in the molecule, i.e., configurational stability of the chiral tin center, as well as for the occurrence of intramolecular Sn–N coordination in solution (13).

An x-ray diffraction study of the diphenyl compound 2 has confirmed the occurrence of intramolecular Sn–N coordination in the solid.   The tin atom has distorted trigonal bipyramidal geometry with the 2-$Me_2NCH_2C_6H_4$ ligand spanning one equatorial and one axial site and the bromine ligand residing at the other axial position (see Figures 2a and 2b).

NMR spectroscopy cannot distinguish between a tetracoordinate or pentacoordinate structure for 2 (and 3) in solution, because the prochiral groupings $SnMe_2$ (in 3), NMe_2, and CH_2 possess molecular symmetry planes so that the CH_2 protons and NMe groups are enantiotopic and thus isochronous (see A).   However, this distinction is possible as soon as the tin atom becomes a chiral center as is the case in 6.

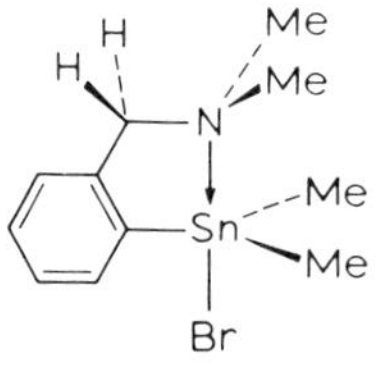

(A)

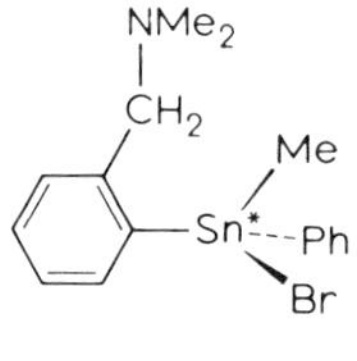

(B)

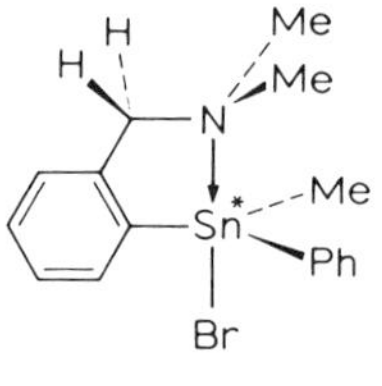

(C)

The tin atom in 6 may be either four-coordinate (B) or five-coordinate (C).   In situations B and C the carbon atom of the CH_2N group lacks a molecular symmetry plane bisecting the HCH angle.   Hence, the benzylic protons will be diastereotopic and anisochronous as long as inversion of configuration at tin is slow on the NMR time scale.   In B the NMe groups are homotopic and isochronous, because in the absence of Sn–N interaction the NMe_2 group is not a stable prochiral assembly as a result of the inversion process taking place at nitrogen. However, if the N atom is coordinated to Sn as in C, the NMe_2 group constitutes a stable prochiral assembly, and accordingly the NMe groups become diastereotopic and thus anisochronous.

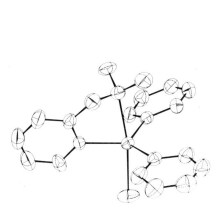

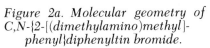

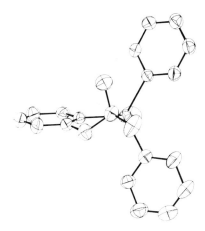

Figure 2a. Molecular geometry of C,N-{2-[(dimethylamino)methyl]-phenyl}diphenyltin bromide.

Figure 2b. Viewed along the N–Sn–Br axis.

The temperature-dependent NMR patterns of **6** can be explained on the basis of the considerations presented above (*see* Figure 3). The benzylic protons (30°C, in toluene-d_8; AB, δ 3.02 and 2.81 ppm J_{AB} 14 Hz) are anisochronous up to 123°C—the highest temperature studied—and thus diastereotopic. Consequently, up to 123°C the rate at which the absolute configuration of the tin atom changes is slow on the NMR time scale. Therefore, **6** is an example of a triorganotin halide with exceptionally high optical stability. The observation at 10°C of two singlets for the NMe protons is interpreted in terms of rate-determining intramolecular Sn–N coordination, because it is only in the pentacoordinate structure that inversion at N is blocked ($k_c \gg k_i$). This makes a stable prochiral assembly of the NMe$_2$ group (*see* C). Coalescence of the NMe resonances indicates that the rate constant for dissociation (k_d) increases with respect to the observation time. The NMe groups become homotopic and isochronous as a result of the inversion process at N in the tetracoordinate Sn-conformer ($\Delta G^{\ddagger}$ for pyramidal inversion at N amounts to $\simeq$6 kcal/mol (*13*) and thus inversion is an extremely fast process over the entire range of temperatures studied).

One further important conclusion can be drawn from the NMR spectra. In the pentacoordinate structure stereomutations resulting in a change of absolute configuration of the tin atom must be very slow or even absent. If rapid inversion of configuration at tin were to take place within the pentacoordinate conformer, this would be reflected in an almost simultaneous coalescence of the CH$_2$ ($\Delta\nu$ = 21 and J_{AB} 14 Hz) and NMe ($\Delta\nu$ = 22 Hz) resonance patterns. However, at 123°C the CH$_2$ protons are still anisochronous, whereas the NMe groups became already isochronous at room temperature.

The five-membered chelate ring in **6** is slightly puckered as a result of the presence of one sp^3 C atom and one sp^3 N atom (*see* Figure 2b). It is obvious

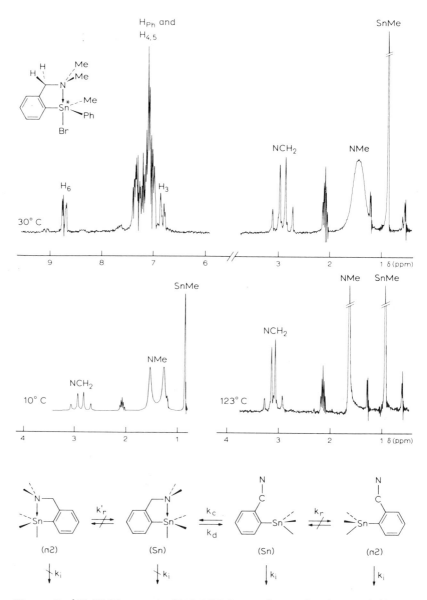

Figure 3. ¹H NMR spectra (100 MHz) in toluene-d₈ of 6 at different temperatures.

that in solution conformational changes in the chelate ring occur. However, the influence on δ is small as appears from the resonance pattern at low temperature.

The naphthyl compound **9** contains a six-membered chelate ring which differs from the five-membered chelate ring in **6** in that one more *sp²* C atom

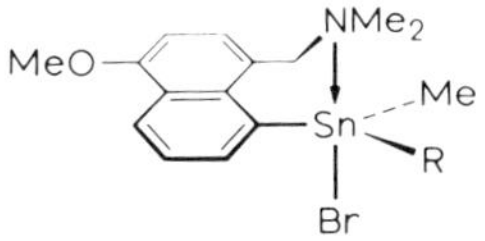

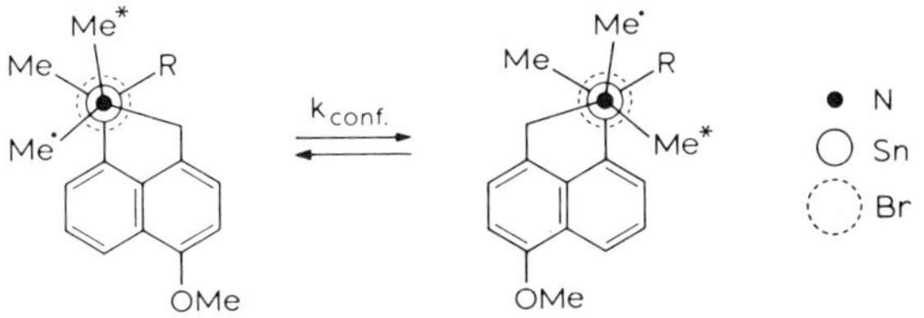

Figure 4. Conformational equilibrium for 8 and 9.

is present. As shown in Figure 4 this six-membered chelate ring exists in two conformations which can be frozen out on the NMR time scale at −30°C. Below this temperature anisochronous signals are observed for the Me_2Sn (two singlets), NMe_2 (two singlets), and CH_2 protons (AB pattern) because in each of the conformations these groups are diastereotopic. At higher temperatures k_{conf} increases with respect to the observation time which renders the Me_2Sn and Me_2N groups as well as CH_2N protons isochronous.

The temperature-dependent behavior of the NMR spectra of the chiral naphthyltin compound **8** reflects the various processes taking place in this molecule, i.e. conformational hanges (limiting spectra are obtained below −30°C with both conformations being observable), Sn–N coordination and dissociation and inversion processes at nitrogen.

Both the benzylamine compound **6** and the naphthylamine compound **8** illustrate the strong influence of the presence of an intramolecular ligand on the optical stability of chiral triorganotin halides. Most probably interaction with external nucleophiles, which is believed to be the first step in the racemization mechanisms (*6*), is blocked by the intramolecular Sn–N coordination. This is supported by the absence of any observable effect on the coalescence pattern of the NMe protons and on the $^2J(Sn–CH_3)$ and $^3J(Sn–H_6)$ coupling constants upon

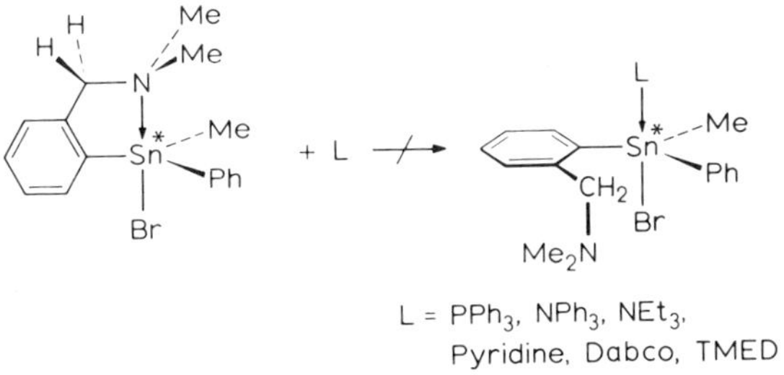

L = PPh$_3$, NPh$_3$, NEt$_3$,
Pyridine, Dabco, TMED

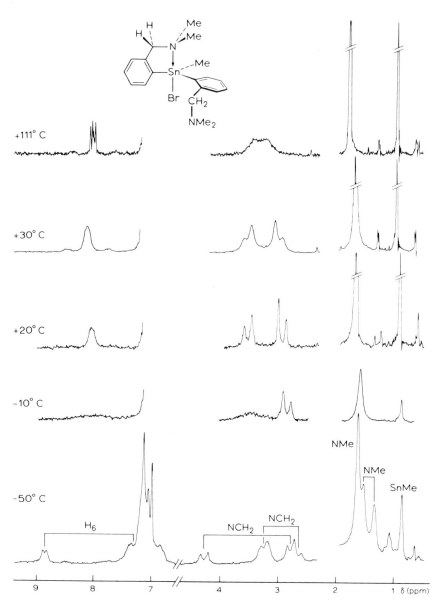

Figure 5. ^1H *NMR spectra (100 MHz) of* **5** *in toluene-d_8 at different temperatures.*

adding external Lewis bases like pyridine and PPh$_3$ to a toluene-d_8 solution of **6**. Only large solvent shifts for high pyridine/**6** ratios were observed as a result of preferential solvent–solute interactions occurring at the site of the intramolecular ligand (*12*).

If pentacoordination is the preferred coordination geometry for the tin atom in **6**, intramolecular ligand exchange might occur in the bis(benzylamine)tin compound **5**. In this compound the two CH$_2$NMe$_2$ ligands can compete for the fifth coordination site at tin.

In Figure 5 the NMR spectrum of **5** in toluene-d_8 is shown. At −50°C two AB patterns for two sets of diastereotopic benzylic protons ($\Delta\delta$, 1.48 ppm, irradiation at 2.7 ppm results in a singlet at 4.27 ppm; $\Delta\delta_2$, 0.58 ppm) are observed. Furthermore, the spectrum shows two multiplets for H$_6$ at δ 8.85 and 7.20 ppm and two sets of signals for the NMe protons, i.e., two singlets at δ 1.34 and 1.52 ppm for diastereotopic NMe groups and one singlet at 1.62 ppm for homotopic NMe groups. This pattern is fully compatible with the pentacoordinate structure shown in Figure 5.

When the temperature is raised to +20°C, the two resonance patterns for H$_6$ coalesce to give an averaged signal at 8.12 ppm, the two AB patterns for the benzylic protons similarly coalesce to one AB pattern, and one broadened singlet is now observed for the NMe protons at the position for homotopic NMe groups. The C,N-, and C-bonded ligands, which at low temperature are observed separately, have become equivalent on the NMR time scale by an intramolecular exchange process. When the temperature is further increased, a second type of averaged spectrum is observed involving exchange between the pentacoordinate situation, and the situation in which both ligands are free and the tin is tetracoordinate. This is reflected in the sharpening of the H$_6$ multiplet as well as by the decrease of the $\Delta\nu$ value of the AB pattern for the anisochronous benzylic protons. However, at 110°C a limiting spectrum for the tetracoordinate situation is not yet obtained, as can be concluded from the broadened AB pattern.

This experiment clearly demonstrates that intramolecular ligand exchange in this type of compound does indeed take place. Some possible routes are shown in Figure 6.

As discussed earlier, the high configurational stability of our triorganotin halides is ascribed to the fact that complex formation with external ligands cannot compete with the intramolecular coordination process. An additional but important factor is that stereoisomerization routes in the pentacoordinate conformer, e.g. by a Berry pseudorotation mechanism (*14,15,16,17*), are energetically unfavorable for two reasons. First, in pentacoordinated triorganotin compounds electronegative ligands occupy axial positions (*19*), whereas stereoisomerization of **6** involves pentacoordinated configurations having the Br and N ligand at equatorial sites. Secondly, the five-membered chelate ring is already highly strained (N–Sn–C angle 75°) in the ground state. Stereosiomerization routes resulting in inversion of configuration at tin involve at least some configurations in which the bidentate spans equatorial sites (*14,15,16,17,18*) requiring an opening

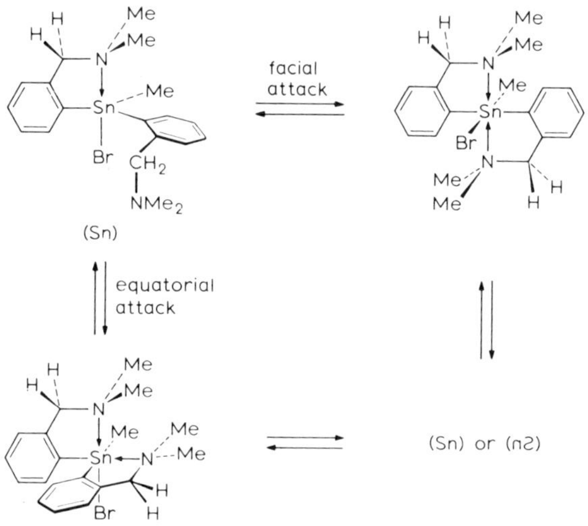

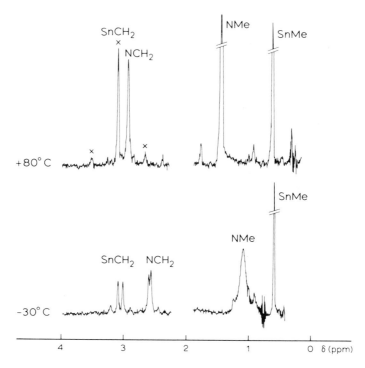

Figure 6. Routes for intramolecular exchange in
5.

Figure 7. ^1H NMR spectra (100 MHz) of **7** *in toluene-d_8 at different temperatures.*

of the N–Sn–C angle in the chelate ring to formally 120°. Obviously this is not a realistic situation in view of the geometric constraints of the five-membered chelate ring.

Study of **7**, where intramolecular Sn–N coordination leads to a six-membered ring with two sp^3 C atoms, lends support to this idea. The NMR spectrum of **7** (*see* Figure 7) shows at 80°C isochronous resonances for the CH$_2$Sn protons [δ 4.07 ppm, ^2J Sn–H (averaged) 86 Hz], the CH$_2$N protons (δ 3.91 ppm) as well as the NMe protons (δ 1.41 ppm).

Upon cooling, the NMe groups ($T_c \simeq -37°C$; $\Delta\delta$ 38 Hz) and the CH$_2$Sn and CH$_2$N protons ($T_c \simeq 0°C$; $\Delta\delta$ 15 and 9 Hz, respectively) become diastereotopic and thus anisochronous at about the same rate. Coalescence of the resonance signals for the diastereotopic SnCH$_2$ and NCH$_2$ protons can be the result only of inversion of configuration at tin, causing loss of dissymmetry on the NMR time scale. Coalescence of the diastereotopic NMe groups may be the result of either rapid inversion of configuration at tin in the pentacoordinated conformer or rapid inversion at N in the four-coordinate conformer. The observation that coalescence of the NMe$_2$, NCH$_2$, and SnCH$_2$ protons occurs at about the same rate points out that stereoisomerization processes resulting in inversion of configuration at tin occur in the pentacoordinated conformer. As a result of the lower geometric constraints of the six-membered ring in **7** which contains two sp^3 C atoms as compared with the five-membered ring in **6**, these processes have a lower energy barrier.

Starting from **6** for the first time the synthesis of an organotin compound containing two optically stable chiral tin centers in one molecule has been realized. Reaction of **6** with NaOH yields the corresponding bis(triorganotin) oxide, **10**, quantitatively.

The NMR spectrum of **10** (Figure 8) show two sets of resonance patterns with a 50/50 intensity ratio. This fully agrees with the fact that from an NMR point of view the SS and RR enantiomeric pair on the one hand and the RS and SR enantiomeric pair on the other hand are chemical-shift equivalent in achiral solvents, but that the two enantiomeric pairs are chemical-shift nonequivalent with respect to each other. The chiral tin centers are most probably tetracoordinated because between $-80°$ and $+100°C$ the NMe groups are homotopic, whereas the anisochronism of the CH$_2$N protons exists over the whole temperature range.

The occurrence of Sn–N coordination in the tetrahedral structure of **6** (A) can be viewed as the first step of an S_N2 type substitution process at tin. Con-

A   C–Sn–N ≈ 109°

tinuation of this process beyond the trigonal bipyramidal configuration would require opening of the N–Sn–C angle from about 75° (*see* Figure 2a) to formally 109°. This is unlikely in view of the geometric constraints of the chelate ring, and accordingly no Sn–Br bond dissociation will occur.

We have now synthesized {2,6-bis[(dimethylamino)methyl]phenyl}phenylmethyltin bromide, **11**, via the organolithium route. **11** has been isolated as a 1/1 adduct with LiBr:

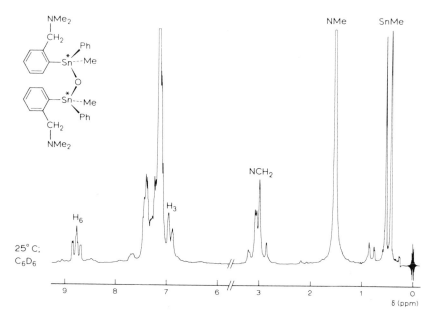

In **11** dissociation of the Sn–Br bond is possible with retention of the pentacoordinated geometry of the tin atom because the second intramolecular ligand can interact with tin thus stabilizing the equatorial R_3Sn unit.

At room temperature the NMR spectrum of **11** shows isochronous signals for the CH_2N and NMe groupings. The CH_2N protons will be diastereotopic if rotation of the equatorial $-Sn{<}^{Me}_{Ph}$ unit around the Sn–C(aryl) axis is sufficiently slow with respect to the NMR time scale. Hindered rotation requires rate-lim-

Figure 8. ^1H *NMR spectrum (100 MHz) of* **10** *in benzene-d_6 at 25°C.*

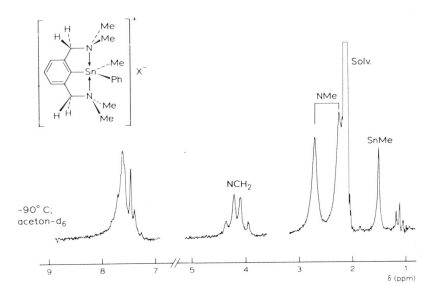

Figure 9. ^{1}H *NMR spectrum* $(100\ MHz)$ *of* **11** *in acetone-d_6 at* $-90°C$.

iting Sn–N coordination which, moreover, results in the NMe groups becoming diastereotopic. This situation is reached at $-90°C$ when anisochronous NCH$_2$ and NMe proton signals are observed in the NMR spectrum (*see* Figure 9).

Experimental

The organocopper and -lithium compounds were prepared by published methods (*20,21*). Solvents were carefully purified, dried, and distilled under nitrogen before use. All reactions were carried out under N$_2$. ^{1}H NMR spectra were recorded on a Varian Associates HA-100 NMR spectrometer. IR spectra were recorded on a Perkin-Elmer 577 grating infrared spectrometer. Elemental analyses were carried out under the supervision of W. J. Buis in the Analytical Department of this institute.

General Procedure for the Synthesis of the Chiral Tin Compounds 6–8. A solution of the organolithium reagent (40 mmol) in ether/hexane (80/30) was slowly added to a stirred solution of phenylmethyltin dibromide (40 mmol) in ether (100 ml). After stirring (1 hr) the solvents were removed, and the residue was extracted with benzene (200 ml). Evaporation of the benzene afforded almost pure product. Further purification was achievd by recrystallization from benzene (or toluene)/pentane mixture (1/1). When the organocopper route is used for the preparation of these compounds, the order of addition of the reagents is not important. However, addition of the MePhSnBr$_2$ solution to the organocopper solution (or suspension) is recommended (*see* Ref. *12*).

6, 90% yield; mp 121°–125°C; found C, 45.3; H, 4.8; N, 3.5; Br, 18.4; Sn, 28.0%; calc. C, 45.21; H, 4.71; N, 3.30; Br, 18.84; Sn, 27.95%; mol wt (osmometry, in C_6H_6) 432; calc. 425.

7, 30% yield; mp 152°–160°C; found C, 45.9; H, 5.3; N, 3.2; Br, 18.6%; calc. C, 46.52; H, 5.05; N, 3.19; Br, 18.20%; mol wt (ebulliometry, in C_6H_6) 439; calc. 438.

8, recyrstallization from $CHCl_3$/pentane (5/1); mp 190°–194°C; found C, 49.7; H, 4.8; N, 2.9; Sn, 23.4%; calc. C, 49.93; H, 4.76; N, 2.77; Sn, 23.52%; mol wt (ebulliometry, in CH_2Cl_2) 505; calc. 505.

Synthesis of Bis{2-[(dimethylamino)methyl]phenyl}Methyltin Bromide 5. A solution of Me_2SnBr_2 (40 mmol) in ether (40 ml) was added to a stirred suspension of the organolithium compound **2.** After stirring (2.5 hr), the reaction mixture was concentrated, yielding a brown oil. Addition of benzene (125 ml) produced a white precipitate (LiBr) which was filtered off. Benzene was removed by evaporation, and the resulting oil was heated at 45°C in vacuum (0.4 mm) for 2 hr (removal of Me_3SnBr). The resulting oil was recrystallized from ether (60% yield). Mp 142°–144°C; found C, 47.2; H, 5.5; N, 5.7; Br, 16.4%; calc. C, 47.34; H, 5.65; N, 5.81; Br, 16.58%; mol wt (osmometry, in C_6H_6) 484; calc. 481.

Acknowledgment

The authors are grateful to A. L. Spek for the x-ray structure determination of **2** and to J. T. B. H. Jastrzebski and E. Natan for experimental assistance.

Literature Cited

1. Pope J., Peachey, S. J., *Proc. Chem. Soc. London* (1900) **16,** 42, 116.
2. E. G. Peddle, G. J. D., Redl, G., *J. Chem. Soc.,* Chem. Commun. (1968) 626.
3. Holloway, C. E., Kandil, S. A., Walker, I. M., *J. Am. Chem. Soc.* (1972) **94,** 4027.
4. Gielen, M., Jamai, H. M-., *Bull. Soc. Chim. Belg.* (1975) **84,** 197.
5. Lequan, M., Meganem, F., *J. Organomet. Chem.* (1975) **94,** C 1.
6. Gielen, M., Barthels, M. R., de Clercq, M., Dehouck, C., Mayence, G., *J. Organomet. Chem.* (1972) **34,** 315.
7. Gielen, M., Jamai, H. M-., *J. Organomet. Chem.* (1975) **91,** C33.
8. Gielen, M., Jamai, H. M-., *Bull. Soc. Chim. Belg.* (1975) **84,** 1037.
9. Abbas, S. Z., Poller, R. C., *J. Chem. Soc.,* Dalton Trans. (1974) 1969; and
10. *J. Organomet. Chem.* (1976) **104,** 187.
11. Kuivila, H. G., Dixon, J. E., Maxfield, P. L., Scarpa, N. M., Topka, T. M., Tsai, K. H., Wursthorn, K. R., *J. Organomet. Chem.* (1975) **86,** 89.
12. Van Koten, G., Schaap, C. A., Noltes, J. G. *J. Organomet. Chem.* (1975) **99,** 157.

13. Jennings, W. B., *Chem. Rev.* (1975) **75**, 307.
14. Muetterties, E. L., Schunn, R. A., *Q. Rev. Chem. Soc.* (1966) **20**, 245.
15. Whitesides, G. M., Mitchell, H. L., *J. Am. Chem. Soc.* (1969) **91**, 5384.
16. Westheimer, F. H., *Acc. Chem. Res.* (1968) **1**, 70.
17. Holmes, R. R., Ibid. (1972) **5**, 296.
18. Shapley, J. R., Osborn, J. A., Ibid. (1973) **6**, 305.
19. Ho, B. Y. K., Zuckerman, J. J., *J. Organomet. Chem.* (1973) **49**, 1.
20. Van Koten, G., Noltes, J. G., *J. Organomet. Chem.* (1975) **84**, 117, 419.
21. Ibid. (1975) **85**, 105.

RECEIVED May 3, 1976.

INDEX

A

Abstraction, intramolecular
  proton 48
Acaricidal activity 9
Acaricides 173, 204
Acidity, Lewis 182
Addition of the thiol R'SH
  to alkenes 179
ADP 206
Agricultural applications
  of organotins 8, 167
Aldehydes 100
Aliphatic diynes, ring
  compounds from 18
Alkali metal derivatives
  of organotins 41
Alkene elimination 120
Alkenes, addition of the
  thiol R'SH to '............ 179
Alkoxides and amines,
  organotin 82, 83
α-Alkoxyketones, synthesis of 90
Alkoxyl radical 35, 36
Alkoxyoxiranes, organotin 89
Alkylaluminum process 3
Alkylation of carbonyl compounds. 91
Alkyl dihalides 99
Alkyl halides 31, 43
Alkyl iodides 117
Alkyltin halides, synthesis
  of novel substituted 123
n-Alkyltins 199
Alkyltins with halogens,
  reactions of 114
Alkyltins, sulfur substituted 113
Amido ligands 72
μ-Amido bridging 72
Amines, organotin 82, 96
Amino acids 235
Angle, C—Sn—C ring 227
Animal toxicity of
  tetraalkyltins 243
Anion, haloaryl radical 50
Anion–hydroxide exchange,
  properties of 221
Anionoids 47
Anion, trichloromethyl 52
Anisochronism 254

Antibacterial properties of

Antibacterial properties of
  organotins, fugicidal
  and 168
Antifeedant effect 168
Antifoulants 7, 184, 204
Applicants of organotin
  compounds 5
Arsines, stibines, and
  bismuthines, organotin
  phosphines 57
Aryl bromide 49
Arylcopper compounds 276
Aryl halides, reactions of
  organostannylanionoids
  with 46
Aryl radicals 47
Aryltin bond cleavage 114
Atom, highly shielded
  tetracoordinate carbon 52
ATP hydrolysis 191
ATP synthesis by trialkyltins,
  inhibition of 188
ATPase(s) 213
  bacterial membrane 214
  cation translocating 219
  inhibitors of ion
    translocating 204
  proton translocating 210
  Ca²⁺-translocating 220
  Na⁺-K⁺ translocating 219

B

Bacteria 213
Bacterial membrane ATPase 214
Bactericide activity 170
Benzalacetone 104
Benzalacetophenone 104
Bis(benzylamine) tin 283
Benzyl chloride 42
Benzyne 51
Bibenzyl 42
Bilharzia disease, tropical 8
Binding energy 228
  of tin compounds 243
Binding, inhibition of 192
Binding to mitochondria 189
Binding of triethyltin to
  cat hemoglobin 193

Bimolecular electrophilic
    substitutions 249
Bimolecular homolytic
    substitution 35
Bimolecular nucleophilic
    substitutions 249
Biochemical activities of
    triorganotins 223
Biocidal organotins 183, 184
    applications of organotins 168
Biocides, tributyltins as 6
Biological activity of
    organotins 172, 197, 204
Bioorganotin chemistry 197
Bipyramidal geometry,
    trans-trigonal 263
Bismuth derivatives, organotin . . . 58
Bismuthines, organotin
    phosphines, arsines,
    stibines, and 57
p-Bistrimethylstannylbenzene . . . 46
Bistrimethylstannydi-
    chloromethane 53
1,4-Bistriphenylsilylbutane 42
Bond dissociation energies 27
Bonding in organotins by
    gamma resonance
    spectroscopy, structure
    and 155
Bonds, coordination of tin by Sn–S 235
Bottles 143
Brestan 168
Bromide, aryl 49
Bromine radicals 36, 37
Bromocyclopropane 44
7-Bromonorbornenes 44
3-Bromonortricyclene 43
n-Butyl halides with
    trimethylstannylsodium
    in THF, reactions of 45

C

Calendered films, rigid 142
Carbanionoids 41
Carbon atom, highly
    shielded tetracoordinate 52
Carbon bonds, metal– 26
Carbon–hydrogen bonds,
    hydroxylation of 197
Carbon hydroxylation 202
α-Carbon hydroxylation 197
β-Carbon methoxyethyltin
    trichloride 126
Carbon tetrachloride,
    reaction of
    trimethylstannylsodium
    with 51

Carbonyl compounds,
    alkylation of 91
Carbonyl derivatives,
    oxidation of organotin
    alkoxides to 93
Catalysts 10, 57, 65
Cation translocating ATPases 219
(CH₃)₂SnS₃ 158
Chain or a non-chain process 27
Charge distribution 228
Charge transfer 117
Chelating ligands 264
CHELEQ 228
Chemical shift effects 229
Chiral naphthyltin 281
Chiral pentacoordinate
    triorganotin halides 275
Chiral tetraorganotin
    compound, optically pure . . . 252
Chiral tin centers, two
    optically stable 285
Chiral tin compounds,
    synthesis of 287
Chloride-free medium, effects
    of triethyltin in a 192
Chloride–hydroxide exchange . . 207, 217
Chloride, sensitivity of
    mitochondrial functions
    to 191
Chloro(di-tert-
    butylphosphino)tin 65
Chloroplasts 213, 215
Chlorosilanes 84
Chlorostannaes 124
Chlorostannylphosphines 65
CIDNP 29
Cinnamaldehyde 103
Clear extruded sheet 139
Cleavage, aryl–tin bond 114
Cleavage, phenyl–tin bond 115
Cleavage reactions 113
CNMR spectral parameters 54
Coalescence 285
    of diastereotopic groups 252
Complexes between oximes
    and oxaziridines and
    praseodynium salts 232
Configurational stability 254
    and structure of
    organotins 54
Coordination of tin by
    SN–S bonds 235
Coupling membranes 206
Covalent organometallic
    compounds, molecular
    weight of 155
Cryoscopy 74
Crystal data 263

Crystal structures   117
Crystallography, x-ray 258, 259
Cycloalkyl halides   43
Cyclohexyltins   200
Cyclopentanes   36
Cyclostannanes   236
Cystein   235
Cytochrome, reduced   189
Cytochrome P-450
    monooxygenase enzyme
    system   199

D

Degradation of triorganotins   175
Depolymerization   163
Destannylation 197, 199, 201
Dialkylaminyl radicals   29
Dialyklstannylenes   29
Dialkyltin dichloride   182
Dialkyltins 29, 177, 232
o-Dibromobenzene   51
1,2-Dibromobutane   42
Dibutyltin dilaurate 10, 134
p-Dichlorobenzene   46
Dichlorocarbene as an
    intermediate   52
Dichlorostannane   125
Dicyclohexylphenyltin
    hydroxide   173
Dicyclopentadienyltin(II)   273
Diethylpyrocarbonate   193
Diethyltin   199
Diffusion   49
1,2-Dimethoxyethane   44
3,3-Dimethyl-3-benzo-
    stannepin   16
Bis-2-[(dimethylamino)-
    methyl]phenylmethyltin
    bromide   288
Dimethyltin bis(fluorosulfonate) .   264
Dimethyltin oxide   273
Dimethyltins   6
    six-coordinated   264
Dioctyltins   5
Diorganotins   186
    affinity for dithiols   186
    inhibition of the
        oxidation of
        α-ketoacids   186
Diphenyltin bis(isooctythio-
    glycolate)   181
Disinfectants   8
Disproportionation, photo-
    induced   76
Disproportionation reactions   128
Dithiols, diorganotin
    affinity for   186

O-Divinylbenzene and
    O-diethynylbenzene,
    ring compounds from   15
Dypnone   105

E

Effective vibrating mass
    (EVM) model   155
Electrophilic reagents   113
Electrolytes   209
    transport of   207
Electron transport
    phosphorylation   205
Electronic effects of substituents   114
Electrophiles   115
Electrophilic substitutions,
    bimolecular   249
Electropositive nature
    of tin   82
Enamine preparations   106
Energies, bond dissociation   27
Energy-using processes   206
α-Enones   103
Enteromorpha   184
Enzyme system, cytochrome
    P-450 monooxygenase   199
ESCA spectroscopy 72, 230, 232
ESR spectroscopy 29, 33
    characterization of
        trivalent tin amides   72
Estertin stabilizers in PVC   134
EVM model 155, 156
Exchange of chloride and
    hydroxide ions
    across membranes   207
Exchange reactions   84
Extruded rigid foil   143
Extrusion   138

F

Foam, flexible polyurethane   10
Foamed profile, rigid   140
Foil, extruded rigid   143
Food packaging   5
Free SH groups   235
Functionally substituted
    organotins 113, 129
Fungi   184
Fungicidal and antibacterial
    properties of organotins   168
Fungicide 8, 168, 204

G

Gamma resonance spectroscopy,
    structure and bonding
    in organotins by   155

Germanium compounds, tin– 254
Germanium, S_E2 reactions at
     silicon or 249
Glass, surface treatment of 10
Grignard process 3

H

Halides, alkyl and cycloalkyl 43
Halides, organic 42
Halides, reactions of
   organostannylanionoids
   with aryl 46
Halides, synthesis of
   novel substituted
   alkyltin 123
α-Haloketones 97
Haloaryl radical anion 50
Halogenoalkoxides, organotin 85
Halogen-substituted organotin
   phosphines 65
Halogens, reactions of
   alkyltins with 114
He(I) photoelectron spectra 76
Hemoglobin, binding of
   triethyltin to cat 193
Hexaalkylditins 29
Hexamethyldisilazane 73, 77
Hexamethyldistannane 46
Histidine 193
HMPT 42
Homoconjugation 34, 39
Homoleptic tin(II) and
   tin(III) amides 70
Homolytic reactions 26
Homolytic substitution,
   bimolecular 35
Hydrogen bonds, hydroxylation
   of carbon– 197
Hydrogen chloride, reaction
   between the stabilizer
   and 181
Hydrolysis, ATP 191
Hydrostannation 12, 14, 33
Hydrostannolysis 12, 19
Hydroxide exchange,
   chloride– 207, 217
Hydroxide exchange, properties
   of anion– 221
Hydroxide transporting action
   of triethyltins 207
β-Hydroxybutyrate, oxidation
   of 191
Hydroxylation of carbon–
   hydrogen bonds 197
Hyperconjugate 34, 39
Hyperfine coupling 30, 33

I

Imidazole polymers, triethyltin– .. 193
Influence of organotin
   compounds on
   mitochondrial functions 186
Inhibition
   of ATP synthesis by
     trialkyltins 188
   of binding 192
   of α-oxoglutarate
     oxidation 187
   by triorganotin
     compounds 205
Inhibitors of ion
   translocating ATPases 204
Injection-molded fittings,
   rigid 150
Intercalation compounds 157
Intramolecular ligand
   exchange 283
Intramolecular nucleophilic
   substitution 85, 90
Intramolecular proton
   abstraction 48
Inversion 44
Iodine charge transfer
   complexes 117
Ion translocating ATPases,
   inhibitors of 204
Ion transport 205
Ionic lattices 157
Ionization potentials 241
Ionophores 204, 225
Iron compounds, tin– 255
Iron–oxygen complex 198
Isotope effect, kinetic 48

K

α-Ketoacid oxidation 232
α-Ketoacids, inhibition of
   the oxidation of 186
Ketone triplets 37
Ketones 100
4-Ketopentyl compounds 182
Kinetic isotope effect 48

L

Lewis acidity 182
Lewis acids 60
Lewis base 76
Ligand exchange, intramolecular 283
Ligands, chelating 264
Light stability 140
Lipid membranes 208
Liposomes 207
Lithium aluminum hydride 11

M

Macromolecules which do not
  bind triethyltin 188
Maneb 9
Manufacture of organotins 3
Marine antifouling paints 7
Membrane(s)
  ATPase, bacterial 214
  coupling 206
  exchange of chloride and
    hydroxide ions across 207
  lipid 208
  proton gradient across 192
Mercapto-2-benizimidazole 235
Mesomeric effect 239
Metabolism of organotin
  compounds 199
Metal–carbon bonds 26
Metals in proteins 233
α-Methyl cinnamaldehyde 104
β-Methyl cinnamaldehyde 104
Mitochondria 213
  binding to 189
  effects of triorganotins on 194
Mitochondrial functions 191
  to chloride, sensitivity
    of 191
  effects of triethyltin on 192
  influence of organotin
    compounds on 186
Mitchondrial oxidative
  phosphorylation 205
Miticide 169
Molecular weight of covalent
  organometallic compounds .. 155
Molluscicides 204
Monooxygenase enzyme systems .. 197
  cytochrome P-450 199
Monooxygenase reactions with
  organotin compounds 201
Mössbauer spectroscopy ..155, 158, 227, 230, 258
  tin 119m 269
Multiple screw extrusion
  of water pipe 136
Mutagenicity 153

N

Naphthyltin, chiral 281
NMR spectroscopy 278
2-Norbornen-5-yltrimethyl-
  stannane 43
Nuclear quadrupole resonance, ·
  UV–PES and 238
Nucleophilic displacement 97
Nucleophilic substitution,
  bimolecular 249

Nucleophilic substitution,
  intermolecular 90
Nucleophilic substitution,
  intramolecular 85

O

Octylthiotins 142
Optical instability 275
Optical stability of
  organotins 249
Optically pure chiral
  tetraorganotin compound ... 252
Optically stable chiral
  tin centers, two 285
Optically stable triorganotin
  compounds 250
Oral toxicity 152, 175
Organic halides 42
Organoarsenic compounds 167
Organofunctional groups 113
Organomercurials 7, 167
Organostannylanionoids 43
  with aryl halides,
    reactions of 46
Organostannylanionoin
  chemistry 41
Organotin(s)
  in agriculture 8, 167
  alkali metal derivatives
    of 41
  alkoxides 82, 83, 93
  alkoxyoxiranes 89
  amines 96
  applications of 5
  biocides 168, 183
  biological oxidation of 197
  bismuth derivatives 58
  configurtion and structures
    of 54
  enolates 105, 106
  functionally substituted 113
  fungicidal and antibacterial
    properties of 168
  halides, functionally
    substituted 129
  halogenoalkoxides 85
  hydrides 2, 11, 124
  manufacture of 3
  metabolism of 199
  on mitochondrial functions,
    influence of 186
  monooxygenase reactions
    with 201
  oxygen-bonded 258
  phosphines, arsines, stibenes,
    and bismuthines 57, 58, 65
  radicals 28

Organotin(s) (*Continued*)
research 1
structure and bonding in 155
-substituted phosphines,
  oxidation of 62
α-substituted sulfur 115
sucrose compounds 184
toxicity 3, 232
Oxaziridines and praseodynium
  salts, complexes between
  oximes and 232
Oxidation
  of α-hydroxybutyrate 191
  inhibition of β-oxoglutarate 187
  α-ketoacid 186, 232
  of organotin alkoxides to
    carbonyl derivatives 93
  of organotin compounds,
    biological 197
  of organotin-substituted
    phosphines 62
  pyruvate 191
  state 230
Oxidative addition 79, 117
Oxidative phosphorylation,
  mitochondrial 205
Oxidative reaction 4
Oximes and oxaziridines and
  praseodynium salts,
  complexes between 232
α-Oxoglutarate oxidation,
  inhibition of 187
Oxygen-bonded compounds,
  structural chemistry
  of organotin 258
Oxygen complex, iron– 198

P

Paint preservation 8
Paints, antifouling 204
Paper mills, slime control in 7
Pentacoordinate triorganotin
  halides, chiral 275
Pentacoordination 283
PES, UV– 227
PES, x-ray– 227
Pesticides 167, 175, 184
Phenylarsenious acid 187
3-Phenyl-3-benzoborepin 16
Phenyl–tin bond cleavage 115
PhMeSnBr₂ 277
Phosphines, arsines, stibenes,
  and bismuthines, organotin .. 57
Phosphines, chlorostannyl 65
Phosphines, oxidation of
  organotin-substituted 62
Phosphoesterase activity 220

Phosphorus bond, tin– 59
Phosphorylation, electron
  transport 205
Photoelectron spectroscopy
  (PES) 227
  UV– 235
Photo-induced
  disproportionation 76
Photolysis experiments 77, 80
Photolysis of new amides 72
Phytotoxicity 171
Pipe, multiple screw
  extrusion of 136
Platinum–tin cluster
  compounds 230
Plictran 169
Plumbyllithiums 42
Point-charge approximation 269
Point-charge potential model 228
Polymers, triethyltin–
  imidazole 193
Polyurethane foams, flexible 10
Praseodynium salts, complexes
  between oximes and
  oxaziridines and 232
Preparation of the stannylene
  complexes 79
Preparation of the tin(II)amides .. 79
Preservation, paint 8
Preservation, wood 7
Properties of anion–
  hydroxide exchange 221
Propyl radicals, *tert*-
  butoxyl and 27
Proteins, metals in 233
Proteins to which
  triethyltin binds 193
Proton
  abstraction, intramolecular 48
  gradient across membranes 192
  permeability 215
  pumping 205
  translocating ATPases 210
PVC stabilizers 4, 123, 177
  estertin 134
Pyruvate oxidation 191

Q

Quadrupole hyperfine interaction 163

R

Racemization 250
Radical, alkoxyl 35, 36
Radical anion, haloaryl 50
Radicals
  aryl 47
  bromine 36, 37

Radicals (*Continued*)
   tert-butoxyl 37
   dialkylaminyl 29
   organotin 28
   β-stannylalkyl 33
   succinimidyl 36
   trialkyltin 30
Radioactivity studies 178
Raman data 163
Rate constants 32, 38
Reaction between the
   stabilizer and
   hydrogen chloride 181
Reactions
   of alkyltins with
     halogens 114
   of *n*-butyl halides with
     trimethylstannylsodium
     in THF 45
   cleavage 113
   disproportionation 128
   homolytic 26
   hydrostannation 12
   hydrostannolysis 12
   not involving a bond to tin 27
   of organostannylanionoids
     with aryl halides ....:.... 46
   oxidative addition 79
   producing organotin radicals ... 27
   at tin centers 189
   of trimethylstannylsodium
     with carbon
     tetrachloride 51
Reduced cytochrome 189
Reversed roll coated
   wall covering 148
Rigid calendered film 142
Rigid foamed profile 140
Rigid injection-molded
   fittings 150
Ring compounds from
   aliphatic diynes 18
Ring compounds from
   O-divinylbenzene and from
   O-diethynylbenzene 15
R'SH to alkenes, addition
   of the thiol 179
Rubbers, silicone 10

S

Salvarsan 167
S_E2 reactions at silicon
   or germanium 249
S_H2 ....36, 37
SH groups, free 235
Silicon or germanium, S_E2
   reactions at 249

Silicone rubbers 10
Six-coordinate dimethyltins 264
Skin and eye irritation 150
Slime control in paper mills 7
$Sn(CH(SiMe_3)_2)_2$ 76
$Sn(N\text{-}i\text{-}Pr_2)(N(SiMe_3)_2)$ 80
$Sn(N(SiMe_3)_2)_2$ 76
$Sn(NR_2)_2$ 78
$(Sn(NR_2)_2)_x$ 80
$Sn(SCM_2CH_2S)_{2n}$ 161
$Sn(SPh)_2)_n$ 77
Solvent effects 114
Spectroscopy
   ESCA 232
   Mössbauer 227, 258
   tin 119*m* 269
   NMR 278
   photoelectron 227
   structure and bonding
     in organotins by
     gamma resonance 155
   UV–photoelectron 235
Stability, configurational 254
Stability, light 140
Stabilizer $R_2Sn(SR')_2$ 179
Stabilizer and hydrogen
   chloride, reaction
   between the 181
Stabilizers in PVC 4, 123,
                          134, 177
Stabilizers, structure of 183
Stanclere 135
Stannacycloalkanes 237
Stannacyclopentanes 36, 38
Stannazanes 102
β-Stannylalkyl radicals 33
Stannylene complexes 79
Stereochemistry 44, 264, 269
Steric factors 114
Structural chemistry of
   organotin oxygen-bonded
   compounds 258
Structure and bonding in
   organotins by gamma
   resonance spectroscopy 155
Structure of organotin
   compounds, configurations
   and 54
Structure of stabilizers 183
Sub-mitochondrial particles 215
α-Substituted sulfur organotins ... 115
Substitution, S_N2 285
Substitution at tin, reactions 250
Succinate oxidase 192
Succinimidyl radicals 36
Sucrose 183
   organotin– 184
Sulfenyl halides 118

β-Sulfides 118
Sulfonium ion intermediate 118
Sulfonium salt intermediate 121
Sulfur organotins, α-substituted .113, 115
Surface treatment of glass 10
Swelling 208
Synthesis of α-alkoxyketones 90
Synthesis of chiral tin
  compounds 287

T

Tetraalkyltins 36
  animal toxicity of 243
Tetracarbonyl nickel 61
Tetracoordinate carbon atom,
  highly shielded 52
Tetraethylene glycol dimethyl
  ether.................... 44
Tetraethyltin 186
Tetraglyme 42
Tetra(manganesepentacarbonyl)-
  ditindihydride 23
Tetraorganotins 186, 200, 252, 275
THF, reactions of n-butyl halides
  with trimethylstannylsodium
  in 45
Tin
  bond cleavage, aryl– 114
  bond cleavage, phenyl– 115
  centers, reactions at 27
  centers, two optically
    stable chiral 285
  cluster compounds,
    platinum– 230
  compounds, binding
    energies of 243
  compounds, synthesis
    of chiral 287
  electropositive nature of 82
  –germanium compounds 254
  –iron compounds 255
  119m Mössbauer spectroscopy .. 269
  phosphorus bond 59
  reactions not involving
    a bond to 27
  by Sn–S bonds,
    coordination of 235
  substitution reactions at 250
  –transition metal compounds .. 254
Tin(II) alkyl 77
Tin(II) amides 70, 72, 73, 76
  disproportionation by
    irradiation 70
  insertion into SnN by PhNCO .. 70
  Lewis base properties 70
  ligand exchange 70
  metathesis with protic reagents . 70

Tin(II) Amides (Continued)
  oxidative addition 70
  preparation of 79
  trimethylsilyl-substituted 70
Tin(II) benzenethiolate 77
Tin(II) diamides 74
Toxicity of organotins 3, 153, 174,
  ....186, 204, 232
  oral 152, 175
Toxicity of tetraalkyltins,
  animal 243
Transition metal carbonyl
  complexes 61
Transition metal compounds,
  tin- 254
Translocating ATPases 219
Trans-trigonal bipyramidal
  geometry 263
Trialkyltins(s) 218, 232
  hydroxides 223
  inhibition of ATP
    synthesis by 188
  radicals 30
Tributyltin(s) 6, 194, 199, 214, 221
  acetate 199
  biocides 6
  chloride 108, 209, 213
  fluoride 7
  oxide 184
Trichloromethyl anion 52
Trichlorostannane 125
Tricyclohexyltin hydroxide
  ....9, 169, 173, 200
Triethylgermylpotassium 42
Triethyltin 186, 189, 191,
  .... 192, 199, 204
  to cat hemoglobin,
    binding of 193
  in a chloride-free medium,
    effects of 192
  hydroxide transporting
    action of 207
  imidazole polymers 193
  macromolecules which
    do not bind 188
  on mitchondrial functions,
    effects of 192
  proteins to which it
    binds 193
Trimerization reactions 65
Trimethylbenzylstannane 42
Trimethylsilylsodium 42
Trimethylsilyl-substituted
  tin amides 70
Trimethylstannylanionoids
  with 7-bromonorbornenes,
  stereochemistry of
  the reaction 44

Trimethylstannylsodium 42, 46, 52
  with carbon tetrachloride,
    reaction of   51
  in THF, reactions of *n*-butyl
    halides with   45
Trimethyltin   189
  chloride   211
  diphenylarsine   58
Triorganostannylamines,
  -phosphines, and -arsines   253
Triorganotin(s) ...7, 171, 204, 207, 215,
                       216, 219, 285
  biochemical activities of   223
  degradation of   175
  halides 250, 275, 283
    chiral pentacoordinate   275
  hydrides   253
  inhibition by   205
  on mitochondria, effects of   194
  optically stable   250
Triphenylgermylsodium   42
Triphenyltin(s) 9, 201, 221
  chloride   214
Tris(neophyl)tin oxide   9
Tris(triphenyltin)arsine   58
Trivalent tin amides, ESR
  characterization of   72
Tropical Bilharzia disease   8

U

Uncoupling 205, 206, 218
UV–photolectron spectroscopy
  (PES) 227, 235
  and nuclear quadrupole
    resonance   238

V

Volatility   152

W

Wall covering, reversed
  roll coated   148
Waterproofing agents   11
Wood preservation   7
Würtz process   3

X

X-ray crystallography 258, 259
X-ray diffraction   278
X-ray–PES   227

The text of this book is set in 10 point Highland with two points of leading. The chapter numerals are set in 26 point Times Roman; the chapter titles are set in 18 point Baskerville Bold.

The book is printed in offset on Text White Opaque, 50-pound. The cover is Joanna Book Binding blue linen.

Jacket design by Linda Mattingly.
Editing by Spencer Lockson.
Production by Barbara Allen.

The book was composed by Mack Printing Co., Easton Pa., and by Service Composition Co., Baltimore, Md., printed and bound by The Maple Press Co., York, Pa.